Sommer

ABOUT THE BOOK

The primary motivation for this book was to provide material that would bridge the gap produced by texts too mathematical and those that attain simplicity by omitting important concepts. This is an applications oriented text which blends the competing Taguchi, Shainin, and classical approaches to designed experiments. Included are over 200 pages of actual industrial case studies plus software.

ABOUT THE AUTHORS

Stephen R. Schmidt, Lt Col (Ret), served for twenty years in the U.S. Air Force both as an instructor pilot and as a Tenure Professor at the U. S. Air Force Academy. In addition to training over 4,000 engineers, managers, and college students, he has personally been a part of over 250 industrial experiments for a wide variety of Fortune 500 companies and their suppliers. Because of his vast experience, he now serves as president of Schmidt/Launsby Consulting and as an adjunct faculty member to the Six Sigma Institute, Motorola University. He earned a B.S. in Math at the U.S. Air Force Academy, an M.S. in Operations Research at the University of Texas and a Ph.D. in Applied Statistics at the University of Northern Colorado. Dr. Schmidt is also the co-author of two other recent texts: *Basic Statistics: Tools for Continuous Improvement* and *Total Quality: A Textbook of Strategic Quality Leadership and Planning*.

Robert G. Launsby, M.S. in Engineering, is Executive Vice President of Schmidt/Launsby Consulting. He has over 20 years of industrial experience in management, engineering, training, and applications of problem solving approaches. Over the last 12 years he has led over 300 applications of statistical methods in a wide variety of industries. Possessing strong presentation and facilitation skills, he has instructed over two thousand persons in the use of statistical problem solving tools. Universities Bob has taught at include University of Colorado, University of Northern Colorado, University of Nebraska, University of Phoenix, Drexel University and Washington State University.

SHORT COURSES FOR INDUSTRY

Several short courses as well as on-site consulting are available. The focus of these activities is on empowering people with the techniques they need and can readily use. Available courses are:

- Basic Statistics (3 days)
- Advanced Topics in Design of Experiments (4 days)
- Introduction to Multiple Factor Designs & Taguchi Techniques (2 days)
- Statistical Process Control (2 days)
- Total Quality Management (3 days)
- Management Overviews (2-4 hours)
- Quality Function Deployment (QFD) (3 days)

OTHER TEXTS BY AIR ACADEMY PRESS

Total Quality: A Textbook of Strategic Quality Leadership and Planning (1991) by
K.D. Lam, F.D. Watson, and S.R. Schmidt. ISBN 0-9622176-9-7
This text is a comprehensive introduction to the philosophy, principles, tools and techniques
of total quality. An excellent teaching tool for the university and community college level,
it includes the basics of total quality as well as advanced topics such as QFD, experimental
design, and total quality applied to software and education.

Basic Statistics: Tools for Continuous Improvement 2nd Edition (1991) by M.J.
Kiemele and S.R. Schmidt. ISBN 0-9622176-8-9
The primary motivation for this book is to educate the readers (managers, engineers,
researchers, analysts, practitioners, and statisticians) in the application of statistical tools to
achieve continuous improvement in how they do business. It is designed for those taking a
first course in statistics, as well as those who may have already taken a statistics course but
need a refresher. The book uses applied examples to illustrate the use of statistical tools
for continuous process improvement. It is easily read (calculus not needed) and there are
no proofs. A statistical applications package, as well as other software, is included with the
text.

Surviving the '90s and Winning! (1990) by B.D. Wellmon. ISBN 0-9622176-3-8
This book has been written specifically for everyone who has to work for a living–from the
worker on the floor to the CEO or owner of the company. Whether your company has 91
people or 91,000; whether your business is manufacturing, service, or government; whether
your customer base is local or global; this handbook should not only be read but used as a
primer on your quest to survive the 1990's. This management survival handbook is designed
as a practical reference guide and implementation tool for anyone in the business world.
There are no theorems or claims of instant answers to all of your problems. There are no
slogans or hidden secrets to be found. There are just three basic steps to be followed to
guide you down the long and confusing path to improved quality. This book will help you
master a system which is unique to your situation and which will contribute to your survival.

Total Quality Management: A Resource Guide (1990) by K.D. Lam.
This resource guide is designed to save you time and money. In it you will find hundreds
of references to help you improve your TQM capabilities. In a handsome looseleaf binder,
this well organized, comprehensive reference guide is designed for both TQM experts and
novices. Those new to TQM will benefit from a chapter on "Getting Started," steps to take
to start the TQM journey. It includes your role in the TQM process, your customers, key
first steps, important tools and techniques you need to learn, and potential pitfalls to avoid.

UNDERSTANDING INDUSTRIAL DESIGNED EXPERIMENTS

Third Edition

Stephen R. Schmidt   **Robert G. Launsby**

Air Academy Press
Colorado Springs, Colorado

Library of Congress Catalog Card Number: 88-92885

ISBN 0-9622176-2-X

The authors recognize that perfection is unattainable without continuous improvement. Therefore, we solicit comments as to how to improve this text. To relay your comments or to obtain further information, contact:

AIR ACADEMY PRESS
1155 Kelly Johnson Blvd. Suite 105
Colorado Springs, CO 80920
PHONE (719) 531-0777
FAX (719) 531-0778

Preface

The time is quickly approaching when you will not be considered a competent engineer or technical manager without a working knowledge of Experimental Design. The demands of increased efficiency of processes, lower product cost, and shortened development cycles will dictate that we use simple, but powerful tools to get the most out of our experiments. No longer do we have the luxury of running one-factor-at-a-time experiments or experiments with excessive sample sizes. In competitive environments, only those groups which **apply** experimental design approaches efficiently and effectively will survive.

Once engineers decide to start using orthogonal or nearly orthogonal designs, they are quickly faced with a plethora of competing design and analysis strategies. Full factorial, Plackett-Burman, Taguchi, and CCD are just some of the competing design types. What is the strength and weakness of each? What is the niche that each strategy best fills? How about analysis approaches? Should the engineer use "Pick the Winner", plots of marginal means, ANOVA, Normal Probability Plots, or Regression Analysis? In which situations do the competing analysis strategies best fit? Should the objective of the experiment influence the design type and analysis approach to be used? Is the objective of my experiment Screening, Modeling, or "Robust" Design? Is the Brainstorming session critical? Who should be involved?

Industry needs a handbook which answers the above questions! The following chapters offer what we believe are appropriate answers to the questions raised. Our approach is not to compulsively follow one specific philosophy of experimentation. Rather, the authors attempt to select the best from competing strategies, combining seemingly dichotomous techniques so as to produce a synergism which leads to an improved approach.

Acknowledgements

Writing a book is very similar to other major undertakings; it requires a team effort to produce a successful product. The magnitude of the team effort required becomes even greater when one considers the first edition of this product hit the market in early 1988. Now, only a couple of years later, the third edition has become a reality.

Many people have assisted in the continuous improvement of this text. A whole host of students from seminars and past university courses have been most helpful in providing suggestions. Several engineers have provided case studies or example problems. Still others have provided editorial support, suggestions, and encouragement. A special thanks must go to Debbie and Derek Brown for the original typesetting and editorial efforts on the first and second editions. Also, a special thanks goes to Diane Launsby for reformatting the entire text into its current form. Only she knows the incredible hours required to format, rework, and edit a text of this size. Thanks also to Ronda Churchill for editing and correcting the latest third edition manuscript. The following is a partial list of those who have helped make this text a success.

Willis Adcock
Alan Arnholt
Maureen Beaty
Ron Berdine
Tom Bingham
Mike Bishop
James Boudot
Randy Boudreau
James Bricknell
Hans Carter
Sandy Claudell
Karen Cornwell
Richard Counts
Thomas Curry
Rodney Davis
Steve Dziuban
Larry Erwin
Gary Fenton
Tom Gardner
Jan Gaudin
Jay Gould

Rafael Gutierrez
Dave Hallowell
Mike Howard
Mike Jackson
Bruce Johnson
Bradley Jones
S.E. Jones
Joe Juarez
Ragu Kacker
Robert Kaliski
Mark Kiemele
Peter Knepell
Kenneth Knox
James Kogler
Jayme Lahey
Robert Lawson
William Lesso
Dan Litwhiler
Jim Merritt
Ed Midkiff
Bill Miller

Rick Mills
Dwight Mitchell
Douglas Montgomery
William Motley
Raymond Myers
Dale Owens
Jack Reece
Mike Reeder
James Riggs
Jim Rutledge
Doug Sheldon
Wen Shih
S. Young Shin
Steve Smith
John Tomick
John Toomey
Barbara Warsavage
Rita Whitely
Don Williams
Don Wilson
Barb Yost

Contents

Chapter 2 Conducting Simple Experimental Designs and Analysis

Chapter 3 Design Types

Chapter 1

Foundations

1.1   Intended Audiences for this Text

This text was written for managers and engineers. It is intended to bridge the gap produced by texts that are too mathematical and those that attain simplicity by omitting important concepts. Mathematical and statistical theory are not emphasized; rather, the presentation is intended to provide conceptual understanding of designed experiments and related topics in statistical quality control. For years statisticians have provided us with textbooks written for understanding statistical theory, written primarily for those well versed in statistical jargon and mathematics. We recommend those texts for anyone seeking more depth into these subjects.

The presentation of material in this text is arranged to include a foundation for understanding experimental design (Chapter 1), an introduction to design and analysis (Chapter 2), a discussion of design types (Chapter 3), a review of basic statistical concepts and techniques (Chapter 4), an analysis of experimental data, variance reduction, and robust designs (Chapter 5), a presentation of Taguchi philosophy and analysis techniques (Chapter 6), an introduction to optimization and response surface methods (Chapter 7), and numerous case studies from industry (Chapter 8).

1.2   Quality and the Role of Statistical Techniques

The American Society for Quality Control has defined quality as "the totality of features and characteristics of a product or service that bear on its ability to satisfy stated or implied needs." A simple interpretation of this definition is that quality is a measure of how well the user's needs are met. The user's interpretation of quality focuses on whether the product functions as advertised with no variation in performance. There are several aspects of quality: the quality of the product design, the quality of the process design, the quality of incoming parts, and the quality of the manufactured product. In the design phase of the product or process, it is important to test which factors affect the quality of the design and determine the settings of these factors that optimize the desired output. In this phase, experimental designs and response surface techniques are used to develop a process which will produce a quality product. The quality of incoming parts and the quality of the manufacturing process are typically achieved through the use of statistical process control techniques; however, on-line experimental designs have also produced excellent results in these areas.

This text will focus on quality improvements through the applications of experimental design and present both sides to the controversy between Taguchi methods and classical techniques. The authors are not committed to either side, but advocate an approach which finds the "best" method for the practitioner to solve the problem at hand. Included are new ideas on how to combine certain aspects of both approaches in order to produce a synergism leading to improved methods of experimental design. Emphasis is on quality improvements and not on selling one approach or another. Also included are some important contributions from Shainin [11].

1.3   What is Experimental Design

Experimental design consists of purposeful changes of the inputs (factors) to a process in order to observe the corresponding changes in the outputs (responses). The process is defined as some combination of *machines, materials, methods, people, environment,* and *measurement* which when used together perform a service, produce a product, or complete a task. Thus, experimental design is a scientific approach which allows the researcher to better understand a process and to determine how the inputs affect the response.

Graphically, a process would appear as shown in Figure 1.1. Some examples of processes are shown in Figures 1.2 through 1.7. Obviously, there exist many different kinds of processes; the ones provided are from various applications of designed experiments.

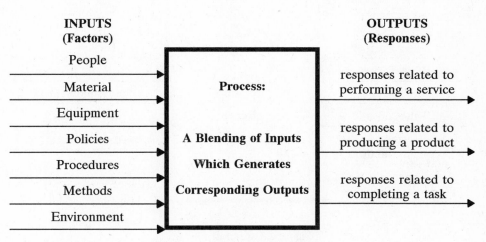

Figure 1.1 Illustration of a Process

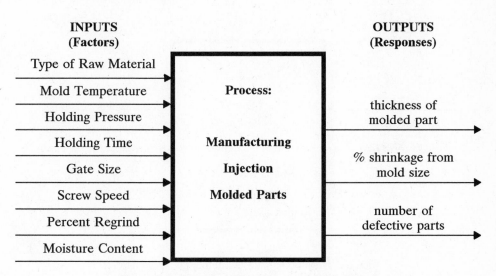

Figure 1.2 Manufacturing Injection Molded Parts

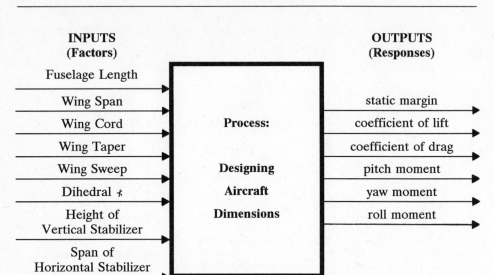

Figure 1.3 Aircraft Design

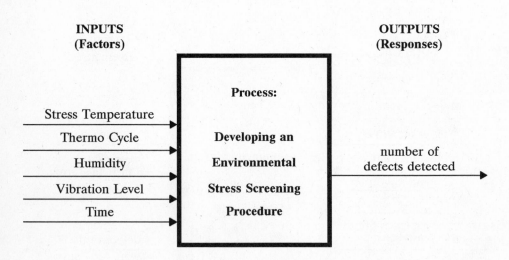

Figure 1.4 Developing an Environmental Stress Screen

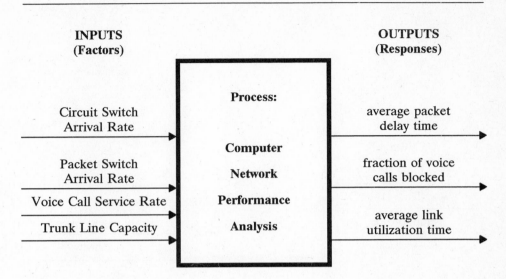

Figure 1.5 Computer Network Performance Analysis

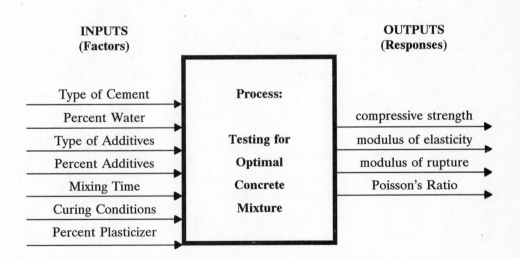

Figure 1.6 Testing for Optimal Concrete Mixture

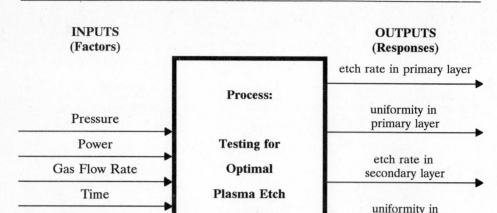

Figure 1.7 Testing for Optimal Plasma Etch Conditions

1.4   Why Use Experimental Design

Historical use of one-factor-at-a-time experimentation and/or arbitrary selection of incomplete full factorial designs has resulted in very inefficient and ineffective attempts to understand and optimize product designs and processes. No longer can these archaic methods be tolerated if a company intends to keep up in a highly competitive market.

The manager should be interested in experimental design to achieve the following: (1) improved performance characteristics; (2) reduced costs; and (3) shortened product development and production time. Improved performance characteristics result from the identification of the critical factor levels that optimize the mean response and minimize response variability. This improved performance also leads to the reduction of scrap and rework, which greatly reduces costs. Understanding which factor levels are critical to improve performance allows for controlling only the important factors while relaxing tolerances on benign factors and, when appropriate, selection of less expensive materials. Reduced time to market is accomplished by understanding what the customer really wants and using efficient testing procedures to optimize the design of the product and process.

For the engineer who tests different design strategies of a new product or trouble shoots problems in an on-line process, experimental designs are used as: (1) efficient methods for gaining an understanding of the relationship between the input factors and the response; (2) a means of determining the settings of the input factors which optimize the response; and (3) a method for building a mathematical model relating the response to the input factors, which is often referred to as process/product characterization.

Perhaps the easiest way to convey to the reader the importance of learning experimental design is to review some alternatives and discuss their weaknesses. The reader is presented three cases which sequentially illustrate the pitfalls associated with historical data, one-at-a-time designs, and full factorials. In addition, Case 4 reveals the need for improvement beyond traditional techniques in the pursuance of robust designs.

Case 1: A plant manager wishes to know if changes in temperature during a production process are related to the number of defects of the product. Since there was an abundance of historical or "happen-stance" data available it was decided to use simple linear regression (see Chapter 4 for a detailed discussion) to correlate temperature and defect rate. The regression results indicated that a high positive correlation existed between these two variables, implying that as you increase the temperature the defect rate would increase (see the scatter diagram in Figure 1.8). Since the goal was to minimize defects, the engineer directed that temperature be maintained at some appropriate low level. What was not revealed from the data analysis was undocumented intervention in the production process by the operator. If the defect rate started to increase (for any reason), the operator (based on years of experience) would raise the temperature to minimize this increase. After the number of defects started to decline, he would reduce the temperature back to the standard operating settings. A low temperature had originally been specified to keep costs down; however, no experimentation had taken place to accurately examine the relationship between temperature and defect rate. Analyzing this non-experimental data results in a confounding of temperature and operator manipulation which makes it impossible to separate the two inputs to the process. The apparent positive correlation of temperature and defect rate did not represent the true relationship, but rather a relationship induced by the operator. Historical data, although often valuable, can lead to this type of faulty conclusion because it probably does not provide sufficient manipulation of input factors and it frequently contains factors which are highly interrelated.

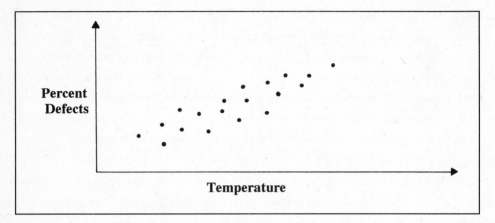

Figure 1.8 Scatter Diagram from Historical Data

Case 2: The plant manager is now convinced that historical data will not always provide a clear understanding of the process so he has decided to design an experiment to test temperature and pressure effects on defect rate. The design consists of the following: (1) select two levels of temperature, T_1 and T_2, and two levels of pressure, P_1 and P_2; and (2) holding pressure constant at level P_1, test the product at both T_1 and T_2 as shown in Table 1.1.

Pressure	Temperature	Response
P_1	T_1	.03
P_1	T_2	.015

Table 1.1 One-At-A-Time Approach (part 1)

Since a more desirable defect rate occurred for T_2, set the temperature at T_2 and test the product at pressure P_1 and P_2. The results are displayed in Table 1.2.

Pressure	Temperature	Response
P_1	T_2	.015
P_2	T_2	.010

Table 1.2 One-At-A-Time Approach (part 2)

At first glance you would say that the ideal settings of temperature and pressure are T_2 and P_2, resulting in a 1 percent defect rate; however, this design, referred to as a "one-at-a-time" design, did not allow for testing the interaction effect of temperature and pressure. Suppose that the untested combination P_2T_1 would produce an interaction as displayed in Figure 1.9. The interaction effect of temperature and pressure is interpreted as follows: as temperature changes from T_1 to T_2, the change in the defect rate is different depending on the pressure settings of P_1 and P_2 (i.e., the lines in the graph are non-parallel).

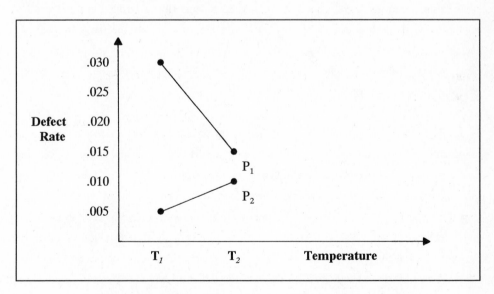

Figure 1.9 Graph of Temperature × Pressure Interaction

A more appropriate design would be to test all four combinations of temperature and pressure, referred to as a full factorial design (discussed further in Chapters 2 and 3).

Case 3: After understanding the pitfall discovered in Case 2 and brainstorming which other factors could potentially affect the defect rate, the manager requests further testing of temperature, pressure, and ten other input factors of interest. If all combinations are to be tested for each factor measured at two levels, there will be 2^{12} or 4096 experimental conditions in the full factorial design. This presents a new problem because of resource limitations in conducting the experiment. Fortunately, there is a solution to the problem. The material in this text includes fractional factorials as one efficient alternative to full factorial designs. Obviously, you cannot get something for nothing. You'll see in Chapter 3 what you give up when conducting a fractional factorial of 32 experimental combinations versus the full factorial of 4096.

Case 4: The manager and his engineering staff have given the 12 factors in Case 3 some more thought and they agree that 3 of the 12 factors cannot be controlled in the operating environment of the product. What is needed is a design to determine the 9 controllable factor settings such that the defect rate and the variability of defect rate across the 3 uncontrollable factors are minimized. The solution to their problem is found in the study of robust designs which are described in Chapters 5 and 6.

These four cases demonstrate that unless the researcher is well informed on all the statistical tools available and his experiments are well designed, results can be misleading and large amounts of resources can be consumed unnecessarily. An even more compelling reason for educating American industrial managers and engineers in the use of designed experiments is based on the need to compete with countries who have successfully adopted these techniques. In 1964, American industries were basking in a 6 billion dollar trade surplus, not realizing they were under attack by aggressive overseas companies. By 1984, the U.S. was looking at a 123 billion dollar trade deficit. Over this 20 year period, U.S. productivity increased a mere 35% as compared to about 60% for some European countries and an amazing 120% for Japan. The same country that had a reputation for making "junk" in the 50's and 60's was threatening the survival of U.S. industry in the 70's and 80's by producing high quality products at a lower price. How did this happen? It is becoming well known that much of this remarkable change in the quality of Japanese products is attributed to designing quality into the product through the use of designed experiments and monitoring quality during production by way of statistical quality control. Ironically, these techniques were developed mostly in the U.S. and Europe over 60 years ago, but U.S. companies failed to implement them on a continuous basis. In the 1950's American statisticians, having been unsuccessful at convincing most of industry to adopt these methods, soon directed their efforts towards new theory instead of industrial implementation. U.S. industries became less and less interested in statistics primarily because there was little post World War II competition to warrant a quality improvement focus. Thus, industry in this country entered a stagnant mode with regard to quality. After World War II, the Japanese were very receptive to the ideas of two American statisticians, W. Edwards Deming and Joseph M. Juran. As a result, the Japanese soon became the quality experts and their companies prospered. The few American companies that did implement statistical techniques did not do so with the intensity seen in Japan. The Japanese managers view the use of statistical methods as very important to their companies' success. Thus, everyone in the company, from management to on-line workers, are taught these techniques in training sessions and they all use them not just to monitor quality in production, but to steadily improve quality throughout the production process. This philosophy was initiated through Deming's 14 points which serve as a guide for management on how to improve quality and productivity.

The following 14 points of Deming are taken from Hogg and Ledolter [3] and can be further studied in Deming's book *Out of the Crisis* [2].

(1) Create a constancy of purpose toward the improvement of product and service. Consistently aim to improve the design of your products. Innovation, money spent on research and education, and maintenance of equipment will pay off in the long run.

(2) Adopt a new philosophy of rejecting defective products, poor workmanship,

and inattentive service. Defective items are a terrible drain on a company; the total cost to produce and dispose of a defective item exceeds the cost to produce a good one, and defectives do not generate revenues.

(3) Do not depend on mass inspection because it is usually too late, too costly, and ineffective. Realize that quality does not come from inspection, but from improvements on the process.

(4) Do not award business on price tag alone, but consider quality as well. Price is only a meaningful criterion if it is set in relation to a measure of quality. The strategy of awarding work to the lowest bidder has the tendency to drive good vendors and good service out of business. Preference should be given to reliable suppliers that use modern methods of statistical quality control to assess the quality of their production.

(5) Constantly improve the system of production and service. Involve workers in this process, but also use statistical experts who can separate special causes of poor quality from common ones.

(6) Institute modern training methods. Instructions to employees must be clear and precise. Workers should be well trained.

(7) Institute modern methods of supervision. Supervision should not be viewed as passive "surveillance", but as active participation aimed at helping the employee make a better product.

(8) Drive out fear. Great economic loss is usually associated with fear when workers are afraid to ask a question or to take a position. A secure worker will report equipment out of order, will ask for clarifying instructions, and will point to conditions that impair quality and production.

(9) Break down the barriers between functional areas. Teamwork among the different departments is needed.

(10) Eliminate numerical goals for your work force. Eliminate targets and slogans. Setting the goals for other people without providing a plan on how to reach these goals is often counterproductive. It is far better to explain what management is doing to improve the system.

(11) Eliminate work standards and numerical quotas. Work standards are usually

without reference to produced quality. Work standards, piece work, and quotas are manifestations of the inability to understand and provide supervision. Quality must be built in.

(12) Remove barriers that discourage the hourly worker from doing his or her job. Management should listen to hourly workers and try to understand their complaints, comments, and suggestions. Management should treat their workers as important participants in the production process and not as opponents across a bargaining table.

(13) Institute a vigorous program of training and education. Education in simple, but powerful, statistical techniques should be required of all employees. Statistical quality control charts should be made routinely and they should be displayed in a place where everyone can see them. Such charts document the quality of a process over time. Employees who are aware of the current level of quality are more likely to investigate the reasons for poor quality and find ways of improving the process. Ultimately, such investigations result in better products.

(14) Create a structure in top management that will vigorously advocate these 13 points.

Other philosophies such as those of Juran, Crosby, and Taguchi may vary slightly from Deming, but all appear to center on four main aspects which lead to improved quality: **customer satisfaction, teamwork, statistical methods,** and **continuous improvement**.

There are many reasons why most of U.S. industry has been slow to implement the techniques and philosophies described above. The authors suggest the following 6 reasons:

(1) Failure of management to recognize the value of these techniques.

(2) Failure of statisticians to teach these techniques in an easy to understand fashion.

(3) Misperception that statistically designed experiments are costly, time-consuming, and impractical.

(4) Systems of reward which provide financial incentives primarily for short term

planning and decision making.

(5) Misperception that quality is achieved by merely meeting specifications.

(6) Until recent years, a lack of competition which imparted an air of complacency within U. S. industry.

1.5 Variation and Its Impact on Quality

It should not be a surprise to the reader that, even in well controlled production processes, there will be variation in the final product. The variation is either caused by uncontrolled factors or production noise. Too much variation degrades the quality of the product and causes a loss to the company. If large numbers of the product are produced outside specifications, and if products are not inspected before they are shipped, then (1) complaints will increase, (2) extra resources will be expended to repair items under warranty, and (3) eventually customers will become discouraged and seek a more reliable product. Historically, the approach to this problem has been to set up specification limits and perform inspections of finished products to ensure zero defects out the back door. This approach will attempt to dichotomize the quality aspect of any product into either acceptable (within spec) or unacceptable (out of spec). Loss to the company is based on being out of spec as shown in Figure 1.10.

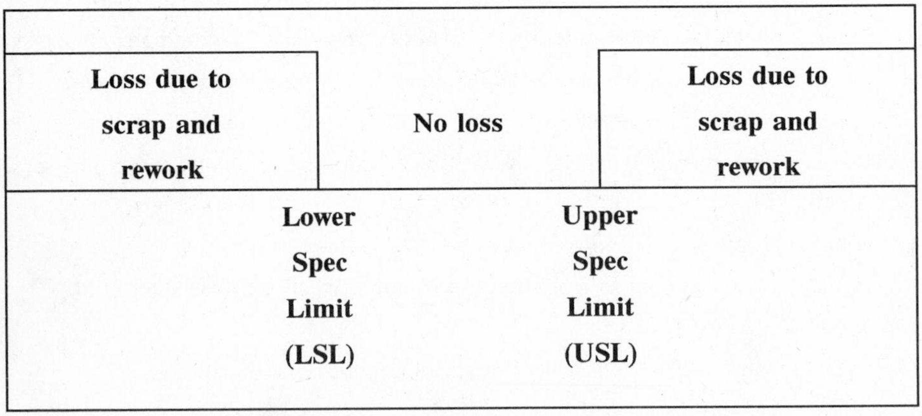

Loss due to scrap and rework	No loss	Loss due to scrap and rework
Lower Spec Limit (LSL)		Upper Spec Limit (USL)

Figure 1.10 Old Philosophy of Quality

The optimal solution to this problem is not to increase inspections at the end of the production line. According to Juran, these inspections are only 80% effective. Furthermore, this approach requires more manpower and results in rework and scrap costs that can become prohibitive. In addition, maintaining profits under these conditions will require sales prices to increase, resulting in a continued decrease in consumer satisfaction. It is also unreasonable to assume that a product just inside a specification limit results in no loss to the company. For example, consider building window glass and window frames. If the glass thickness is at the lower limit and the frame width is at the upper limit (or vice versa) the quality of the product is reduced.

To develop a quality product at a reduced cost, quality must be designed into the process [5] and focus must be on the target (nominal) value instead of just being "in spec" as depicted in Figure 1.11. Deviations from the target result in loss to the producer in the form of scrap, rework, warranty costs, increased cycle time, tied up capital, increased inventories, etc. Losses to the purchaser are due to degraded performance and reliability, as well as increased maintenance costs. Eventually the purchaser (customer) will find a better producer (supplier) and, thus, the ultimate loss to the producer is reduced market share.

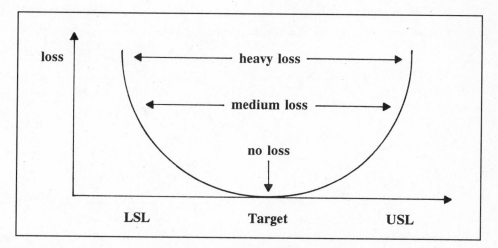

Figure 1.11 New Philosophy of Quality

To minimize the loss incurred due to deviation from the target, the variability of the product about the target must be reduced as shown in Figure 1.12. Product B has a much higher proportion of measured values close to the target. Therefore, Product B will result in less overall loss.

Whether you are in research, development, design, manufacturing, packaging or any other process related job, variance reduction is best achieved through a thorough understanding of the relationships between the output variables (responses or quality characteristics) and the input variables (factors or parameters). The simplest and most important tool to begin with is the cause-and-effect diagram [10]. If properly developed, a cause-and-effect diagram will contain a list of all possible sources (causes) of variation in the response (effect). All identified sources of variation that can be easily removed at low cost should be dealt with prior to any experimentation in order to bring the process in an approximate state of control, (i.e., consistency).

A designed experiment will provide the most efficient set of testing conditions to:

1) Develop simple models that represent the relationship of the outputs with the inputs,

2) Determine the critical input factors,

3) Determine optimal factor settings.

A properly conducted cause-and-effect diagram with a subsequent designed experiment will typically result in a large amount of variance reduction and thus improved quality.

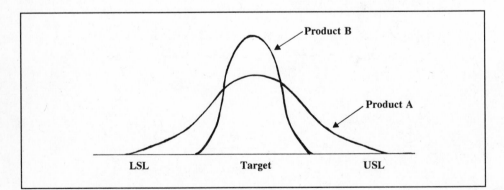

Figure 1.12 Higher Quality Implies Variance Reduction Around the Target

1.6   Measurements of Quality

After a process has been brought into statistical control or stabilized (i.e., all special causes of variation have been detected and eliminated), we can assess the capability of the process through the use of metrics such as the C_p or C_{pk} indices. The C_p index (referred to as the process potential) was designed to compare the variability of some quality characteristic (y) based upon the lower specification limit (LSL) and the upper specification limit (USL) while assuming the average of the characteristic ($\bar{y}$) is equal to the desired target value (T). The formula for C_p is shown in Equation 1.1.

$$C_p = \frac{specification\ width}{process\ width} = \frac{USL - LSL}{6\hat{\sigma}}$$

$$where \quad \hat{\sigma} = \sqrt{\frac{\sum_{1}^{n}(y_i - \bar{y})^2}{n-1}} \quad and \quad \bar{y} = \frac{\sum_{1}^{n} y_i}{n}$$

(Throughout this text, the symbols $\hat{\sigma}$ and s are both used to represent sample standard deviation. $\hat{\sigma}^2$ or s^2 are used to represent variance.)

Equation 1.1 Computing the C_p Index

According to acceptable standards in industry, C_p values less than 1.00 are unacceptable, values between 1.0 and 1.33 are marginally acceptable, and values greater than 1.33 are desired. Many quality oriented companies such as Ford Motor Company and Motorola are now requiring C_p values greater than 2.0.

Since the C_p index does not take into account $\bar{y}$ deviations from the target, T, a more realistic estimate of your actual state of quality is the C_{pk} index computed as shown in Equation 1.2. Interpretation of C_{pk} values is similar to that described for C_p. Notice that once the process average is on the target, the only possible way to improve C_{pk} or C_p is through variance reduction.

$$C_{pk} = minimum \ of \ (\frac{USL-\bar{y}}{3\hat{\sigma}}) \quad or \quad (\frac{\bar{y}-LSL}{3\hat{\sigma}})$$

Equation 1.2 Computing the C_{pk} Index

A third measure of product quality is the number of defects per million (dpm). Obviously we need to minimize dpm in order to insure high levels of quality. For attribute (pass/no pass) type quality characteristics, one would simply count the number of defects per number of opportunities and multiply times 10^6 to obtain an estimate of dpm. For quality characteristics that are quantitative, we can estimate dpm from our previous capability index, C_{pk}. The relationship between C_{pk} and dpm is shown in Table 1.3.

C_{pk}	dpm
.50	133,600
.67	44,400
.80	16,400
.90	6,900
1.00	2,700
1.33	66
1.67	<1
2.00	.0018
3.00	<<1
4.00	<1 defect per billion

This information was obtained from an R&M 2000 Variability Reduction Process slide prepared by HQ USAF/LE-RD.

Table 1.3 Relationship Between C_{pk} and dpm

Taguchi has developed a loss function which allows the researcher to associate a dollar value to the current state of process/product quality. This dollar amount can be used to identify key processes/products to be improved and to evaluate the amount of improvement. The quadratic loss function shown in Figure 1.13 will, in many cases, approximate the true loss due to deviations from the target. Actual losses result from a combination of scrap, rework, poor performance, lack of customer satisfaction, increased cycle time, tied-up capital, inventories, loss of market share, etc.

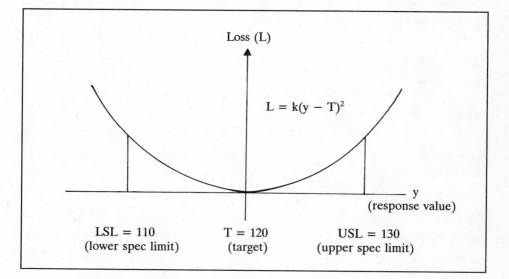

Figure 1.13 Taguchi Loss Function

The measured loss for a single product can be estimated using $L = k(y - T)^2$ where y is the response value, T is the target value, and k is the monetary constant. To determine k, we need to estimate the loss for any specific value of y. For example, if the estimated loss for y = 130 is $100.00, then $k = 100 \div (130 - 120)^2 = 1.0$. The loss function $L = 1.0(y - T)^2$ can now be used to estimate the loss associated with other y values.

Another use of the quadratic loss function is the determination of average loss per product. Using expected loss, E(L), to depict average loss, it is shown in Chapter 7 that $E(L) = k(\sigma_y^2 + (\bar{y} - T)^2)$.

The route to improved C_p, C_{pk}, dpm, or average loss is the same. Improvements will occur as we reduce the variance, σ_y^2, and reduce the deviation of the response average from the designated target, i.e., reduce $|\bar{y} - T|$. The use of designed experiments can assist the practitioner in determining which inputs (factors) shift the average of the response, shift the response variability, shift both the average and the variability, or have no effect on either the average or the variability of the response. This type of knowledge at the research, development or design phase allows for a technology transfer which is manufacturable. In the manufacturing phase, the knowledge of input factor influence on the response can be used for achieving continuous process improvement.

1.7 Organizing the Experiment

The reason for any designed experiment is to provide data that can be used to answer well thought out questions about the process. Failure to allocate sufficient time and thought in formulating these questions often results in wasting resources throughout the remainder of the experimental process. In many situations, up to 50% of the overall effort should go into the planning phase of the design.   It is important that all of the key players (management, engineers, experts in the area, operators, and analysts) be involved in the planning phase along with inputs from internal and/or external customers. The goals should be specific and statements of how the goals will be attained (objectives) must be complete.

One of the most effective approaches to the planning phase is the use of the "brainstorming" technique. This technique consists of an uninhibited method for creating ideas that pertain to the goals, objectives, and variables involved. A list of inputs (factors) and outputs (responses) must be obtained. In addition, how the inputs are measured and how many levels per input must be decided. Through the use of brainstorming and a team effort, there is a higher likelihood that the planning phase will be a success. If the planning phase fails, there is little chance of salvaging the experiment and further efforts could be a waste of resources. Barker's text, *Quality by Experimental Design* [1], presents the guidelines for brainstorming given in Figure 1.14. While the brainstorming process appears simple, there are a few additional concerns to ensure success. Brainstorming groups should be small, i.e., 4 to 10 people. Someone should record the ideas and everyone should

GUIDELINES FOR BRAINSTORMING

TEAM MAKEUP

Experts

"Semi" experts

Implementers

Analysts

Technical staff who will run the experiment

Operators

DISCUSSION RULES

Suspend judgement

Strive for quantity

Generate wild ideas

Build on the ideas of others

LEADER'S RULES FOR BRAINSTORMING

Be enthusiastic

Capture all the ideas

Make sure you have a good skills mix

Push for quantity

Strictly enforce the rules

Keep intensity high

Get participation from everybody

Figure 1.14 Guidelines for Brainstorming

participate. Ideas are not to be challenged until after the brainstorming session is over. When the brainstorming is complete the list can be critiqued and reduced to a workable number of ideas. For a more structured brainstorming session, you can use fishbone diagrams [4] or the nominal group technique (NGT) [9].

At the completion of the brainstorming session, a form similar to the one shown in Figures 1.15 and 1.16 should be initiated and continued throughout the remainder of the experiment. The form is an expanded version from that of Barker [1]. The actual form you use may be a modification of this one; however, it is imperative that a similar document be used to facilitate an organized flow of events and provide a common basis for documentation and communication.

It is important that any experimental effort be aimed at design or production problems where there is the most need. For example, assume a product consists of components A, B, C, D, and E. Failures of this product are the "company's" number one concern. Management directs the use of experimental design to solve the problem. You could attempt to design an experiment with parameters from all components, or you could use historical data presented in a Pareto diagram as shown in Figure 1.17. Thus, historical data can sometimes be used to identify critical problem areas. Figure 1.17 reveals that component D contributes to 60% of the defects, component C contributes to 20% of the defects, etc. Since component D contributes the most to product defects, the best approach would be to build an experimental design to (1) determine the cause of defects within component D, and (2) improve the quality of component D, thereby improving the product quality. After completing the work on product D, the product quality is again monitored and defects categorized as to component contributions. If necessary, further experimentation can be conducted. This process of continuous improvement will lead to a high quality item. Other Pareto diagrams on estimated cost of correcting components or estimated return from an improvement effort might also prove valuable. In summary, you want to attack that aspect of your product where the most gains can be made.

STEPS FOR EXPERIMENTAL DESIGN

I.   STATEMENT OF THE PROBLEM: _____
_____
_____

(During this step you should estimate your current level of quality by way of C_{pk}, dpm, or total loss.   This estimate will then be compared with improvements found after Step XII.)

II.   OBJECTIVE OF THE EXPERIMENT: _____
_____
_____

III.   START DATE: _____   **END DATE:** _____

IV.   SELECT QUALITY CHARACTERISTICS (also known as responses, dependent variables, or output variables). These characteristics should be related to customer needs and expectations.

RESPONSE	TYPE	ANTICIPATED RANGE	How will you measure the response? Is the measurement method accurate and precise?
1.			
2.			
3.			

V.   SELECT FACTORS (also known as parameters or input variables) which are anticipated to have an effect on the response.

FACTOR	TYPE	CONTROLLABLE OR NOISE	RANGE OF INTEREST	LEVELS	ANTICIPATED INTERACTIONS WITH
1.					
2.					
3.					
4.					
5.					

Figure 1.15 Part I of Steps for Experimental Design

STEPS FOR EXPERIMENTAL DESIGN (CONT)

VI.   DETERMINE THE NUMBER OF RESOURCES TO BE USED IN THE EXPERIMENT. (Consider the desired number, the cost per resource, time per experimental trial, and the maximum allowable number of resources.)

VII.   WHICH DESIGN TYPES AND ANALYSIS STRATEGIES ARE APPROPRIATE? (Discuss advantages and disadvantages of each.)

VIII.   SELECT THE BEST DESIGN TYPE AND ANALYSIS STRATEGY TO SUIT YOUR NEEDS.

IX.   CAN ALL THE RUNS BE RANDOMIZED? _____
WHICH FACTORS ARE MOST DIFFICULT TO RANDOMIZE? _____

X.   CONDUCT THE EXPERIMENT AND RECORD THE DATA. (Monitor both of these events for accuracy.)

XI.   ANALYZE THE DATA, DRAW CONCLUSIONS, MAKE PREDICTIONS, AND DO CONFIRMATORY TESTS.

XII.   ASSESS RESULTS AND MAKE DECISIONS. (Evaluate your new state of quality and compare with the quality level prior to the improvement effort. If necessary, conduct more experimentation.)

Figure 1.16 Part II of Steps for Experimental Design

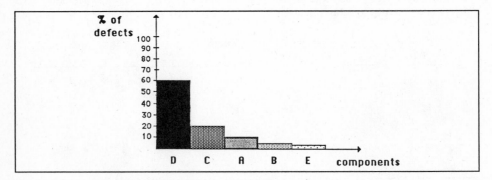

Figure 1.17 Pareto Diagram of the Contribution of each
Component to the % of Defectives

1.8 Selecting the Quality Characteristic(s)

Selecting the appropriate quality characteristic(s), also referred to as the response(s), is critical to good experimentation. For example, continuous responses will typically provide more information and require fewer resources than non-continuous or categorical type responses. If the response comes from a set of ordered categories (e.g., poor, fair, good, very good, excellent), you can assign the categories ordered integers (e.g., 1, 2, 3, 4, 5) and conduct the analysis in the same manner described in Chapters 2, 4, and 5. However, if binary responses (e.g. defective/non-defective) are used, you will need large numbers of replications of each experimental condition in order to avoid ambiguous results. The authors have found that, on occasion, continuous variables related to the binary response can be used in the analysis, resulting in large savings of experimental resources. For example, the vibration of a device during processing can be highly related to whether the device will be defective or non-defective. Therefore, finding the optimal input values which minimize vibration will also minimize defects, but with fewer experimental resources. Some interesting guidelines [7] in the selection of the proper response are as follows:

(1) The quality characteristic should be related as closely as possible to the basic engineering mechanism of the technology.

(2) Use continuous responses if possible.

(3) Use quality characteristics which can be measured precisely, accurately, and with stability. (The authors have found this can be a major stumbling block in effective experimentation with leading-edge technologies. Further brainstorming may be needed to resolve this problem.)

(4) Ensure the quality characteristics cover the important dimensions of the ideal function of the technology with respect to customer needs and expectations.

(5) Exercise a liberal blending of technical engineering knowledge, information on customer requirements, statistical prowess, and common sense.

(6) For complex systems, select responses at the subsystem level and run experiments at this level before attempting overall system optimization.

1.9 Conducting the Experiment

The importance of exercising extreme discipline in the actual conduct of the experiment cannot be overstressed. During the brainstorming process, it is recommended that a wide open, creative environment be encouraged in which the free flow of wild ideas is welcomed. Conducting the experiment, however, requires a very disciplined approach. The authors find that groups who are not accustomed to functioning in a disciplined environment will lower the likelihood of success. Factor levels must be set as close as possible to the specific design settings. Samples must be prepared in exactly the same way each time. Data sheets need to be clearly prepared in advance. All parties involved in the conduct and analysis of the experiment must be well versed in the intimate details of how the experiment is conducted and not accept any deviations from the plan. You, as the champion of experimental design within your area, must communicate and enforce the need for discipline in the conduct of the experiment without alienating the other key team members. All too often, a lack of communication, leadership, or control can doom an otherwise brilliantly conceived experimental scheme.

1.10 Causality versus Correlation

When testing the effects of input factors on the output, it is desired to be able to ascertain causality, not just correlational relationships. Just as pointed out in Section 1.4, a high correlation does not necessarily relate to the cause of the problem. To ascertain the *causality* of an input factor, it is important to ensure that changes in the output occurred because of changes in the input and not because of some other related factor.   To accomplish this task, experiments must be designed such that input factors are uncorrelated and all *nuisance variables* (variables not controlled) have their effects averaged out over all experimental conditions. This procedure is referred to as *randomization* and consists of running experimental conditions in a random sequence. The advantage of randomization is that it averages out the effects of all nuisance variables. This is especially important when the brainstorming process fails to identify all factors or some factors cannot be measured or controlled in the experiment.

As an example demonstrating the need for randomization, consider the following: A new product is being tested using two different bonding agents. Assume that there are three batches of raw material being used to make these products and the batch factor was uncontrolled in the design. After testing bond strength, the data appear as shown in Table 1.4.

Measured Bond Strength for Bonding Agents A and B				
	A	Batch #	B	Batch #
	2.4	3	2.7	3
	2.5	3	3.3	1
	2.1	3	3.1	1
	3.1	1	3.7	2
	2.7	3	3.5	2
	2.9	3	3.6	1
	3.0	2	2.8	2
Mean ($\bar{y}$)	2.670		3.243	
STD Dev ($\hat{\sigma}$)	.359		.391	

Table 1.4 Example Demonstrating the Need for Randomization

Initially, batch information was not considered and thus it appears that bonding agent B provides a higher mean bond strength. Therefore, it is concluded that product B is the better bonding agent. However, consider the consequences of batch #3 being defective. Since more observations using bonding agent A were conducted with batch #3 than any other, the results are now suspect. Therefore, we cannot state unequivocally that bonding agent differences caused substantial differences in bond strength because the batch effect is confounded with bonding agents. To avoid this problem, we could have randomly assigned materials to bonding agents in an attempt to average out the effects of the defective batch over the two treatments. It is also possible to control the different batches beyond the randomization process by including them as an input factor. More than likely, the best design in this case will be one that provides information on which bonding agent will maximize strength and simultaneously provide insensitivity (robustness) for batch-to-batch variations. Robust designs are discussed further in Chapters 5 and 6.

1.11   A Quality Improvement Example

Suppose you are responsible for vehicle maintenance in a large organization. As an initial attempt to implement continuous process improvement (CPI), you would probably form a team to assist you in defining your mission and developing a process flow diagram for vehicle repair. The diagram, shown in Figure 1.18, can be used to improve quality by identifying non-value added steps to be modified or eliminated. An obvious CPI objective would be to ensure that "the right repairs are done right the first time." Just imagine how many steps could be eliminated from the process flow diagram if you could educate, train, and motivate your people and suppliers to perform in this manner.

After reviewing the process flow diagram, assume you decide to implement a gas mileage improvement strategy. To benchmark your current state of vehicle quality, you would probably direct that gas mileage data be recorded on each vehicle. The best way to display this data is shown in Figure 1.19.

As you can see, gas mileage does vary from week to week. Your concern is whether this variability is natural or due to specific causes. The use of control charts can assist you in answering this question.

To present a simple overview of control charts, we will assume that the 20 weeks of data in Figure 1.19 are from a vehicle whose performance is currently known to be stable. Computing the average gas mileage over the 20 weeks will give us a feel for the vehicle's expected gas mileage. This average is found as shown below.

$$\bar{y} = \frac{\sum_{1}^{20} y_i}{20} = \frac{[21+19+20+...+22]}{20} = \frac{400}{20} = 20$$

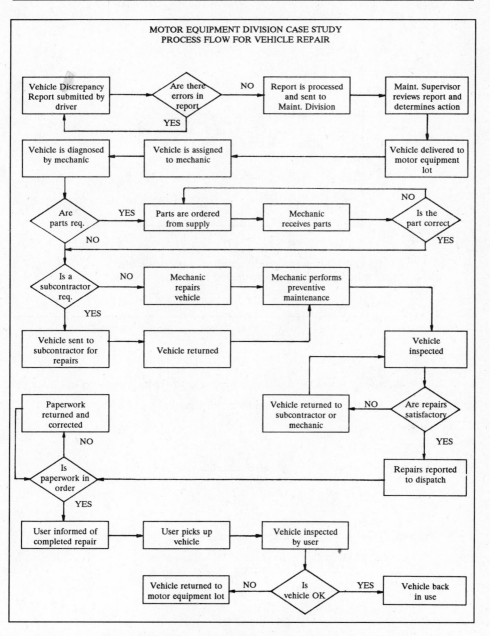

MOTOR EQUIPMENT DIVISION CASE STUDY
PROCESS FLOW FOR VEHICLE REPAIR

Figure 1.18 Process Flow Diagram for Vehicle Repair

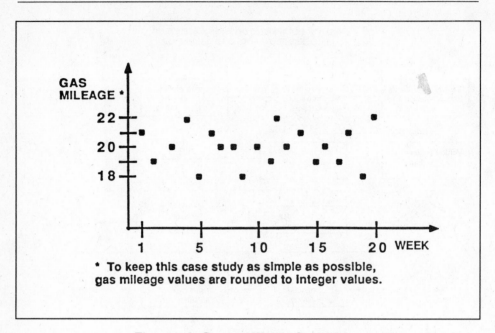

Figure 1.19 Graph of Weekly Gas Mileage

To obtain a measure of the variability in vehicle gas mileage, we can calculate the sample variance, $\hat{\sigma}^2$, of the 20 mileage values. The equation for $\hat{\sigma}^2$ is:

$$\hat{\sigma}^2 = \frac{\sum (y-\bar{y})^2}{n-1}$$

which can be described as the sum of the squared deviations from the average, divided by $(n-1)$. The deviations from the average and the variance calculations are displayed in Figure 1.20.

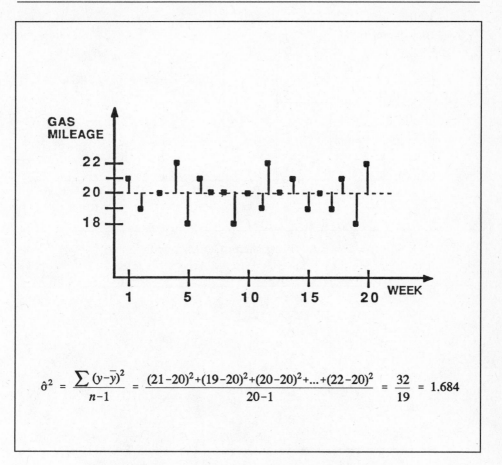

Figure 1.20 Graph of Deviations from the Mean Gas Mileage

A common technique which is a simple, yet powerful, way to get a visual feel for the way data is distributed is to draw a histogram. A histogram is a graph which displays how frequently a given outcome occurs. The histogram for our data is shown in Figure 1.21. If a smooth curved line were used to outline the histogram, it would appear to be symmetric and bell shaped, much the same as a normal curve. Assuming, then, that our gas mileage values are approximately normally distributed allows us to make use of the empirical rule shown in Figure 1.22.

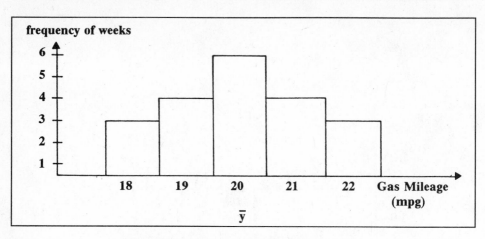

Figure 1.21 Histogram for Gas Mileage Data

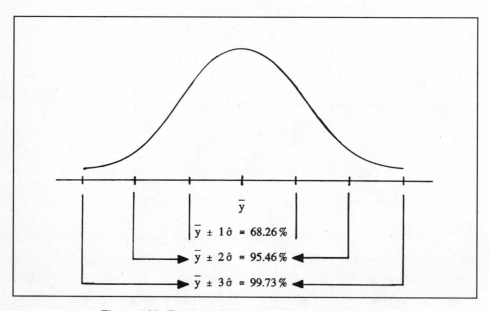

Figure 1.22 Empirical Rule of the Normal Distribution

The normal distribution of gas mileage values and the randomness of the data in Figure 1.19 supports our assumption that the vehicle's performance is stable. We can now use the empirical rule to assist us in developing control charts. For example, $\bar{y} \pm 3\hat{\sigma}$ (where $\hat{\sigma}$ is the standard deviation, $\sqrt{\hat{\sigma}^2}$) results in a lower control limit (LCL) of 16.1 and an upper control limit (UCL) of 23.9 (see Figure 1.23). The empirical rule states that approximately 99.73% of all future gas mileage values for this vehicle should fall within these limits. Therefore, we can use these limits to evaluate the future performance of this specific vehicle. For instance, if week 21 produced a gas mileage of 23, chances are that nothing specific has caused this value; however, a value of 16 is outside the control limits (i.e., outside the natural variability) and indicates that there is a 99.73% chance this value was caused by some specific change in the vehicle. Other control chart rules exist for determining out of control conditions due to trends or increases in variability [10],[12].

Once an out of control condition is determined, the obvious next step is to determine its cause. To obtain this information, a brainstorming session with the appropriate members should be conducted. The results of such a session might resemble the cause and effect diagram displayed in Figure 1.24. After completing the brainstorming, the group should try to reduce the information to those inputs most likely to affect the gas mileage. You now have a set of input factors to the process which can be investigated each time the mileage control chart detects a change.

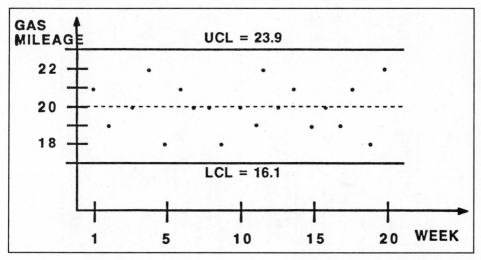

Figure 1.23 Control Chart Based on Gas Mileage Data

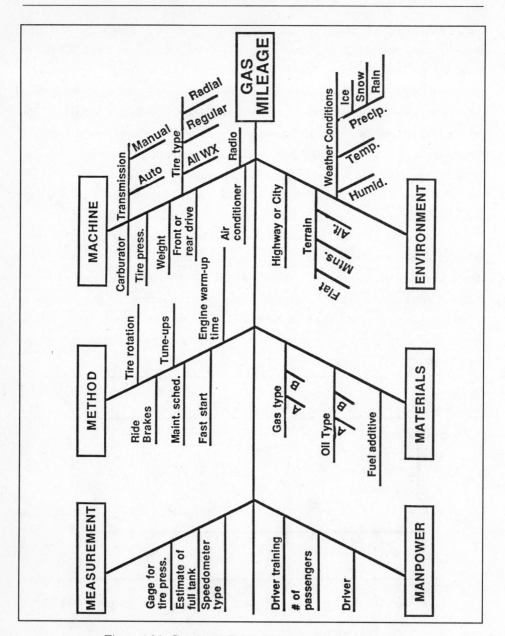

Figure 1.24 Cause and Effect Diagram for Gas Mileage

Assume your investigation revealed an extremely low tire pressure which you determine to be the cause of the out of control mileage value. You would obviously correct the low tire pressure problem and possibly begin to chart the tire pressures on a weekly basis. In this way, you are now controlling a critical input to the process instead of waiting for a substantial change in the process output. Thus, you have now entered a problem prevention mode versus detecting poor quality through inspection of output values.

Control charting is an excellent process monitoring tool; however, engineers are not always sure which inputs are the important ones (vital few) and which ones are unimportant (trivial many). Furthermore, using control charting to gain an in-depth understanding of your process can be a long drawn out procedure. What is needed is a faster method for achieving "profound knowledge" of the gas mileage process. The statistical tool which provides this type of in-depth knowledge of a process in a timely manner is experimental design. The basic idea is to characterize the process by determining which inputs have a critical impact on the response. The design of experiments approach can be used effectively after control charts have been implemented; however, it is advantageous to characterize the process before production begins. This strategy allows for quality to be designed into the product from the beginning of the design engineering phase. Before the actual experiment is conducted, you should review the cause-and-effect diagram to remove unnecessary sources of variation. For example, it would be wise to develop training and/or a standard operating procedure to prevent operators from riding the brakes and making fast starts.

Assuming the brainstorming resulted in four input factors to be used in the experiment, the remaining of this section will provide overview of a designed experiment. To investigate all possible combinations of four factors at two levels would require the 16 tests (runs) shown in Table 1.5. The ($-$) values are shorthand for -1 and indicate a factor set at its low value. The ($+$) values are shorthand for $+1$ and represent a high factor setting. Notice that the design columns are all balanced vertically, i.e., there are an equal number of ($+$) and ($-$) values in each column. That is, the sum of all the numbers in each column is zero. Furthermore, these columns are also balanced horizontally for each level of each column. For example, when A is at a ($-$), the remaining columns have a balanced number of ($+$) and ($-$) values, the same balancing occurs for A at ($+$) and all other columns evaluated at their ($+$) or ($-$) settings. These balancing properties result in design orthogonality which allows us to estimate the effects of each factor independent of the others.

| RUN | FACTORS | | | |
	A	B	C	D
1	−	−	−	−
2	−	−	−	+
3	−	−	+	−
4	−	−	+	+
5	−	+	−	−
6	−	+	−	+
7	−	+	+	−
8	−	+	+	+
9	+	−	−	−
10	+	−	−	+
11	+	−	+	−
12	+	−	+	+
13	+	+	−	−
14	+	+	−	+
15	+	+	+	−
16	+	+	+	+

Table 1.5 Full Factorial Low (−) and High (+) Settings
for a Four Factor Experiment

The design in Table 1.5 is referred to as a full factorial design. It is capable of estimating the effect of all possible factor combinations on the response. In this case, there are 15 possible "effects." There are 4 "linear" or "main" effects (A, B, C, D); 6 "two-way" interactions (AB, AC, AD, BC, BD, CD); 4 "three-way" interactions (ABC, ABD, ACD, BCD); and only 1 "four-way" interaction (ABCD). Experimenters infrequently find interactions beyond two ways to be significant, and thus time and resources can be conserved by the use of a fraction of the full factorial design. For the gas mileage experiment, it was decided to use a fractional factorial design with eight runs or rows. (Fractional factorial designs are discussed in depth in Chapter 3). The resulting design matrix is shown in Table 1.6. Notice that this eight-run design also has the balancing properties required for orthogonality. The column for factor D was formed by using the elements of the ABC interaction column, that is, multiplying the corresponding row values in columns A, B, and C together to get the value in column D. The result is that the settings for factor D are "aliased" (or identical) with the settings for the ABC interaction. This, in turn, will produce other "aliasings" as shown in Table 1.7. Terms that are aliased have the same (+) and (−) sequence in their columns. The result of aliasing is that any two terms cannot be evaluated separately. This loss of information is a result of saving resources by conducting fewer tests than the number required for a full factorial. The experimenter should assess the amount of aliasing in any design which is less than the full factorial. As long as the amount of aliasing is acceptable he/she should continue; else chose a different design matrix. More will be said regarding aliasing once we have analyzed the data.

	FACTORS			
RUN	**A**	**B**	**C**	**D**
1	−	−	−	−
2	−	−	+	+
3	−	+	−	+
4	−	+	+	−
5	+	−	−	+
6	+	−	+	−
7	+	+	−	−
8	+	+	+	+

Table 1.6   Half Fraction Design Settings for Four Factors,
Each at 2 Levels (Low and High)

ALIAS PATTERN	
effect	**alias**
A	BCD
B	ACD
C	ABD
D	ABC
AB	CD
AC	BD
AD	BC
BC	AD
BD	AC
CD	AB
ABC	D
ABD	C
ACD	B
BCD	A

Table 1.7   Alias Pattern for the Half Fraction Design in Table 1.6

· The four factors which were brainstormed to be used in the experiment and the low (−) and high (+) levels of each are shown in Table 1.8.

FACTOR	LEVELS
A: TIRE PRESSURE	28, 35
B: TIMING SETTING	LOW, HIGH
C: TYPE OF OIL	1, 2
D: TYPE OF GAS	1, 2

Table 1.8 Factors for the Gas Mileage Case Study

The completed experimental matrix with the response values (taken over three weeks for each run) are displayed in Table 1.9.

RUN	A	B	AB	C	AC	BC	D	y_1	y_2	y_3	$\bar{Y}$	s*
1	−	−	+	−	+	+	−	22.27	21.12	21.37	21.59	.60
2	−	−	+	+	−	−	+	14.22	15.40	10.46	13.36	2.58
3	−	+	−	−	+	−	+	22.49	23.15	22.08	22.57	.54
4	−	+	−	+	−	+	−	9.96	13.80	11.92	11.89	1.92
5	+	−	−	−	−	+	+	17.35	18.60	17.97	17.98	.62
6	+	−	−	+	+	−	−	27.08	24.54	24.57	25.40	1.46
7	+	+	+	−	−	−	−	18.36	17.63	17.04	17.68	.66
8	+	+	+	+	+	+	+	22.78	26.97	27.14	25.63	2.47

Table 1.9 Complete Experimental Matrix with Response Values
for Gas Mileage Case Study

A graphical analysis for each effect in the design matrix is obtained by plotting the average response at the (−) and (+) settings and then connecting these two dots (see Figure 1.25). The * in Figure 1.25 was obtained by averaging the response values when factor A is set at its high (+), that is, the average response for runs 5, 6, 7, and 8. The other dots are found in a similar fashion. The steepness of the slope reflects the importance of each effect. From the graphical results in Figure 1.25, it is apparent that the most important

---

* s represents the response standard deviation for each run, the symbols s and $\hat{\sigma}$ are used interchangeably throughout the book.

effects are factor A and the AC or BD interaction. Since AC is aliased with BD (see Table 1.7), we must go back to the factors to try and determine which interaction is most likely responsible for the steep slope in Figure 1.25. Since it is highly unlikely to have a "tire pressure × type of oil" interaction (AC), it is decided that the steep slope in Figure 1.25 is due to the BD, or "timing × type of gas" interaction. The BD interaction graph is shown in Figure 1.26 where the ★ was determined from the response average of the runs with B at (−) and D at (−).

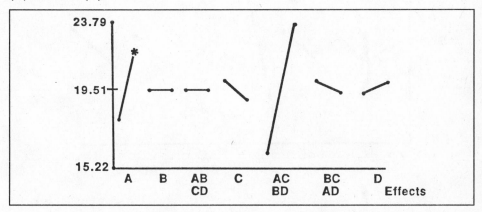

Figure 1.25 Marginal Mean Graphs for Gas Mileage Case Study

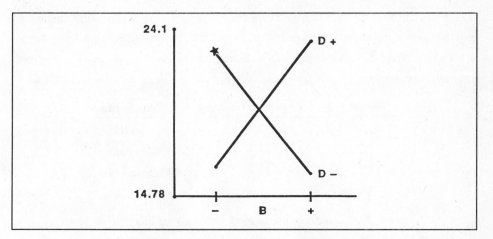

Figure 1.26 Interaction Graph for Factors B and D

Another graphical analysis can be used to determine if any of the factors shift the variability in the gas mileage. Figure 1.27 indicates that factor C has a big effect on variability.

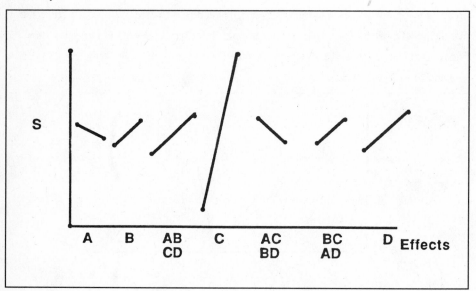

Figure 1.27 Marginal Average Graphs for Standard Deviation

To determine the best factor settings for maximizing gas mileage while minimizing variability, we must return to Figures 1.25, 1.26, and 1.27. Based on Figure 1.25, set factor A at (+). Figure 1.26 indicates that factors B and D should both be set at (+). Factor C is set at (−) to minimize variability. Thus, the final factor setting decision is as follows:

Factor	(−) or (+) Setting	True Setting
A	+	35 psi
B	+	high timing
C	−	type 1 oil
D	+	type 2 gas

At this point, one should feel pretty good about solving a quality problem and thus satisfying the customer. Be careful, though, because we made the assumption that the customer wanted high gas mileage. When a survey was conducted in Madison, Wisconsin [8] as to what the customer (the driver and/or passenger) thought was important, the primary response was not mileage, but safety. The point is that designing experiments to solve problems which the customer may not think important will not always be a productive strategy. One should first survey the customer through the use of a procedure such as Quality Function Deployment (QFD). It is not enough to simply determine customer needs and expectations. To satisfy the customer, processes should be designed to meet customer needs and expectations so that customers will purchase the process outputs (products or services). To accomplish this task, crossfunctional teams from marketing, purchasing, engineering, and others must work closely together from the time the need is first conceived until the need is met. QFD provides a framework within which the crossfunctional teams can work.

To accomplish the QFD procedure you must answer the following questions: (Total Quality Management: A guide for Implementation, DoD 5000.51.6):

(1) What do customers really expect?

(2) Are all the customer expectations equally important? If not, rank order them.

(3) Will meeting customer expectations enable you to obtain a competitive advantage?

(4) What are the engineering characteristics that are related to customer expectations?

(5) How does each engineering characteristic affect each customer expectation?

(6) How does each engineering change affect other characteristics?

For more information on QFD, see *Basic Statistics: Tools for Continuous Improvement* [10].

After assessing the customer needs and translating these needs into product/process design, then the designed experiments approach provides the type of profound knowledge we should really be striving for.

1.12 Taguchi Approach to Designed Experiments

The objective of parameter (Robust) design is to determine factor settings that achieve desired response targets (min., max., or specific value) while minimizing the variability due to uncontrolled or noise factors. To illustrate this concept in the gas mileage example, consider weather conditions which are outside the vehicle operator's control. It might be desirable to find the best settings of factors A, B, C, and D that maximize gas mileage while also making mileage insensitive (as much as possible) to weather conditions. In this case, we would make specific replicates due to various weather conditions. In other words, the first replicate (first recorded column for the response) would all be obtained under rainy conditions (w_1), the second replicate obtained during snowy conditions (w_2), and the third replicate obtained under dry conditions (w_3). Thus, the test matrix would appear as shown in Table 9.9. The only difference between this strategy and that of previous sections is the specific condition of the replicated response. By minimizing the response variability over the noise factor levels we have found a condition for the process that makes it insensitive—or *robust*—to environmental conditions. Robustness can also be sought for process parameters which are difficult or expensive to control such as material types, different lots of material, other environmental conditions, etc. (see Chapters 5 and 6).

| | Factors & Interactions | | | | | | | Replicated Responses | | | | |
Run	A	B	AB	C	AC	BC	D	w_1	w_2	w_3	$\bar{w}$	s
1	−1	−1	1	−1	1	1	−1					
2	−1	−1	1	1	−1	−1	1					
3	−1	1	−1	−1	1	−1	1					
4	−1	1	−1	1	−1	1	−1					
5	1	−1	−1	−1	−1	1	1					
6	1	−1	−1	1	1	−1	−1					
7	1	1	1	−1	−1	−1	−1					
8	1	1	1	1	1	1	1					

Table 9.9 Replicated Responses Based on Weather Conditions

1.13 Steps for Quality Improvement: Getting Started

Across America, purchasers of products and services are demanding that their suppliers provide (1) **better** quality, (2) **faster** response times, and (3) **cheaper** prices. To meet these demands, suppliers must be able to use a host of statistical and non-statistical tools. Other texts that address some of the tools that are not included in this book are *Total*

Quality — A Textbook of Strategic Quality Leadership & Planning (1991) by Lam, K.D., Watson, F. and Schmidt, S. R. [9] and *Basic Statistics: Tools for Continuous Improvement*, 2nd Edition (1991) by Kiemele, M. J. and Schmidt, S. R. [10]. (For more information on these texts, contact Air Academy Press at (719) 531-0777.) To improve quality, it is important to be familiar with all available tools; however, equally important is knowing when to use them. The following steps are suggested as an example of how to get started in the quality improvement cycle. We highly recommend that you use them as an outline for your company and that you make modifications as necessary.

QUALITY IMPROVEMENT STEPS

WHAT	WHO	HOW
1. Commitment	Everyone, from the top to the on-line worker.	Cultural Change Strategic Planning Awareness Training
2. Define products/ processes.	Crossfunctional Teams	Process Flow Diagrams Brainstorming
3. Mission Statement	Crossfunctional Teams	Brainstorming Nominal Group Technique(NGT)
4. Identify customers and suppliers (Internal/ External).	Crossfunctional Teams	Brainstorming NGT
5. Identify customer expectations and what you expect from your suppliers.	Crossfunctional Teams including Customer and Supplier Representatives	Surveys Brainstorming
6. Translate customer expectations into key quality characteristics.	Crossfunctional Teams and the Customer	QFD Brainstorming NGT
7. Determine current state of quality for all products/processes.	Process Action Teams (PATs) and the Customer	Benchmarking Loss Function Process Capability (C_{pk}) Defective Parts per Million (dpm) Surveys

8. Identify the best opportunity for improvement.	PATs and the Customer	Loss Function Cost Analysis Pareto Charts
9. For the best opportunity, identify causes of degraded quality.	PATs	Brainstorming Cause and Effect Diagrams Historical Data
10. Characterize product/ process, i.e., find relationship of outputs (effects) and inputs (inputs).	PATs	Designed Experiments
11. Find optimal settings for input factors.	PATs	Designed Experiment Analysis Empirical Models
12. Determine input tolerances.	PATs	Use empirical models in (11) as simulators for Designed Experiments using available tolerances.
13. Set up control charts on critical inputs and outputs.	PATs	Statistical Process Control (SPC)

14. Recycle the steps, implementing the philosophy of continuous process improvement (CPI).

1.14 U.S. Air Force Approach to Total Quality Management (TQM)

Total Quality Management is a phrase that most industry people are becoming quite familiar with. In order to compete in the 21st century, a company must embrace TQM as a new way of life. Numerous TQM success stories have been documented for all types of industry [10]. Several large companies such as Motorola, Boeing, IBM, Xerox, Ford Motor Company, and General Motors have instituted strong TQM type improvement strategies within their companies and for their suppliers. The U.S. Air Force has also implemented many aspects of TQM as shown in the following article on R&M 2000.

R&M 2000 VARIABILITY REDUCTION PROCESS

BRUCE A. JOHNSON, CAPTAIN, USAF

Office of the Special Assistant for Reliability and Maintainability
Headquarters, United States Air Force

THE PROCESS

Improving combat capability is a major Air Force objective. This is becoming increasingly difficult in the face of constrained manpower and fiscal resources. However, there is a solution. Substantial increases in combat capability are achievable through more reliable and maintainable weapon systems. Such systems are able to complete more missions with less spares, support equipment, facilities and maintenance personnel.

Weapon systems fail for many reasons. Some components, like tires, wear out. But most systems fail because of poor design, the use of defective parts and materials, or poor workmanship. The cause of these failures is *variability* in the design and manufacturing processes. The problem variability presents is that it exists in nearly all processes and it results in marginal or non-conforming products. The variability comes from the fact that conditions under which these items are produced change. Variability reflects the differences in raw material, machines, their operators and the manufacturing conditions. When process variation increases, the product's physical properties or functional performance can degrade, and the number of product defects increases. The significance variation has on a product's reliability and quality depends on the criticality of the manufacturing process and part characteristics.

There are two ways to reduce variability. Traditionally, the approach has been to tighten design tolerances and increase inspections. Costs climb as scrap and rework increase, and productivity drops. Inspections and tighter tolerances only treat the symptoms and do not resolve the actual problem.

The preferred method is to reduce the variability by improving the process. This can be done by eliminating the causes of variation through statistical techniques, and by developing more robust products which are insensitive to the causes of variation. The methods of reducing variability is aptly named the Variability Reduction Process (VRP).

VRP is a proven set of practices and technologies which yield more reliable and nearly defect-free products at lower cost. It is a structured, disciplined design and manufacturing approach aimed at meeting customer expectations and improving the development and manufacturing process while minimizing acquisition time and cost (Figure 1).

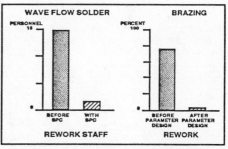

Figure 1. Process Improvements from VRP.

The objectives of VRP are to design robust products which are insensitive to the causes of failure; to achieve capable manufacturing processes that produce nearly defect-free products; and to adopt the managerial attitude of continuously improving all processes. The basic tools are teamwork, statistical process control (SPC), loss function, design of experiment (DOE), parameter design and quality function deployment (QFD). VRP must span all of engineering, manufacturing and management, and include the suppliers (Figure 2).

PURPOSE: MEET CUSTOMER EXPECTATIONS, IN MINIMUM TIME, AT LOWEST COST		
OBJECTIVES	ROBUST DESIGNS	CAPABLE PROCESSES	CONTINUOUS IMPROVEMENT
PRIMARY RESPONSIBILITY	ENGINEERS	MANUFACTURING	MANAGEMENT
TEAMS	INTERDISCIPLINARY	IMPROVEMENT	MULTI-FUNCTIONAL
TOOLS / TECHNOLOGIES	QFD / DOE / PARAMETER DESIGN	DOE / SPC	LOSS FUNCTION

Figure 2. The Elements of VRP.

CAPABLE MANUFACTURING PROCESSES

Capable manufacturing processes can only be achieved when the critical parameters are known, and the causes of variation are eliminated or minimized. For most processes, SPC is highly effective (Figure 3). It allows the operator to observe the process and distinguish between patterns of random and abnormal variation. It assists the operator in making timely decisions such as adjusting or shutting down the process before defects are produced. When combined with other statistical tools and problem solving techniques (Figure 4), the worker can isolate and remove the causes of abnormal variations.

When the abnormal variations are removed from the process, the process is said to be under statistical control. In many processes, this will not be sufficient. The random variations alone can result in

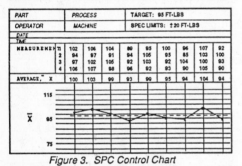

PART		PROCESS		TARGET: 95 FT·LBS				
OPERATOR		MACHINE		SPEC LIMITS: ±20 FT·LBS				
DATE								
TIME								
MEASUREMENT 1	102	106	104	89	95	100	96	107	92
2	94	97	91	94	105	95	85	103	100
3	97	102	105	92	103	92	104	100	93
4	106	107	98	96	92	93	90	105	90
AVERAGE, $\bar{X}$	100	103	99	93	99	95	94	104	94

Figure 3. SPC Control Chart

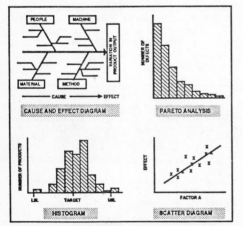

Figure 4. Tools Workers Can Use to Identify the Causes of Variation.

defective products, and their causes should be identified and removed until the process is capable of producing near defect-free products. However, causes of random variation are more difficult to identify, are usually systemic and normally require management action to remove.

The difference between VRP and traditional methods of quality control is that improvements in quality are achieved through improvements in the manufacturing processes. No longer is better quality to be achieved through tightening specifications and more inspections. In the case of SPC, the manufacturing processes are improved by eliminating the causes of variability. Usually, the process can be centered around the design target and variation reduced well within the specifications (Figure 5).

When implemented correctly, the results can be impressive. For Parlex Nevada Inc, a circuit card manufacturer, SPC was used to cut scrap cost by 90 percent in one year, and changed the company's losses into profits. Boeing used SPC to resolve a rivet flushness problem on the nose section of the 737 aircraft. The improvements saved a half-million dollars a year.

A more powerful method of resolving difficult or complex industrial problems is the statistical design of experiments (DOE). DOE methods have been around for 60 years and have been extensively used by the agricultural, pharmaceutical and chemical industries to advance their products. These techniques can greatly accelerate the rate of improving product designs and manufacturing processes. Such statistical experiments will aid the engineer in identifying the critical parameters for SPC, isolating the causes of variation, and improving the product's technical or operational characteristics.

DOE works by measuring the effects that different inputs have on a process. This is done by identifying a prospective set of input factors, varying

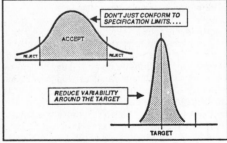

Figure 5. Design and Build to Target Values, Not Specification Limits.

the inputs over a series of experiments, collecting the data and analyzing the results. An input may be varied over a range of values such as can be done with an oven's curing temperature or conditionally, such as the decision to add or withhold a curing additive. The methods, whether they employ a full-factor, fractional-factor, orthogonal array or surface response technique, use a statistical approach that ensures accuracy and validity.

Well-planned experiments can have dramatic results. For example, a government-owned-company-operated (GOCO) munitions plant had a serious problem in producing the ADAM mine. Although SPC was in use and 12 of 13 processes were within their tolerances, 19 out of 25 lots were rejected. Aerojet Ordinance, the plant operator, decided to apply a Taguchi experiment to identify the critical parameters. They selected the 13 parameters used in the SPC program and tested parameters at three different levels. Only 27 experiments were conducted, firing 6 rounds each. The results were profound. Four parameters were found to be critical, and when set at their best levels, the process produced good lots without any rejects. The other nine parameters were less important and their tolerances can be relaxed. Results: production schedule met while achieving significant cost savings.

ROBUST DESIGNS

Having a capable manufacturing process is not enough. It may not be economical to remove or control some of the causes of variation. Therefore, it is necessary to develop robust manufacturing processes which are insensitive to the manufacturing conditions, materials, machines and operators. In most cases, *the greatest improvements come from robust designs.* These improvements are achieved through parameter design, a technique of selecting the optimum conditions (i.e. determining the ideal parameter settings) that minimize the variability without removing the causes of variation.

During parameter design, a set of parameters is identified to enhance the product, and a series of experiments is conducted to observe the effects of the parameters on the desired part characteristics. The results

could identify new parameter settings that improve the product and increase yield. For example, the problem an engineer may want to solve is the variability of ceramic parts. The source of variation is the uneven temperatures in the kiln. Because modifying the kiln is too expensive, the engineer conducts several experiments to identify a way to minimize the effects of the uneven temperature. For the experiment, he selects as parameters the amounts of the ingredients, their textures, blending procedures and firing temperatures. For validity, the engineer should use DOE procedures for conducting the experiment. An orthogonal array may be used to minimize experimental time and cost. The results will enable the engineer to fine tune new parameter settings that minimize the effects of uneven firing temperatures.

The problem with most design approaches is that parameter design is rarely done. Most engineers focus on the system design to develop the product, and immediately transition to tolerance design to establish the specification limits. Often, the results is a inferior product which is sensitive to variations in the manufacturing process. Parameter design should be done before tolerance design.

Parameter design can also be used to design and produce a more robust product that will perform better over a wider range of operating conditions and environments. It can be used to enhance a desired customer's need such as a smooth automobile ride, or to enhance an engineering requirement such as to lower the susceptibility of corrosion.

The success of parameter design and SPC hinge on the engineers' understanding of the customers' needs. Quality Function Deployment (QFD)

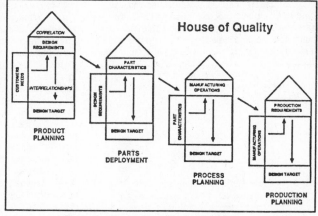

Figure 6. Quality Function Deployment.

is a systematic approach for developing and translating the customers' needs into the critical part characteristics and production requirements. The QFD requirements matrices are designed to minimize the chance of starting the design process with incomplete or erroneous requirements. They provide a methodology which assures an orderly translation of the customers' requirements throughout the product development process (Figure 6).

The basic approach used in QFD is conceptually similar to the practice employed by most companies. The difference lies in its structure. It compels the different disciplines and departments to communicate. QFD starts by defining the customers' requirements in the customers' terminology and translates these requirements into engineering requirements. These engineering requirements become the product characteristics which should be measurable and given target values. If properly executed, the product should fulfill the expectations of the customer.

The other matrices translate the engineering requirements into part characteristics, required manufacturing operations, and production requirements. Each matrix identifies the design targets, interrelationships and priorities. The end result should be a set of operating procedures which the factory can follow to consistently produce the critical part characteristics.

The design environment best suited to produce robust products is concurrent engineering (also referred to as simultaneous engineering). Concurrent engineering addresses all the customer, design and manufacturing issues up-front starting with concept exploration. The process employs good design practices, interdisciplinary teams and a structured requirements process, such as QFD, to concurrently develop the product and manufacturing processes. Its practice encourages communication between the design, product and production engineers. Concurrent engineering replaces the typical "sequential" approach to product design, which is more costly and time consuming. The effects concurrent engineering can have may be summed up by the following example. Using sequential engineering practices, the Allison Transmission Division estimated in 1982 it would cost $100 million in capital investment and $75,000 per unit to replace the transmission in the M-113 Armored Personnel Carrier. In 1987, using concurrent engineering, Allison's estimate dropped to $20 million for capital investment and $50,000 per unit (Figure 7).

CONTINUOUS IMPROVEMENT

For VRP to succeed, management from the top down must adopt new attitudes about reliability and quality, and must become *directly involved* in continuously improving the design and manufacturing processes. They must implement programs to foster improvement, teardown the barriers that inhibit change, instill teamwork, establish goals for improvement, and provide education and training for successful implementation.

Management's primary objective should be to satisfy the customer and serve the customer's needs if a company is to stay in business and make a profit. Reliability and quality must come first — not profit. If done smartly, reliability and quality will reduce cost and increase profit. For example, Hewlett-Packard's Yokogawa plant implemented many of the VRP techniques and, after eight years, they achieved 240 percent increase in profit, 120 percent increase in productivity, 19 percent increase in market share, 79 percent decrease in failure rate, and 42 percent decrease in manufacturing costs.

Management must become *process-oriented* and stimulate efforts to improve the way employees do the job. Teamwork is the foundation for continuous improvement. An important part of team building is the assignment of people to multifunctional management teams, interdisciplinary design teams and process improvement teams. Everyone should be involved in process improvement.

Management should implement programs to foster continuous improvement, use education to change attitudes and provide training. Change will be gradual and will require a long-term outlook. In Japan, most of the small improvements come from the workers' suggestion system called *Kaizen Teian*. Kawasaki Heavy Industries Aircraft Works has one of the more impressive programs. In 1987, each employee submitted an average of 229 suggestions

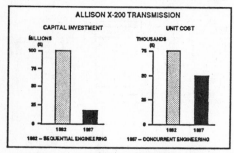

Figure 7. Concurrent Engineering.

and 92 percent were adopted. Savings were esti-mated at $35 million. At a Texas Instruments plant, they introduced an enhancement program, and over the past five years, the program has reduced defects by 2,300 percent (Figure 8).

Management must take responsibility for process improvement while giving the workers the responsibility of maintaining the process. Without the ability to maintain the existing process, there can be no improvement. This means management must give the worker ownership of his processes and allow the worker to improve or stop the process as necessary. In many progressive companies, work-ers are involved in the development of their own operating procedures, and in some cases, they write their procedures.

Management must change the accounting procedures. The notion that loss only occurs when the product is outside the specification limits is obso-lete. Loss includes not only the cost of scrap and rework, but also the cost of warranties, excess inven-tory and capital investment, customer dissatisfac-tion, and eventual loss of market share. The tradi-tional go/no-go approach to quality should be re-placed with a powerful monetary loss function to better account for loss (Figure 9). A quadratic loss function allows management to better assess the true cost of production processes and the benefits derived from process improvements (Figure 10). Most important, the loss function supports continu-

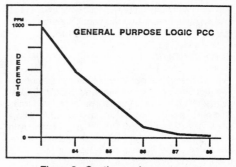

Figure 8. Continuous Improvement.

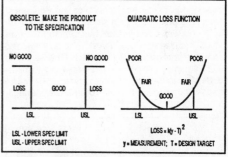

Figure 9. Quality Loss Functions.

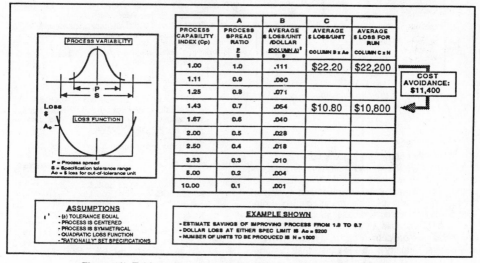

Figure 10. Table for Evaluating the Benefits from Process Improvement.

ous process improvement because minimizing the loss equates to reducing the variability around the target.

IMPLEMENTATION

The adoption of VRP begins with the conviction that change is necessary and beneficial. Implementation is an evolutionary process. VRP may start out in a single product line at one factory with a single group of suppliers. But it must be grown with the long-range goal that it will encompass the entire enterprise. The Air Force strategy for VRP is to encourage defense contractors and suppliers to (1) foster top-level commitment to VRP, (2) involve all levels, departments and vendors, (3) apply the VRP in a systematic approach, and (4) create a culture of continuous process improvement. Within the Air Force, the Vice Chief of Staff has directed all commands involved in weapon system acquisition and support to implement VRP by 1993. The Air Force acquisition regulations have been rewritten to incorporate VRP in the acquisition process. VRP will be an essential part of Air Force's Total Quality Management program (Figure 11).

	TQM TOOLS	R&M 2000 VRP	
		CAPABLE PROCESSES	ROBUST DESIGN
TEAM WORK		X	X
SPC		X	
LOSS FUNCTION		X	X
DESIGN OF EXPERIMENTS		X	X
PARAMETER DESIGN			X
HOUSE OF QUALITY (QFD)			X

(left side vertical labels: INCREASING COMPLEXITY)

Figure 11. TQM / VRP Relationhhip

SUMMARY

The Variability Reduction Process makes two seemingly contradictory goals compatible: to produce highly reliable and maintainable weapon systems while reducing development time and costs. The method is to design robust systems, produce them with capable manufacturing processes, and achieve continuous improvement.

VRP provides a win-win situation. The Air Force obtains more combat capability with the available dollars. Industry is able to satisfy their customers, improve productivity and lower costs.

REFERENCES

Statistical Process Control:

E. L. Grant and R. S. Leavenworth, Statistical Process Control (5th ed), McGraw Hill, New York, 1979.

J. S. Oakland, Statistical Process Control, Wiley, New York, 1986.

D. J. Wheeler and D. S. Chambers, Understanding Statistical Process Control, Statistical Process Controls, Inc., Knoxville, TN, 1986.

Design of Experiment/Parameter Design:

G. C. P. Box, W. G. Hunter and J. S. Hunter, Statistics for Experimenters, Wiley, New York, 1978.

G. C. P. Box, S. Bisgaard and C. Fung, "An Explanation and Critique of Taguchi's Contribution to Quality Engineering," Quality and Reliability Engineering International, Vol. 4, 123-131, (1988).

C. Daniel, Applications of Statistics to Industrial Experimentation, Wiley, New York, 1976.

R. V. Hogg and J. Ledolter, Engineering Statistics, MacMillan Publishing, New York, 1987.

S. R. Schmidt and R. G. Launsby, Understanding Industrial Design of Experiments, Department of Mathematical Sciences, USAF Academy, CO, (pending publication, Summer 1989).

G. Taguchi and Y. Wu, Introduction to Off-Line Quality Control, Central Japan Quality Control Association, Nagoya, Japan, 1980.

G. Taguchi, System of Experimental Design — Engineering Methods to Optimize Quality and Minimize Costs, UNIPUB/Kraus International Publications, White Plains, NY, 1987.

Quality Function Deployment:

J. R. Hauser and D. Clausing, "The House of Quality," Harvard Business Review, (May-June 1988)

B. King, Better Designs in Half the Time: Implementing QFD Quality Function Deployment in America, GOAL/QPC, Methuen, MA, 1987.

L. P. Sullivan, "Quality Function Deployment," Quality Progress, (June 1986), pp 39-50.

Concurrent (Simultaneous) Engineering:

R. I Winner, J. P. Pennell, H. E. Bertrand and M. Slusarczuk, The Role of Concurrent Engineering in Weapon System Acquisition (IDA Report R-338), Institute for Defense Analyses, Alexandria VA, 1988.

Continuous Improvement:

M. Imai, Kaizen, Random House Business Division, New York, 1986.

W. W. Scherkenbach, The Deming Route to Quality and Productivity — Road Maps and Roadblocks, Mercury Press/ Fairchild Publications, Rockville, MD, 1986.

Chapter 1 Problem Set

1.   What is experimental design?

2.   Discuss how the use of experimental design can result in higher quality products at lower costs.

3.   List the disadvantages of one factor at a time experiments.

4.   List the disadvantages of full factorial (all possible combinations) experiments.

5.   What is variability in the response and how can it be reduced?

6.   Describe what is meant by a robust design.

7.   Assuming your process is in control, use the following information to calculate C_p and C_{pk}. Interpret these values and discuss the proper use of each.

$$
\begin{aligned}
\text{Upper Spec} &= 2100 \\
\text{Target} &= 2000 \\
\text{Lower Spec} &= 1900 \\
\\
\bar{y} &= 2040 \\
\hat{\sigma} &= 20
\end{aligned}
$$

8.   What is meant by randomization and why is it important?

9.   How important is the brainstorming session? How much time should be devoted to brainstorming?

10.   Discuss the difference in the following two philosophies:
      (1)   Manufacture the product to meet specifications.

      (2)   Manufacture the product to the designed target value.

      What are the advantages/disadvantages of each philosophy?

11.   Your travel time to work tends to vary by several minutes. As a result of this variation problem, you have been late on several occasions. Your boss is concerned and has asked that you brainstorm the factors that contribute to travel time variation. Use the fishbone diagram below to organize the brainstorming process.

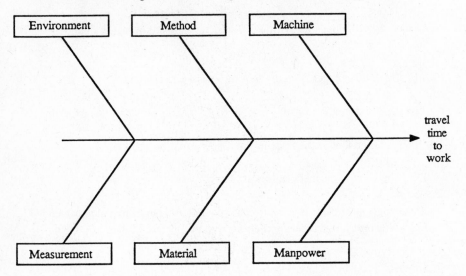

12.   Processed material thickness is a critical dimension with specifications of [0.49, 0.51] and a designed target value of 0.5. Your internal customer has indicated that any time the thickness deviates from the design target by more than ±0.004, they are required to complete an $8.00 extra processing step.

    (1)   Compute the monetary constant, k.

    (2)   Compute the average loss if your process has a $\bar{y}$ = 0.499 and s = .008.

    (3)   Find the total loss for one year if n = 100,000 per year.

    (4)   If reducing s to .004 is estimated to cost $700,000, will this improvement be a good investment? (Assume $\bar{y}$ remains .499.)

Chapter 1 Bibliography

1. Barker, Thomas B. (1985), *Quality by Experimental Design*, Marcel Dekker, Inc.

2. Deming, W. E. (1982), *Out of the Crisis*, Cambridge, Mass: MIT Center for Advanced Engineering Study.

3. Hogg, Robert V. and Ledolter, Johannes (1987), *Engineering Statistics*, MacMillan Publishing Company, New York.

4. Prince, George M. (1972), *The Practice of Creativity*, Collier Books.

5. Taguchi, G. (1986), *Introduction to Quality Engineering*, Asian Productivity Organization.

6. Ott, Ellis R. (1975), *Process Quality Control*, McGraw-Hill, Inc., New York.

7. Phadke, M. S. (1989), *Quality Engineering Using Robust Design*, Prentice-Hall, New Jersey.

8. Hunter, W., O'Neill, J., and Wallen, C. (1988), *Doing More with Less in the Public Sector: A progress report from Madison, Wisconsin*, Center for Quality and Productivity Improvement, University of Wisconsin.

9. Lam, K.D., Watson, Frank D. and Schmidt, S. R. (1991), *Total Quality – A Textbook of Strategic Quality Leadership and Planning*, Air Academy Press, Colorado Springs, Colorado.

10. Kiemele, M. J. and Schmidt, S. R. (1991), *Basic Statistics: Tools for Continuous Improvement*, 2nd Edition, Air Academy Press, Colorado Springs, Colorado.

11. Bhote, Keki R. (1988), *World Class Quality*, American Management Association, New York.

12. Montgomery, D. (1991), *Introduction to Statistical Quality Control*, 2nd Edition, John Wiley and Sons, Inc.

Chapter 2

Conducting Simple Experimental Designs and Analysis

2.1 Introduction

Before beginning an experimental design, it is important that the following preparatory steps be taken:

(1) Management must establish an environment suitable for action. Any statistical method will ultimately fail unless management has prepared a responsive environment where employees are motivated toward quality improvement, willing to work in teams, and unafraid to be candid with their inputs [1].

(2) Employees must be well trained in their jobs. Whether an employee is a supervisor, operator, engineer, or analyst, the experiment is likely to be conducted improperly unless everyone knows his/her job.

(3) Establish a team consisting of people from different backgrounds. An ideal team would have engineers, supervisors, analysts, operators, and possibly someone from maintenance, materials, and marketing. Many well designed experiments have been unsuccessful because someone did not provide the necessary support. Support is easier to obtain from someone if you make him/her part of the team.

(4) Define the process (use a process flow diagram) and ensure standard operating procedures are being adhered to.

(5) Complete a thorough Cause-and-Effect diagram analysis and remove all unnecessary sources of variation.

2.2   Experimental Design Example (Full Factorial)

Perhaps the simplest way to discuss designs and analysis is to take an example and walk through the 12 step process described in Chapter 1. Let us consider a nickel plating process.

Step I: Statement of the Problem

In the manufacture of high-technology disks for computer disk drives, several layers of materials are typically plated by electrolytic processes. The base material used in these products is aluminum. A pictorial cross-section of the product is shown below:

Chromium Layer

Nickel Layer

Aluminum Base Material

The customer of the nickel plated layer is unsatisfied with product performance. A process action team has identified the problem as inability to achieve the target plating thickness resulting in unsatisfactory process capability.

Step II: Objective of the Experiment

The objective of the experiment is to determine a way to increase process capability. To achieve this objective, it is necessary to characterize the process (i.e., determine which are the important factors and how they are related to plating thickness). Figure 2.1 depicts the types of factors we will try to identify.

The 1's and 2's in Figure 2.1 relate to the settings or levels for the various factors. As shown in example (i), changing factor A from its high setting (level 2) to its low setting (level 1) shifts the average of the response, but not the variability. Factor B, however, shifts the variability, but not the average. Factor C shifts both the response average and variability, while Factor D appears to be a benign factor, i.e., it does not shift either the response average or variability. (Note: Because of the introductory nature of this chapter, our example will only discuss identifying factors that shift the response average. Variability shifting factors are discussed in Chapter 5.)

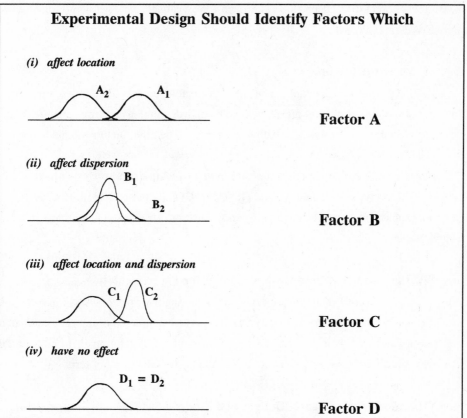

Experimental Design Should Identify Factors Which

(i) affect location

A_2 A_1

Factor A

(ii) affect dispersion

B_1

B_2

Factor B

(iii) affect location and dispersion

C_1 C_2

Factor C

(iv) have no effect

$D_1 = D_2$

Factor D

Figure 2.1 Objectives of an Experimental Design

Step III: Start Date - May 7
        End Date - June 1

Step IV: Select the Response

The response or quality characteristic to be used in the experiment should be measurable and related to the customer's needs and expectations. Plating thickness variability has been determined to be related to customer dissatisfaction. The thickness measurement process has been studied through a gage capability study [2] and found to be precise, accurate, and stable.

Step V: Select the Factors

Although numerous factors can be tested in a plating process (e.g., bath temperature, time, ph, phosphorous, percent nickel, bath volume, hydrogen ion concentration, etc.), this first example will test only time and bath temperature. The other factors are held constant at predetermined settings.

Bath temperature and plating time are both quantitative factors whose low and high experimental levels are determined to be (16°C, 32°C) and (4 sec, 12 sec) respectively. In addition, engineering knowledge and experience suggests a strong possibility that these two factors interact.

Step VI: Determine the Number of Resources for Experimentation

The process is available for experimentation during the week of 14 - 18 May. Each experimental test will take approximately 50 minutes (including setup time). The estimated cost per test is $400. Due to cost and time constraints, it is determined to allocate 20 resources (product) for experimentation and another 5 for confirmatory tests.

Step VII: Which Experimental Designs and Analysis Strategies Are Appropriate

A wide variety of design types and analysis strategies are available. The advantages/disadvantages of each design type are presented in Chapter 3 and the advantages/disadvantages of each analysis strategy are presented in Chapter 5. With only two factors selected for testing in this example, some of the design type possibilities are:

(i)   2 factor (2 levels) full factorial

(ii)   2 factor (3 levels) full factorial

(iii)   central composite design

Numerous data analysis approaches are available. You have a choice ranging from simple graphs to more complex statistical techniques such as Analysis of Variance (ANOVA) or regression.

Step VIII: Select the Best Design Type and Analysis Strategy to Suit Your Needs

Assuming that quadratic effects are not important, the best design type for 2 factors (2 levels) that will provide information on both linear effects and their interaction is the 2 factor (2 levels) full factorial. The number of test combinations can be enumerated through the use of the tree diagram displayed in Figure 2.2. Notice that we are now using a (+1) to indicate the high factor setting and a (−1) to indicate the low factor setting. This coded notation will allow us to perform more powerful analysis than the previously used 1 and 2 for low and high settings.

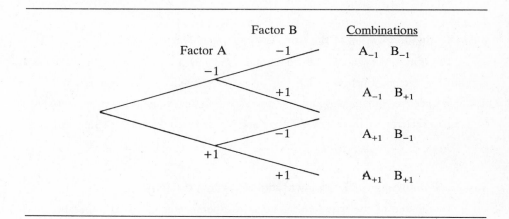

Figure 2.2 Enumerating All Combinations for 2 Factors (2 levels)
in a Tree Diagram

For our 2 factor experiment we will let factor A be bath temperature where (−1) = 16°C and (+1) = 32°C. Similarly, factor B will represent plating time where (−1) = 4 seconds and (+1) = 12 seconds.

The combinations produced in the tree diagram can be represented in a design (or test) matrix as shown in Table 2.1. The actual setting design matrix (Table 2.1a) is used to run the experiment and the coded setting matrix (Table 2.1b) is used for analysis purposes.

The analysis strategy selected for this introductory example will be a graphical approach with a prediction equation relating plating thickness to the factors time and temperature.

a)	Actual Settings		b)	Coded Settings	
Run #	A (Temp)	B (Time)	Run #	A (Temp)	B (Time)
1	16°C	4 sec	1	−1	−1
2	16°C	12 sec	2	−1	+1
3	32°C	4 sec	3	+1	−1
4	32°C	12 sec	4	+1	+1

Table 2.1 a) Design Matrix for Actual Settings and
b) Design Matrix for Coded Settings

Step IX: Randomization of the Runs in the Design Matrix

The run order should be conducted randomly to compensate for potential uncontrolled factors which were left out of the experiment. If a factor is difficult to randomize, such as bath temperature, you can randomize runs within each temperature level (sometimes referred to as a blocked design because you have blocked the randomization process).

Step X: Conducting the Experiment and Recording the Data

This is one of the most critical steps in experimentation. It is imperative that everyone involved in the process understand the discipline required in conducting the experimental runs. It is recommended that a very specific test procedure be laid out for the designed experiment. This allows you to do two things: 1) it forces you to think through the details of the experiment, and 2) it helps you and your team communicate on exactly how the experiment is to be conducted. Even after this has been done, things can still go wrong in the conduct of the experiment; therefore, it is also recommended that you be present for the setup and running of the experiment.

Based on the 20 allocated resources for this experiment, we have decided to collect thickness measurements on a sample of size five (i.e., five tested products) for each design combination. Each sample of size five can be either repetitions or replications. With replications every experimental trial represents a total reconfiguration of the test setup. To do five replications would typically result in randomly testing one product for each of the four combinations and then repeating this procedure four more times. Repetitions are conducted by testing the entire sample of the same design combination (or run) before adjusting factor levels and conducting tests on another combination. Replications are best from a statistical standpoint (they provide for a better estimate of experimental error), but tend to be more expensive. See Appendix M for a Rule of Thumb on the desired sample size per combination.

Continuing with our example, we conduct the full factorial design as specified with 5 measures of each experimental run. The design and data collected are shown in Table 2.2.

Run #	Temp	Time	Thickness Readings	Average
1	16°C	4 sec	116.1, 116.9, 112.6, 118.7, 114.9	115.8
2	16°C	12 sec	116.5, 115.5, 119.2, 114.7, 118.3	116.8
3	32°C	4 sec	106.7, 107.5, 105.9, 107.1, 106.5	106.7
4	32°C	12 sec	123.2, 125.1, 124.5, 124.0, 124.7	124.3

Table 2.2 Example Design Matrix with Response Data

Step XI: Analyze Data, Draw Conclusions, Make Predictions, and Do Confirmatory Tests

Our choice of analysis strategy is a graphical approach with a prediction equation relating plating thickness to time and temperature. The steps for this analysis approach are:

(1) Conduct eye-ball analysis.

(2) Calculate the response average at each level for each design column.

(3) Calculate the effects and half effects for each column.

(4) Plot the averages from Step 2.

(5) Generate a Pareto chart of the absolute value of each half effect from Step 3.

(6) Determine the "importance" of factors/interactions.

(7) Generate a prediction equation using "important" factors and interactions.

(8) Based upon the objective, select the best settings for the factors.

(9) Using the prediction equation generated in Step 7, predict the response. This value will be used as a target for the confirmation runs.

Analysis of the Example Data

Each of the nine steps for analysis will be individually discussed for the plating example.

Step (1)   Eye-ball analysis − First review the response values looking for anything that appears unusual. Investigate all unusual response values and correct any discovered errors. Once we are certain the data is accurate, we can continue.

Suppose we wish to maximize the nickel thickness. Run #4 (temperature = 32°C, time = 12 seconds) is the best of the four possible combinations with an average thickness of 124.3 microinches. Using this "pick the winner" technique, each experimental run is evaluated in light of the objective. The run which most closely meets our objective is declared the winner. The obvious advantage of this approach is in its simplicity. Disadvantages lie in its inability to indicate which factors have the greatest effect, possible suboptimization, and/or inability to predict untested combinations (this will be very important with fractional factorial designs discussed in Chapter 3). Another disadvantage of simply picking a design run winner occurs when trying to achieve a target thickness and none of the run averages equal the target.

Steps (2) & (3)   Calculate the average at each level for each column. Calculate the effect and half effect for each column. This technique will first be demonstrated using the temperature column. For the design in Table 2.2, temperature is at the low setting (16°C) for runs #1 and #2. There are ten thickness readings (116.1, 116.9, 112.6, ... , 114.7, 118.3) when temperature is at 16°C. The average of these ten values is 116.3. When temperature is at 32°C, the average reading is 115.5. The difference (average at the high minus the average at the low) between these two averages is −0.8 microinches. This difference is defined as the effect of the temperature factor. On the average, we estimate that changing temperature from 16°C to 32°C decreases the plating thickness by 0.8 microinches. Analysis of the time column gives us an average of 111.3 microinches when time is 4 seconds and 120.6 microinches when time is 12 seconds. The estimated effect for time is 120.6 − 111.3 = 9.3 microinches. Comparing the effect of temperature with the effect of time indicates that the time factor appears to be approximately twelve times more important than temperature (over the tested levels) in effecting plating thickness.

The simple analysis approach used above to estimate the effects of time and temperature is possible because of the nature of the design matrix. All full factorial designs (as well as many other families of designs) are orthogonal designs. These types of designs are simple, but powerful in that each column in the design can be used to independently estimate the effect of the assigned factor. Had our design not been an orthogonal design, simple analysis would not have been applicable. Table 2.3 is a summary of the analysis up to this point.

Run #	Temp	Time	Thickness Readings	Average
1	16°C	4 sec	116.1, 116.9, 112.6, 118.7, 114.9	115.8
2	16°C	12 sec	116.5, 115.5, 119.2, 114.7, 118.3	116.8
3	32°C	4 sec	106.7, 107.5, 105.9, 107.1, 106.5	106.7
4	32°C	12 sec	123.2, 125.1, 124.5, 124.0, 124.7	124.3
	116.3	111.3	Average Thickness at Low	
	115.5	120.6	Average Thickness at High	
	−0.8	9.3	Effect (Δ)	

Table 2.3 Analysis Summary for Factor Effects

In addition to providing independent estimates of the effects of factors, full factorial designs (and other types of designs − see Chapter 3) can be used to provide independent estimates of the factor interaction effects. Interactions refer to dependencies between a factor's effect on the response and the levels of another factor.

Let's leave our example briefly to discuss the concept of interactions and how orthogonal designs can be used to analyze interactions. First look at an example where factors do not interact. Consider two factors, dwell time and cure temperature, where dwell time is being investigated at two levels and cure temperature is being investigated at three levels. The one response being considered is tensile strength. Using a full factorial design, the runs are shown in Table 2.4.

A plot of the data in Table 2.4 with strength on the y-axis, cure temperature on the x-axis, and dwell times displayed as two lines is shown in Figure 2.3.

Run #	Dwell Time	Cure Temp	Tensile Strength (using ASTM D3916)
1	60 min	250°F	100
2	60 min	300°F	150
3	60 min	350°F	200
4	100 min	250°F	125
5	100 min	300°F	175
6	100 min	350°F	225

Table 2.4 Design Matrix and Response Values for an Example with No Interaction Effects

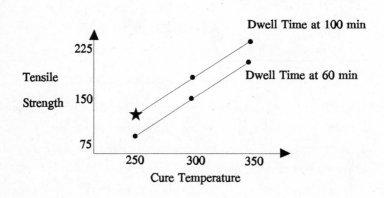

Figure 2.3 Graph of 2 Factors That Do Not Interact

To understand how an interaction graph is generated, verify for yourself that the ★ on Figure 2.3 represents the average response when cure temperature is 250°F and dwell time is 100 minutes, i.e., the average response for run number 4. The other dots in Figure 2.3 can be verified in a similar manner.

Note that the lines in Figure 2.3 are parallel. With this type of plot, parallel or near parallel lines are indicators of little or no interaction. In our Table 2.4 and Figure 2.3 example with non-interacting data, you can quantify the impact of increasing dwell time from 60 minutes to 100 minutes independent of the setting for cure temperature. The effect is an increase of 25 units regardless of the level for cure temperature.

Next, the scenario is modified to look at an example of factors which do interact. In this case, suppose the same full factorial design is conducted for cure temperature and dwell time, but now the data is somewhat different (see Table 2.5 and Figure 2.4).

Run #	Dwell Time	Cure Temp	Tensile Strength
1	60 min	250°F	200
2	60 min	300°F	200
3	60 min	350°F	200
4	100 min	250°F	100
5	100 min	300°F	200
6	100 min	350°F	300

Table 2.5 Design Matrix and Response Values of 2 Factors That Do Interact

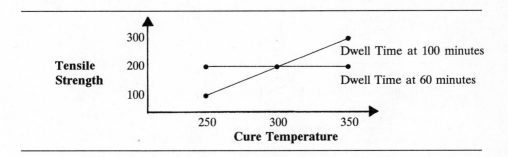

Figure 2.4 Graph of a 2-Factor Interaction

No longer are the lines parallel. In this scenario we can no longer discuss the impact of dwell time without first specifying the level for cure temperature. There now appears to be a dependency between cure temperature's effect on tensile strength and the levels of dwell time.

Returning to our plating example with two factors at two levels, we will plot the interaction graph and then mathematically calculate the effect of the interaction. The interaction plot for temperature and time in the plating example is shown in Figure 2.5. This interaction graph indicates that there may be an interaction between temperature and time. To mathematically evaluate the effect of the time and temperature interaction, our design matrix must be converted from the actual factor settings to the coded values (for 2 level designs, the low level is always a −1 setting and the high level is always a +1 setting). The coded matrix for temperature and time is shown in Table 2.6.

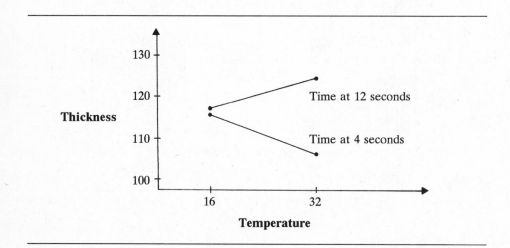

Figure 2.5 Interaction Graph for Temperature and Time in the Plating Example

Run #	A(Temp)	B(Time)
1	−1	−1
2	−1	+1
3	+1	−1
4	+1	+1

Table 2.6 Coded Design Matrix for the Plating Example

In order to quantify the temperature and time interaction effect, a coded column is generated to represent the interaction. An interaction column is generated by simply multiplying together the appropriate column values of temperature and time for each row. For row #1, the value for A × B is +1. This value is obtained by multiplying together −1 (for temperature) and −1 (for time). Continuing this operation for all four rows we obtain the columns shown in Table 2.7.

Run #	A(Temp)	B(Time)	AB
1	−1	−1	+1
2	−1	+1	−1
3	+1	−1	−1
4	+1	+1	+1

Table 2.7 Coded Design Matrix with AB Interaction Column

We can now use the AB* column to mathematically determine the effect of the interaction. The actual calculations are performed in the same manner as for the factor columns. For the AB interaction column, the response average of the −1 values is $\frac{116.8 + 106.7}{2} = 111.75$. The response average of the +1 values is $\frac{115.8 + 124.3}{2} = 120.05$. Thus, the interaction effect is $120.05 - 111.75 = 8.3$.

The next part of analysis step (3) is to calculate the half effects for each column. The half effects will be used in a prediction equation to be presented later. The half effects (as the name implies) are simply one half of the effects (Δ's). For our example, the values are:

	A (Temp)	B (Time)	AB
Δ/2	−0.4	4.65	4.15

---

* Throughout this book you will see interaction terms (such as the one representing the A and B interaction) as A × B or AB.

Step (4) in our 9 Step Data Analysis Procedure is to plot the averages from steps (2) and (3). The graphical analysis of the averages will reveal the relative importance of all of the effects (see Figure 2.6).

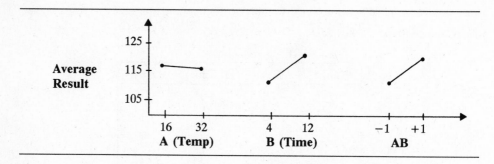

Figure 2.6 Graph of the Averages for Each Effect
(Also referred to as a graph of the Marginal Means.)

Step (5) directs us to generate a Pareto chart of the absolute value of each half effect. The Pareto chart for our example is shown below.

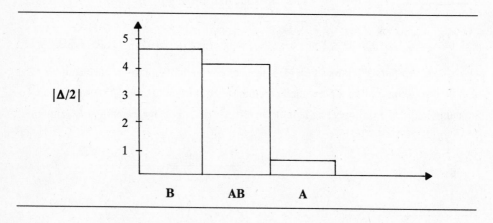

Figure 2.7 Pareto Chart for $\frac{\Delta}{2}$ Values or Half Effects

In **Step (6)** we must determine which factors and interactions are important and which are trivial. The three approaches available to assist in making this decision are:

        (1)   Graphical

        (2)   Analysis of Variance (ANOVA)

        (3)   Regression Analysis

In Chapters 4 and 5, ANOVA and regression approaches are discussed. At this point we will present only the graphical approach. There are two reasons for determining which factors/interactions should be included in our prediction equation: (1) If too many are included, we tend to overpredict; whereas, if too few are included, we tend to underpredict. Neither of these situations is good. (2) Optimal factor levels will be based on the $\Delta/2$ for factor effects that appear in the model and based on cost, convenience, etc. for benign factors whose effects are not strong enough to be in the model. Our objective is to develop a prediction equation which includes only those factor effects and interactions which are important in describing the response. The calculated effect for important factors/interactions would be non-zero; however, even benign factor effects will possibly be non-zero due to sampling error. Using the shape of the Pareto Chart, the size of the half effects, and prior knowledge, we should be able to make a fairly good decision concerning which effect should be in the model and which should not. If in doubt, we will normally place the term in the model.

The shape of the Pareto chart can reveal a separation in important and benign effects. For example, the Pareto chart in Figure 2.7 indicates that effects B and A $\times$ B are important and effect A is potentially benign. The size of the half effect can also be used in helping us make a decision in regard to which factors/interactions should be included in the prediction equation. For example, the calculated half effect for time in our example is 4.65 microinches. Leaving this term out of the model when it is truly important would result in an inaccurate prediction of ± 4.65 microinches. If 4.65 microinches is technically significant, time should be included in our prediction equation. On the other hand, leaving the temperature effect out of the prediction equation would only result in a $\pm.4$ microinch inaccuracy.

Using our technical knowledge of a process can also be useful in understanding what needs to be included in the prediction equation. Suppose it is known beforehand that a particular factor has a positive effect upon the response. In this case the factor will usually be included in our prediction equation regardless of the size of the calculated effect.

If prior technical knowledge suggests that both time and temperature are important factors for the plating example, then the Pareto chart for our process suggests the Time × Temperature interaction (A × B) is important as well.

Step (7) The general form of the prediction equation is:

$$* \quad \hat{y} = \bar{y} + \left(\frac{\Delta_A}{2}\right)A + \left(\frac{\Delta_B}{2}\right)B + \left(\frac{\Delta_{AB}}{2}\right)AB + ... \text{ (other terms for larger designs)}$$

Where:
$\hat{y}$ = the predicted response

$\bar{y}$ = the average of all response values from the experimental data

$\dfrac{\Delta_A}{2}$ = half effect for factor A

$\dfrac{\Delta_B}{2}$ = half effect for factor B

$\dfrac{\Delta_{AB}}{2}$ = half effect for AB interaction

*NOTE:
(1) We only include the important effects in the prediction equation.

(2) This prediction equation applies only to orthogonally-coded factor settings for 2-level designs. See Chapter 4 techniques for other designs.

(3) The origin of this equation is the slope and intercept model, i.e., for only

1 factor $\hat{y}$ = intercept + slope(A) = $\bar{y} + \dfrac{\Delta_A}{2}A$.

Equation 2.1 Prediction Equation for a Coded 2-Level Orthogonal Design

Although the use of coded factor setting values for the prediction equation appears to be an inconvenience, models built on the true factor settings will frequently provide equations whose coefficients are not independent. Therefore, only equations for coded settings will provide a clear understanding of the response/inputs relationship.

The prediction equation for our example is:

$$\hat{y} = 115.93 - \left(\frac{.8}{2}\right)A + \left(\frac{9.3}{2}\right)B + \left(\frac{8.3}{2}\right)AB = 115.93 - (.4)A + (4.65)B + (4.15)AB$$

Notice that the A effect is included in the model even though it was thought to be unimportant. The hierarchy rule for modeling dictates that if a higher order term (quadratic or interaction) is included in the model, then all linear effects represented in the higher order term will be included in the model regardless of their significance. In our example, the inclusion of the AB term as an important effect dictates (using the hierarchy rule) that the linear A and B terms both be included in the model. The prediction equation represents a Taylor Series approximation of the true physical process model for the response and the tested factors. An important note is that the prediction equation is obtained quickly and is based on actual process data.

Step (8) Select the best settings for the factors. If our objective is to minimize the response then our optimum predicted thickness occurs for $A_{+1}B_{-1}$ (i.e., temperature $= 32\,^{\circ}C$ and time $= 4$ seconds). The predicted minimum thickness for $A_{+1}B_{-1}$ is:

$$\hat{y} = 115.93 + - .4(+1) + 4.65(-1) + 4.15(+1)(-1) = 106.73$$

The plot of averages at each level (see Figure 2.6) would have provided the same conclusion as we found with the prediction equation, however, the graphical approach does not provide an accurate prediction value.

If the objective is to maximize thickness, the best settings appear to be temperature at $32\,^{\circ}C$ and time at 12 seconds. The predicted maximum thickness is:

$$\hat{y} = 115.93 - .4(+1) + 4.65(+1) + 4.15(+1)(+1) = 124.33$$

Both of the above analysis approaches agree with the conclusions reached by the eye-ball analysis approach. This is always the case with full factorial designs. Why then do we need to go to all of the work required to create the prediction equation? There are three important reasons: 1) Typically we will not conduct a full factorial design and the prediction equation can be used to estimate the untested combinations in a fractional factorial design, 2) the equation is useful in predicting a target value, and 3) the coded prediction equation is used to complete sensitivity analysis and tolerance design.

Suppose we have a target thickness of 120 microinches. Since we have one prediction equation with 2 variables, suppose we decide to set factor A at 32°C (+1). The prediction equation can then be used as follows:

$$\hat{y} = 115.93 + 4.65(B) - .4(A) + 4.15(A)(B)$$

$$120 = 115.93 + 4.65(B) - .4(+1) + 4.15(+1)(B)$$

$$4.47 = 8.80(B) \Rightarrow B = .51$$

In order for B = .51 to be meaningful, the coded value must be translated to actual units for the factor time.* The following picture indicates what is happening.

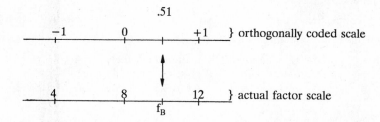

To resolve the above problem, the following equation must be solved:

$$\frac{.51 - 0}{f_B - 8} = \frac{1 - 0}{12 - 8} \Rightarrow .51(4) = f_B - 8 \Rightarrow f_B = 10.04 \text{ sec}$$

where f_B is the actual setting for factor B.

Thus, setting time at 10.04 seconds and temperature at 32°C should give us a thickness of 120 microinches.

Step (9) involves obtaining a predicted best value and conducting a confirmation test. Suppose (as discussed above) our target is 120 microinches. From the prediction equation, setting time at 10.04 seconds and temperature at 32°C should give us 120 microinches.

---

*Normally, this transformation from coded to actual settings (and vise versa) is computed by a software package such as Q-Edge, RS/Discover, Catalyst, DOE/Target, etc.

Confirmation tests are conducted to validate our model (prediction equation). We recommend a confirmatory test sample size between four and twenty depending on what you can afford. The confirmation process consists of repeated trials with time at 10.04 seconds and temperature at 32°C. If the response average of the confirmation is close to the predicted value, you can confirm your previous assumptions and the accuracy of the model. If, on the other hand, the confirmatory response average is not close to the predicted value, you will conclude that one or more of the assumptions made was not valid and/or the model is inaccurate.

For example, suppose we conducted five verification tests for our example and obtained the following results: 120.5, 122.7, 118.4, 125.3, 117.0
Pictorially, the data would appear as follows:

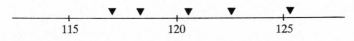

Since the actual values are distributed nicely about 120, it appears reasonable to conclude that our confirmation test verified our assumptions and our model. If, on the other hand, our results had been: 130.6, 125.9, 127.3, 133.3, 119.1.

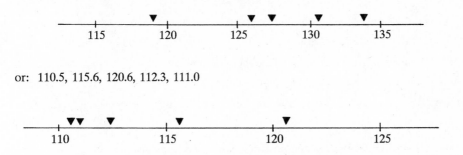

or: 110.5, 115.6, 120.6, 112.3, 111.0

then our conclusions are questionable. The previous discussion is based on a graphical approach for determining if our experiment has been confirmed. Chapter 5 provides a more in depth discussion of confirmation.

Step XII Assess Results, Make Decisions, and If Necessary, Conduct More Experimentation

Let's assume for our example that the results of the confirmation runs closely matched the predicted value with relatively small variability. We would then set up and run our plating process at the predicted best settings and conduct a capability study to re-evaluate our new state of quality. The new C_{pk} would be compared with the old process C_{pk} described in Step I. In Chapter 5, you will also learn how to model the standard deviation, s. Using the $\hat{s}$ prediction model you will be able to optimize and average the standard deviation. If you are a researcher, developer, or designer you can use both the $\hat{y}$ and $\hat{s}$ models to estimate C_{pk} values before the technology transfer. Obviously, if the estimated C_{pk} is unsatisfactory, it is advisable to redesign the product to insure it is manufacturable.

The previously discussed plating example was intended to provide an overview of the steps in setting up a designed experiment, as well as a discussion of simple data analysis. The design analyzed was a full factorial design. Full factorial designs are good in that they provide estimates of all factor effects, as well as the effects of all possible interactions. A disadvantage is that full factorial designs will require a large number of runs when more than a few factors exist. For example, it is not unusual to want to experiment with ten factors at two levels. A full factorial design would require 2^{10} or 1024 experiments. This number is excessive for even the largest experimental budgets. Fortunately, alternatives to full factorial designs exist. One very popular alternative is the fractional factorial design. As the name implies, we run only a portion or fraction of the full factorial. The good news with fractional factorials is that we can dramatically reduce the total number of runs required to estimate the effects of factors. The bad news is that some (or all) information about interactions is lost. See Appendix M for Rules of Thumb on when to use full and fractional factorials.

2.3 Fractional Factorial Designs for 2 Level Factors

Let's discuss how fractional factorial designs can be generated. Suppose we have two factors (A and B) to be tested at two levels. The coded design matrix is shown in Table 2.8.

Run #	A	B
1	−1	−1
2	−1	+1
3	+1	−1
4	+1	+1

Table 2.8 Coded Design Matrix for 2 Factors at 2 Levels

We could then generate the A × B interaction column as demonstrated previously in the plating example. Our coded design matrix would now appear as shown in Table 2.9.

Run #	A	B	AB
1	−1	−1	+1
2	−1	+1	−1
3	+1	−1	−1
4	+1	+1	+1

Table 2.9 Coded Design Matrix with the Interaction Column

The above matrix provides us with three orthogonal columns. Orthogonal designs for 2 level factors will have vertical and horizontal balancing. For example, all three columns have equal numbers of low and high values, hence, vertical balance. To demonstrate horizontal balance, look at only the runs for A at the −1 setting (runs 1 and 2), and notice that column B is balanced (one −1 and one +1). Similarly, when column A is at the +1 level, column B is again balanced. Continuing in this manner for all possible pair-wise comparisons, we find all columns have horizontal balance. Thus, the response average at the (+) or (−) settings for any column has an equal influence from the other columns (i.e., the evaluation of any effect is independent of other column effects).

Now suppose a third factor C exists which we want to include in our experiment. The two choices are: 1) full factorial with eight runs, or 2) a fractional factorial of only four runs. Although 3 factors can be tested in 4 runs, there is a loss of all information on interactions. The mechanics of creating a complex fractional factorial are somewhat burdensome (see Chapter 3 for the details). In this case we simply take the AB column and assign the settings for Factor C to it. Our design for 3 factors in 4 runs is shown in Table 2.10.

Run #	A	B	C=(A × B)
1	−1	−1	+1
2	−1	+1	−1
3	+1	−1	−1
4	+1	+1	+1

Table 2.10 Fractional Factorial Design of 3 Factors (at 2 Levels) in 4 Runs

The tree diagram for three factors at two levels is presented in Figure 2.8.

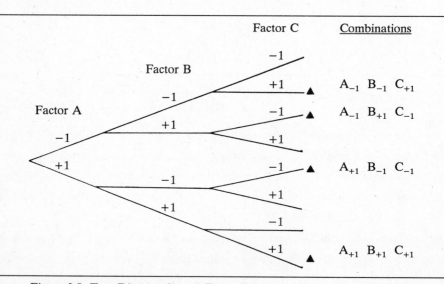

Figure 2.8 Tree Diagram for a 3 Factor Fractional Factorial in 4 Runs

Note that the branches with triangles are the four runs specified in the fractional factorial which is a subset of the full factorial. It is, however, a special subset in that it is an orthogonal subset. If we had chosen the remaining four runs in the above tree diagram, it would have been the complementary fractional factorial which is another orthogonal subset of the full factorial.

Returning to the coded design matrix of Table 2.10, let's generate the possible interaction columns for 3 factors at 2 levels: AB, AC, BC, ABC (see Table 2.11).

Run #	A	B	C	AB	AC	BC	ABC
1	−1	−1	+1	+1	−1	−1	+1
2	−1	+1	−1	−1	+1	−1	+1
3	+1	−1	−1	−1	−1	+1	+1
4	+1	+1	+1	+1	+1	+1	+1

Table 2.11 Coded Matrix for All Terms in a 3 Factor (2 Level) 4 Run
Half Fractional Factorial Design

Notice that column A is identical to column BC. B is identical to AC, while C is identical to AB. A is said to be aliased with BC, B with AC, and C with AB. The impact of aliasing is encountered when we evaluate the effects for each column. Since A is aliased with BC, analysis for these columns will provide the same effect values. The practical implication of A aliased with B × C is that we are unable to determine whether the effect calculated for column A is due to A, due to BC, or due to some combination of the two. Fortunately for industrial experimenters, the effects of interactions are usually small relative to the effects of factors, thus providing justification for the use of fractional factorial designs. In addition, not all fractional factorial designs result in the loss of all interaction information. Chapter 3 provides a detailed discussion of fractional factorial designs, as well as tabled fractional factorial designs. See Appendix M for Rules of Thumb on the use of full and fractional factorials.

Consider another example of testing 4 factors in a half fraction of 8 run combinations. The full factorial for an 8 run matrix is based on only 3 factors (see Table 2.12).

Run #	A	B	C	AB	AC	BC	ABC
1	−1	−1	−1	+1	+1	+1	−1
2	−1	−1	+1	+1	−1	−1	+1
3	−1	+1	−1	−1	+1	−1	+1
4	−1	+1	+1	−1	−1	+1	−1
5	+1	−1	−1	−1	−1	+1	+1
6	+1	−1	+1	−1	+1	−1	−1
7	+1	+1	−1	+1	−1	−1	−1
8	+1	+1	+1	+1	+1	+1	+1

Table 2.12 Three Factor (2 Level) Full Factorial Coded Matrix

Now consider which column in Table 2.12 you would be willing to give up to estimate a fourth factor D without going to the 16 run full factorial for 4 factors. The least likely column in Table 2.10 to produce an important effect is the 3 way interaction A × B × C; therefore, it is the best choice for aliasing with the settings for factor D. The half fraction (8 run matrix versus 16) for 4 factors at 2 levels is displayed in Table 2.13.

Run #	A	B	C	A × B (C × D)	A × C (B × D)	B × C (A × D)	D (A × B × C)
1	−1	−1	−1	+1	+1	+1	−1
2	−1	−1	+1	+1	−1	−1	+1
3	−1	+1	−1	−1	+1	−1	+1
4	−1	+1	+1	−1	−1	+1	−1
5	+1	−1	−1	−1	−1	+1	+1
6	+1	−1	+1	−1	+1	−1	−1
7	+1	+1	−1	+1	−1	−1	−1
8	+1	+1	+1	+1	+1	+1	+1

Table 2.13 A Half Fraction Coded Matrix for 4 Factors at 2 Levels

As you will learn in Chapter 3, aliasing factor D with the A × B × C interaction also created other effect aliasings, i.e., A × B = C × D, A × C = B × D, and B × C = A × D. Thus, if any 2-factor interaction column has a significant effect, it could be due to one or both of the aliased effects for that column. Which effect is actually placed in the prediction equation is based on prior knowledge as to which interaction is most likely to be important. It is important to understand that the term aliased effects does not imply that their physical effect with the response is the same. Rather, aliased effects have identical coded columns and, thus, evaluation of that column cannot separate the two effects. In practice, an experimenter typically chooses a fraction of the full factorial design and evaluates the aliasing patterns before running any tests. If the aliasing patterns are unacceptable, then more runs or fewer factors must be selected. Those experimenters not familiar with the generation of fractional factorial designs could simply refer to the tabled designs in Section 3.7.

2.4   Analysis of a 2-Level Fractional Factorial Design

Suppose you are a process engineer for an aircraft company and are responsible for the painting process. You wish to conduct a designed experiment in order to determine which of 4 factors are most responsible in influencing paint thickness and so that you can hit a target thickness of .5 mm. The factors being tested are shown in Table 2.14.

Factor	Low	High
A - Thinner Amount	10	20
B - Fluid Pressure	15	30
C - Temperature	68	78
D - Vendor	1	2

Table 2.14 Factors and Settings for Aircraft Painting Example

An 8-run fractional factorial design is conducted with 4 replicates of each run. The design and results are displayed in Table 2.15.

2-26 *Understanding Industrial Designed Experiments*

Run #	Thinner Amount	Fluid Pressure	Temperature	Vendor	Paint Thickness	Average	Standard Deviation
1	10	15	68	1	.638, .489, .477, .541	.536	.073
2	10	15	78	2	.496, .509, .513, .495	.503	.009
3	10	30	68	2	.562, .493, .529, .531	.529	.028
4	10	30	78	1	.564, .632, .507, .757	.615	.108
5	20	15	68	2	.659, .623, .636, .584	.626	.031
6	20	15	78	1	.549, .457, .604, .386	.499	.096
7	20	30	68	1	.842, .910, .657, .953	.840	.131
8	20	30	78	2	.910, .898, .913, .875	.899	.017

Table 2.15 Experimental Data for Aircraft Painting Process

Analysis Steps

Step (1) Runs #2 and #6 come closest to meeting our target value of .5 inches.

Steps (2) and (3) The averages, effects and half effects for the applicable factors and interactions are shown in Table 2.16.

Factor/Interaction	Avg−	Avg+	effect (Δ)	Δ/2
A - Thinner amount	.546	.716	.17	.085
B - Fluid pressure	.541	.721	.18	.090
C - Temperature	.633	.629	−.004	−.002
D - Vendor	.623	.639	.016	.008
*AB(CD)	.567	.695	.128	+.064
*AC(BD)	.646	.616	−.030	−.015
*BC(AD)	.593	.669	.076	+.038

Table 2.16 Table of Effects

Step (4) A plot of the averages at the high and low for each column is shown in Figure 2.9.

---

* Since we are evaluating 4 factors in an 8-run design the 2-factor interactions are aliased with each other, i.e., AB(CD) implies AB is aliased with CD.

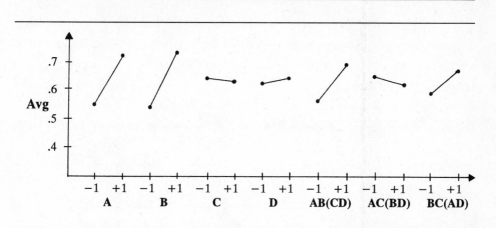

Figure 2.9 Graph of the Averages for Each Effect

Step (5) Generate a Pareto chart of the absolute value of each half effect as shown in Figure 2.10.

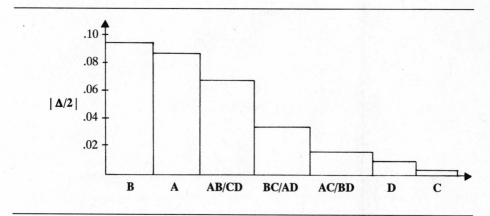

Figure 2.10 Pareto Chart for $|\Delta/2|$ Values of Half Effects

Step (6) Determine the important factors/interactions. Based upon the above Pareto chart and engineering knowledge, we decide that the factors A and B and the interaction AB/CD are the important effects. Since AB and CD are aliased, we are unable to determine whether the associated effect is due to AB or CD. It is, however, unusual for an interaction to be important without one or both of the constituent factors being important. Therefore, the important interaction is probably AB which is plotted in Figure 2.11.

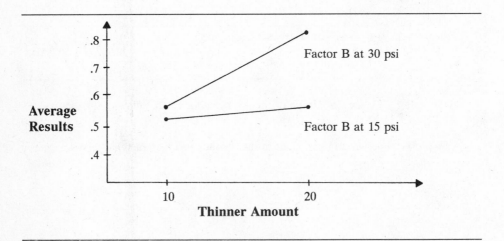

Figure 2.11 Interaction of Thinner Amount and Fluid Pressure

Step (7) Generate the following prediction equation:

$$\hat{y} = .631 + .085(A) + .064(A)(B) + .09(B)$$

Step (8) From our interaction plot and prediction equation, it appears A = −1 and B = −1 will provide the closest predicted value to .50.* Inserting these values into the prediction equation results in $\hat{y}$ = .52 which is not the desired .50, but the best we can do within the experimental region.

---

* Since the lowest possible thinner amount and fluid pressure are 10 and 15 respectively, it does not appear we can do better than a .52 inch thickness.

Step (9) Conduct between four and twenty verification runs. If the results are close to the predicted value, the experiment has verified. If the results of the confirmation runs are not close to the predicted value, the experimental results have not been confirmed thus indicating a need for further investigation of the paint process. See Chapter 5 for a detailed discussion of this topic.

Simple analysis can also be used to determine possible variance reduction factors. The details of these approaches are discussed in Chapter 5. We can, however, visually look for possible factor settings which provide reduced variation. For our example the paint thickness standard deviation for each row of data is:

Run #	Vendor	Standard Deviation
1	1	.073
2	2	.009
3	2	.028
4	1	.108
5	2	.031
6	1	.096
7	1	.131
8	2	.017

A visual inspection of our data indicates the standard deviation is relatively low for runs 2, 3, 5, and 8. These runs were made with vendor 2. Runs 1, 4, 6, and 7 have relatively high standard deviations. For these runs vendor 1 was used. The average standard deviations for vendor 1 and 2 are .102 and .021. Using the rules discussed in Chapter 5, we find that vendor 2 provides a significantly reduced paint thickness variance.

2.5   Example of a Full Factorial Design for 3 Level Factors

Suppose you are a chemical process engineer in charge of rubber compounding formulations for a particular butadiene based formulation. Your objective is to maximize the durometer (hardness) reading. The two factors you choose to experiment with are cure temperature and amount of filler in the formulation. Each factor is to be tested at 3 levels in a full factorial design with all other non-experimental factors held constant. The design and data are shown in Table 2.17.

Run #	Cure Temperature (°F)	Filler (lbs)	Durometer Reading	Average
1	250	50	20, 21, 20	20.3
2	250	60	23, 22, 20	21.7
3	250	70	24, 24, 25	24.3
4	275	50	29, 30, 31	30.0
5	275	60	31, 32, 30	31.0
6	275	70	34, 33, 34	33.7
7	300	50	27, 26, 25	26.0
8	300	60	27, 27, 29	27.7
9	300	70	28, 24, 26	26.0

Table 2.17 Design and Response Data for a 2-Factor (3-Level) Full Factorial

Step (1) The pick the winner approach indicates run #6 provides the maximum durometer reading. The average for this run is 33.7.

Step (2) directs us to calculate the average at each level for each column in the design. The average values for our factors are:

	Cure Temperature	Filler
Avg Low	22.1	25.4
Avg Middle	31.6	26.8
Avg High	26.6	28.0

Step (3) calculates the effects and half effects for each column. The steps to analysis (p. 2-7) break down at this point for 3-level designs. Regression analysis or ANOVA must be used for complete analysis of the data. Some graphical analysis, however, can be calculated with 3-level designs. Plots of averages for each factor (shown in Figure 2.12), as well as interaction plots (see Figure 2.13), can be generated.

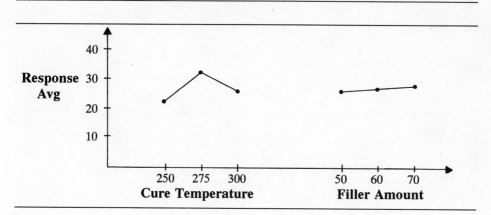

Figure 2.12 Plot of Averages

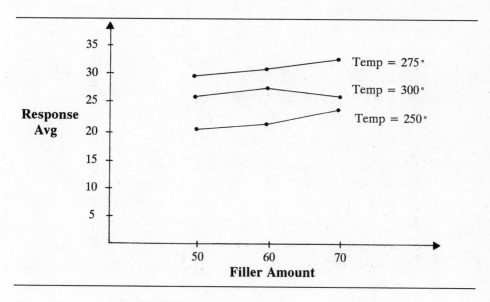

Figure 2.13 Temperature × Filler Interaction Plot

Since our objective is to maximize hardness, it appears from Figure 2.12 that cure temperature = 275°F and filler amount = 70 lbs provides the best settings. The next step would be the confirmation tests.

Chapter 2 Problem Set

1. A company that manufactures can-forming equipment wants to set up an experiment to help understand the factors influencing surface finish on a particular steel subassembly. Three factors were chosen to be varied in the experiments. The factors were:

Factor Name	Low Level	High Level
A - RPM	588	1182
B - Feed Rate	.004	.008
C - Tool Radius	1/64	1/32

    a) Generate a tree diagram to represent the possible combinations with a full factorial at two levels.

    b) Generate a design matrix representing the possible combinations.

    c) Suppose changing the tool radius requires an 8 hour teardown procedure. Discuss strategies for limiting the number of times tool radius must be changed.

    d) Suppose the part being machined is 4130 preheated steel. The part is a 1" diameter rod which is 4" long. Should the part be measured at one point or several points? Discuss various approaches for making multiple measures on the part.

2. The company in the previous example decides to test an 8-run full factorial design with the 3 factors at 2 levels. A brainstorming session conducted with the operator, supervisor, and engineer on the response resulted in the finished part being measured at 4 places. (Two measures were taken on each end 180° from each other.) The design and resultant response values were:

Run #	RPM	Feed Rate	Tool Radius	Surface Finish
1	588	.004	1/64	50, 50, 55, 50
2	588	.004	1/32	145, 150, 100, 110
3	588	.008	1/64	160, 160, 155, 160
4	588	.008	1/32	180, 200, 190, 195
5	1182	.004	1/64	60, 60, 60, 55
6	1182	.004	1/32	25, 35, 35, 30
7	1182	.008	1/64	160, 160, 160, 160
8	1182	.008	1/32	80, 70, 70, 80

a) Calculate the averages, effects, and half effects for all factors and interactions. Plot the average values for each factor.

b) Generate a Pareto chart of the absolute value of each half effect. Which effects appear to be most important?

3.   For the previous problem,

a) Generate an interaction plot for any interactions with large effects.

b) Suppose the objective in Problem 2 is to minimize the response. What are the best settings for the factors to achieve this objective?

4.   For a process you work with, pull together a cross functional team and begin brainstorming the setup of a designed experiment. Be sure and use the check sheet from Chapter 1 as a guide in assisting you through the proper steps.

5. a) Using *Factsim* on the disk in the back of the book, select 3 factors at 2 levels, and set up a designed experiment (full factorial).

b) Conduct the experiment using four repeated measures per run.

c) Calculate the average response at each level for all factors and interactions.

d) Generate a Pareto chart of the absolute value for each half effect. Decide which factors and interactions appear to be most important in influencing the response.

e) Build the prediction equation.

f) Determine one set of factor settings which can achieve a target of .5 inches.

Chapter 2 Bibliography

1.   Kiemele, Mark J. and Schmidt, S. R. (1991), *Basic Statistics: Tools for Continuous Improvement*, 2nd Edition, Air Academy Press, Colorado Springs, CO.

2.   Ott, Ellis R., (1975), *Process Quality Control*, McGraw-Hill, Inc., New York.

Chapter 3

Design Types

3.1 Introduction*

A design matrix displays the levels of each factor for all of the run combinations in the experiment. For example, Table 3.1 contains 3 factors and 4 runs where the first run is made with factors A, B, and C set at values of 20, 800, and 50, respectively.

Run	Factors		
	A	B	C
1	20	800	50
2	20	600	30
3	10	800	30
4	10	600	50

Table 3.1 Actual Settings for 3 Factors in a 4-Run Design Matrix

Although the actual settings are used to run the experiment, for analysis purposes the researchers will code the actual factor values using the transformation in Equation 3.1. Since $\overline{f}_j$ and d_j are known, it is easy to untransform x_j back to f_j whenever necessary. Transforming the actual values in Table 3.1 to coded values results in the design matrix shown in Table 3.2a. For convenience, researchers will sometimes abbreviate the "1" and "−1" with "+" and "−" as shown in Table 3.2b.

---

*This chapter shows the reader the mechanics required for developing good experimental designs. To avoid the complexity contained in this chapter please refer to Appendix M and a software package such as Q-Edge, RS/Discover, Catalyst, DOE/Target, etc.

$$x_j = 2 \frac{f_j - \bar{f}_j}{d_j}$$

where x_j is the coded setting for factor j
            f_j is the actual setting for factor j
            $\bar{f}_j$ is the average of all the actual settings for factor j
            d_j is the distance between the largest and smallest actual settings of factor j

Equation 3.1 Transforming Actual Factor Settings to Coded Values

(a)	Factors			(b)	Factors		
Run	A	B	C	Run	A	B	C
1	+1	+1	+1	1	+	+	+
2	+1	−1	−1	2	+	−	−
3	−1	+1	−1	3	−	+	−
4	−1	−1	+1	4	−	−	+

Table 3.2 Coded Values for 3 Factors in a 4-Run Design Matrix
(Note: Column A began with half + and half − which is the opposite of
Chapter 2; either choice can be used with the same end result.)

Each column of factor values is referred to as a vector in the design matrix. A design is considered to be **balanced vertically** when the transformed values, x_j, for each factor column sum to zero, i.e., $\sum_{i=1}^{n} x_{ji} = 0$ for all factors j. Balanced designs are desirable because they simplify the calculations during analysis, and under certain conditions they lend themselves to orthogonal designs. A 2-level design matrix is said to be **orthogonal** if it is balanced vertically and if the sum of every dot product for all possible variable pairs is zero, i.e., $\sum_{i=1}^{n} x_{ji} x_{ui} = 0$ (for all n combinations of factors j and u, j ≠ u). The dot product test for orthogonality is similar to testing horizontal balancing described in Chapter 2. The reader should verify that the design in Table 3.2 is balanced and orthogonal.

    The reason for orthogonality is that it allows the desired effects to be evaluated independently. A high degree of non-orthogonality results in a dependency of effects which is particularly undesirable.* For example, if effects A and B are dependent and only A is important, B may appear important due to the dependency with A. Even worse, if A and B are both important, but one in a positive direction and the other in a negative direction, their dependency may cause both to appear unimportant.

Designs for 2-level factors that are only balanced vertically will not be orthogonal (see Table 3.3). In this case, factors B and C have the same set of values and are referred to as **aliased**. Notice that the B and C dot product in Table 3.3 sums to 4 and not zero.

Run	Factors A	Factors B	Factors C
1	+1	+1	+1
2	+1	−1	−1
3	−1	+1	+1
4	−1	−1	−1

Table 3.3 Non-Orthogonal Matrix with Aliased Columns B and C

Design matrices can be non-orthogonal even without containing aliased columns. For example, the design in Table 3.4 is a vertically balanced, non-orthogonal design without any two vectors (or columns) which are identical. In Table 3.4, the sum of each possible dot product is not equal to zero and the factors are not aliased; instead, all factors are partially correlated (confounded) with each other. In other words, aliased design columns are perfectly correlated (in a positive or negative sense); whereas partially confounded design columns are neither independent nor aliased. (See Chapter 4 for information on how to use regression to determine if any two columns are aliased, partially correlated, or independent.)

The designs in Tables 3.1 through 3.4 all represent factors measured at only two levels. For that reason, they are typically referred to as **2-level designs** which are particularly useful for factor screening and in estimating first order models with or without interactions.

---

*Software packages such as Q-Edge, RS/Discover, Catalyst, DOE/Target, etc. will automatically generate orthogonal or nearly orthogonal designs. Thus, the reader need not be overly concerned about non-orthogonality problems when using these packages.

The inclusion of a few center points in a 2-level design will create a third level to allow for curvature evaluation; however, there exist several different types of 3-level designs, most of which are discussed in detail later in this chapter. For now, an example of an orthogonal 3-level design is shown in Table 3.5. The reader should be able to verify that all 4 columns in Table 3.5 are balanced vertically and horizontally.

Run	Factors			
	A	B	C	D
1	+1	−1	−1	−1
2	−1	+1	−1	−1
3	−1	−1	+1	−1
4	−1	−1	−1	+1
5	+1	+1	+1	+1
6	−1	+1	+1	+1
7	+1	−1	+1	+1
8	+1	+1	−1	+1
9	+1	+1	+1	−1
10	−1	−1	−1	−1

Table 3.4 Non-Orthogonal Matrix without Aliased Columns

Run	Factors			
	A	B	C	D
1	−	−	−	−
2	−	0	0	0
3	−	+	+	+
4	0	−	0	+
5	0	0	+	−
6	0	+	−	0
7	+	−	+	0
8	+	0	−	+
9	+	+	0	−

Table 3.5 Example of an Orthogonal Design for 4 Factors, Each at 3 Levels (Interactions cannot be evaluated independently in this design.)

3.2   Full Factorial Designs

Given that the experimenter desires to examine 3 factors (A, B, and C) each at two levels, the design used to estimate all possible effects would be a **full factorial design**. The set of all possible effects for this example is described in Table 3.6.

Main Effects	2-Way Interactions	3-Way Interactions
A	AB	ABC
B	AC	
C	BC	

Table 3.6 List of All Possible Effects for a Full Factorial Design
for 3 Factors, Each at 2 Levels

The number of combinations or runs (n) for a full factorial design of k factors will be $n = 2^k$. The number of orthogonal columns representing all possible effects will be $n - 1$. In this example, a total of 7 effects can be analyzed in 8 runs. The full factorial design matrix for 3 factors (2 levels) is designated a 2^3 design matrix and is displayed in Table 3.7. Notice the pattern used to generate the A, B, and C columns. (Any similar scheme can be used to obtain vertical and horizontal balancing of these columns.)

Run	A	B	C	AB	AC	BC	ABC
1	−1	−1	−1	+1	+1	+1	−1
2	−1	−1	+1	+1	−1	−1	+1
3	−1	+1	−1	−1	+1	−1	+1
4	−1	+1	+1	−1	−1	+1	−1
5	+1	−1	−1	−1	−1	+1	+1
6	+1	−1	+1	−1	+1	−1	−1
7	+1	+1	−1	+1	−1	−1	−1
8	+1	+1	+1	+1	+1	+1	+1

Table 3.7 Coded Design Matrix for a Full Factorial of 3 Factors
Each at 2 Levels (2^3 Design)

The experimenter will use the actual setting values for columns A, B, and C when running the experiment. The remaining columns are obtained by multiplying the associated coded settings in each of the factor columns and they are used only in the analysis phase of the experiment. Graphically, runs in Table 3.7 will appear as shown in Figure 3.1.

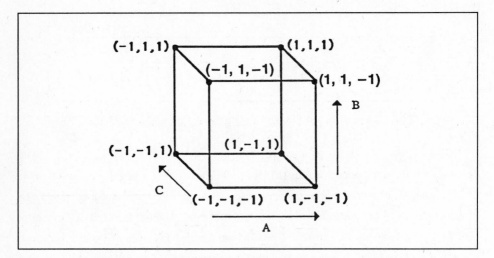

Figure 3.1 Geometric Representation of All Runs in Table 3.7

For 3 factors, each with 3 levels, a full factorial design consists of $3^3 = 27$ runs. The reader should verify the vertical and horizontal balancing of the orthogonal 3 factor (3 level) full factorial design in Table 3.8. The full factorial design in Table 3.8 is graphically displayed in Figure 3.2.

The advantages of full factorial designs are: (i) orthogonality, (ii) no aliasing concerns, and (iii) all main factors and all interactions can be evaluated. The obvious disadvantage is the cost, time, and resources needed to make all the runs required by a full factorial. A full factorial for 7 factors, each at 2 levels, is 2^7 or 128 runs. Three-level factor full factorials are even more cost, time, and resource demanding. For example, a full factorial for 7 factors, each at 3 levels, is 3^7 or 2187 runs. See Appendix M for Rules of Thumb on when to use full factorial designs.

	Factors				Factors				Factors		
Run	A	B	C	Run	A	B	C	Run	A	B	C
1	+	+	+	10	0	+	+	19	−	+	+
2	+	+	0	11	0	+	0	20	−	+	0
3	+	+	−	12	0	+	−	21	−	+	−
4	+	0	+	13	0	0	+	22	−	0	+
5	+	0	0	14	0	0	0	23	−	0	0
6	+	0	−	15	0	0	−	24	−	0	−
7	+	−	+	16	0	−	+	25	−	−	+
8	+	−	0	17	0	−	0	26	−	−	0
9	+	−	−	18	0	−	−	27	−	−	−

Table 3.8 Full Factorial for 3 Factors, Each at 3 Levels
(To save space, interaction and quadratic columns are not displayed.)

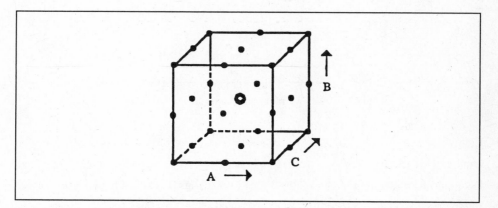

Figure 3.2 Graphical Representation of All Runs in Table 3.8

3.3   Fractional Factorials for 2-Level Designs

Often a researcher does not have the time, funds, or resources to complete a full factorial design. Obviously, it doesn't take too many factors, k, before 2^k or 3^k runs are infeasible (it is not often that you would expect to find 3-way (and higher) interactions to be significant). A type of orthogonal design which allows you to estimate all linear effects and desired interactions while requiring fewer runs is a **fractional factorial**. Given any number of factors, you can build a 2-level fractional factorial design with n = 2^{k-q} runs by completing the following:

(1) Write the full factorial design for k−q factors (q is some positive integer < k).

(2) Alias the extra q factors with higher order interactions which can be assumed to be insignificant. When q > 1 try to use interactions with the same number of letters.

Run	A	B	C	AB	AC	BC	D=ABC
				Factors			
1	−1	−1	−1	+1	+1	+1	−1
2	−1	−1	+1	+1	−1	−1	+1
3	−1	+1	−1	−1	+1	−1	+1
4	−1	+1	+1	−1	−1	+1	−1
5	+1	−1	−1	−1	−1	+1	+1
6	+1	−1	+1	−1	+1	−1	−1
7	+1	+1	−1	+1	−1	−1	−1
8	+1	+1	+1	+1	+1	+1	+1

Table 3.9 A 2^{4-1} Fractional Factorial Design Matrix

Consider the 2^{4-1} design in Table 3.9 as a simple example. The full factorial for k = 4 requires 2^4 = 16 runs. If we can assume an insignificant ABC interaction, then a fractional factorial can be accomplished in 2^{4-1} = 8 runs. This design is referred to as a half fraction where the ABC column is used to generate the coded settings for factor D, i.e., we purposely aliased factor D coded settings with the coded values for ABC. Since each letter in the design matrix represents a column of (+) and (−) elements, aliasing D with ABC implies D = ABC. This does not imply that the D and ABC effects are the same, only that the evaluation of the D and ABC column effects is the same. If the column effect is significant, it is either due to D, ABC, or some combination of the two.

It should be obvious that in orthogonal 2-level designs if any column is multiplied by itself, the result is an identity column, I, which contains all +1 values. Thus, DD or D^2 =

I. Using the **design generator**, $D = ABC$, we can also conclude $D^2 = ABCD = I$. We can now use $I = ABCD$ as a relationship which will define all of the aliasing of effects contained in our fractional factorial design. From this point on, $I = ABCD$ will be referred to as the **defining relation** for the 2^{4-1} fractional factorial design. Since there is only one generator represented in the defining relation for this design, the alias pattern is easily constructed by multiplying both sides of the defining relation by the desired effect. For example, $AB(I) = AB(ABCD) = A^2B^2CD = IICD = CD$; therefore, $AB = CD$. The right side then becomes the alias of the effect on the left (see Table 3.10). Aliasing patterns should be evaluated before experimentation is begun.* If the aliasing of certain effects is undesirable, you should (i) select different generators (ii) make more runs, and/or (iii) hold one or more of the factors constant.

effect	alias
A	BCD
B	ACD
C	ABD
D	ABC
AB	CD
AC	BD
AD	BC
BC	AD
BD	AC
CD	AB
ABC	D
ABD	C
ACD	B
BCD	A
ABCD	I

Table 3.10 Aliasing Pattern for a 2^{4-1} Fractional Factorial Design Where $D = ABC$

---

*Aliasing patterns will usually be obtained from a software package. The intent of this discussion is to show the reader the actual mechanics required.

In the event you run a 2^{4-1} design and later desire to add more runs to overcome the alias problems shown in Table 3.10, you need only run the other half fraction using the negative of the generator ABC = D, i.e., build another 8 runs with factor settings for A, B, and C the same as in the first 8 runs and this time let D = −ABC. Then augment the second set of 8 runs with the previous 8 runs to make a 2^4 full factorial. The D factor will be de-aliased from the ABC interaction; however, the ABCD interaction will now be aliased with any effect produced by differences in the two blocks or sets of 8 runs. If this column has a significant effect, the experimenter would investigate the cause, i.e., changes in materials, manpower, environment, etc. or possible ABCD interactions. More details on overcoming aliasing problems are discussed in the foldover design section.

A more complex example of a fractional factorial would be to have 7 factors each at 2 levels. Not wanting n = 2^7 = 128 runs, you choose to fractionate the design so that n=16, i.e., you want a 2^{7-3} design. The design consists of a full factorial for (7 − 3) factors with 3 higher order interactions as generators for factors E, F, and G.

Run	\multicolumn{7}{c}{Factors}
	A	B	C	D	E=ABC	F=BCD	G=ABD
1	+	+	+	+	+	+	+
2	+	+	+	−	+	−	−
3	+	+	−	+	−	−	+
4	+	+	−	−	−	+	−
5	+	−	+	+	−	−	−
6	+	−	+	−	−	+	+
7	+	−	−	+	+	+	−
8	+	−	−	−	+	−	+
9	−	+	+	+	−	+	−
10	−	+	+	−	−	−	+
11	−	+	−	+	+	−	−
12	−	+	−	−	+	+	+
13	−	−	+	+	+	−	+
14	−	−	+	−	+	+	−
15	−	−	−	+	−	+	+
16	−	−	−	−	−	−	−

Table 3.11 Coded Matrix for a 2^{7-3} Design (Interactions other than generators are not shown for convenience − evaluation of the matrix would include all 15 orthogonal columns.)

The 1/8 fraction for $k = 7$ contains the generators $E = ABC$, $F = BCD$, and $G = ABD$ which result in the following defining words: $I = ABCE = BCDF = ABDG$. Multiplying any 2 defining words at a time results in $I^2 = I = ADEF = CDEG = ACFG$ and multiplying all 3 at one time yields $I^3 = I = BEFG$. Therefore, the complete defining relation consists of:

$$I = ABCE = BCDF = ABDG = ADEF = CDEG = ACFG = BEFG$$

This defining relation will allow the easy computation of the aliasing pattern for the entire design. For example, if you want to determine all effects aliased with factor A, just multiply the defining relation by A which then appears as:

$$A = BCE = ABCDF = BDG = DEF = ACDEG = CFG = ABEFG$$

Similarly, the aliasing pattern for EF is:

$$EF = ABCF = BCDE = ABDEFG = AD = CDFG = ACEG = BG$$

In general, there exists 2^q different fractions of a 2^{k-q} design. For our 2^{7-3} example, there exists $2^q = 8$ different 1/8 fractional designs, represented by "+" or "−" in the previous set of generators, i.e.,

$$E = \pm ABC, \quad F = \pm BCD, \text{ and } G = \pm ABD.$$

To de-alias any particular factor from its 3-way interaction generator, simply add a different 1/8 fraction to the first one. For example, to de-alias factor E from ABC, take two 1/8 fractions, one with $E = ABC$ and the other $E = -ABC$.

At this time the reader is ready to understand the term **resolution** of a design. In general, the resolution of a 2-level design is the length (number) of the shortest word (set of letters) in the defining relation. The previous 2^{7-3} fractional factorial has a resolution of IV (i.e., the 2^{7-3} design is a resolution four design). The meaning of different resolution levels is as follows:

R_V: A design that does not alias the main effects with each other or with 2-way interactions. In addition, 2-way interactions are not aliased with one another. Thus, this design provides for unconfounded main and 2-way interaction effects. R_V designs are excellent for building prediction equations that will typically not have serious interaction concerns.

R_{IV}: A design which does not alias mains with 2-way interactions, but does alias 2-way interactions with other 2-way interactions. R_{IV} designs are used for building prediction equations when resource limitations do not permit R_V designs. Prior knowledge must be used to determine which aliased effect goes into the prediction equation.

R_{III}: A design which does not alias main effects with one another, but does alias mains with 2-way interactions. R_{III} designs are typically used when screening a large number of factors to find only the most important factors for future experimentation.

R_{II}: A design which contains main effects aliased with other main effects.

A R_{II} design is referred to as **supersaturated** and is not recommended. When n is approximately equal to k and 2-way interactions are assumed insignificant or if you are screening a large number of factors, a R_{III} design is used. R_{III} designs are sometimes referred to as **saturated** designs because all or most orthogonal columns are assigned factors. If 2-way interactions are of interest, **unsaturated** designs of R_{IV} or R_V are suggested.

For a few fractional factorial examples, consider the following:

i) 2^{5-2}

Run	A	B	C	AB	AC	D = BC	E = ABC
1	−	−	−	+	+	+	−
2	−	−	+	+	−	−	+
3	−	+	−	−	+	−	+
4	−	+	+	−	−	+	−
5	+	−	−	−	−	+	+
6	+	−	+	−	+	−	−
7	+	+	−	+	−	−	−
8	+	+	+	+	+	+	+

I = BCD = ABCE = ADE
Design is of Resolution III.

ii) 2^{6-3}

Run	A	B	C	AB	D = AC	E = BC	F = ABC
1	−	−	−	+	+	+	−
2	−	−	+	+	−	−	+
3	−	+	−	−	+	−	+
4	−	+	+	−	−	+	−
5	+	−	−	−	−	+	+
6	+	−	+	−	+	−	−
7	+	+	−	+	−	−	−
8	+	+	+	+	+	+	+

I = ACD = BCE = ABCF = ABDE = BDF = AEF = CDEF
Design is of Resolution III.

iii) 2^{5-1}

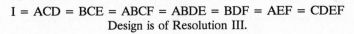

Run	A	B	C	D	AB	AC	AD	BC	BD	CD	ABC	ABD	ACD	BCD	E=ABCD
1	−	−	−	−											
2	−	−	−	+											
3	−	−	+	−											
4	−	−	+	+											
5	−	+	−	−											
6	−	+	−	+											
7	−	+	+	−											
8	−	+	+	+											
9	+	−	−	−											
10	+	−	−	+											
11	+	−	+	−											
12	+	−	+	+											
13	+	+	−	−											
14	+	+	−	+											
15	+	+	+	−											
16	+	+	+	+											

Note: The reader should be able to develop these columns by using appropriate column value multiplication.

I = ABCDE
Design is of Resolution V.

iv) 2^{6-2}

Run	A	B	C	D	AB	AC	AD	BC	BD	CD	ABC	ABD	E=ACD	F=BCD	ABCD
1	−	−	−	−											
2	−	−	−	+											
3	−	−	+	−											
4	−	−	+	+											
5	−	+	−	−											
6	−	+	−	+											
7	−	+	+	−											
8	−	+	+	+											
9	+	−	−	−											
10	+	−	−	+											
11	+	−	+	−											
12	+	−	+	+											
13	+	+	−	−											
14	+	+	−	+											
15	+	+	+	−											
16	+	+	+	+											

Note: The reader should be able to develop these columns by using appropriate column value multiplication.

$$I = ACDE = BCDF = ABEF$$
Design is of Resolution IV.

Note: A rule of thumb for n ≥ 16 is to avoid (when possible) selecting generators from interactions which have consecutive length, i.e., avoid using both a 3-way and a 4-way interaction as generators. The reader should verify that for the 2^{6-2} design above, if E = ABCD, the resolution would have been III instead of IV.

v) 2^{7-3} (see Table 3.11)

vi) 2^{8-4}

Run	A	B	C	D	AB	AC	AD	BC	BD	CD	E=ABC	F=ABD	G=ACD	H=BCD	ABCD
1															
2															
3															
4															
5															
6															
7					Note: By now the reader should be able to										
8					develop the coded values for all										
9					of these columns.										
10															
11															
12															
13															
14															
15															
16															

I = ABCE = ABDF = ACDG = BCDH = CDEF = BDEG = ADEH = BCFG = ACFH
= ABGH = AEFG = BEFH = DFGH = CEGH = ABCDEFGH
This design is of Resolution IV.

In the very recent past, classical statistics and short courses typically spent chapters on the concept of alias patterns, fractionalization, and resolution. With the advent of powerful software, computers can quickly determine what the student was once asked to spend many hours on. Since this is the case, we shall leave you with the previous set of examples and forge on into additional useful design concepts. (Also see Appendix M for Rules of Thumb on when to use fractional factorial designs.)

3.4   Foldover Designs

A **foldover design** of resolution IV consists of two R_{III} designs augmented where the second R_{III} design has all the settings "+1" or "−1" reversed. The new design will have twice as many runs as the original R_{III}. An additional factor can be added by placing a column of "−1" values for the first R_{III} runs and a column of "+1" values for the second R_{III} runs. As an example, consider the 2^{3-1} design of resolution III in Table 3.12.

	Factors		
Run	A	B	C = A × B
1	−	−	+
2	−	+	−
3	+	−	−
4	+	+	+

Table 3.12 Coded Matrix for a 2^{3-1} R$_{III}$ Design

The defining relation for the design in Table 3.12 is I = ABC. Therefore, the 2^{3-1} design is of resolution III. By folding over this R$_{III}$ design, you obtain the 8 runs displayed in Table 3.13. For both sets of 4 runs, D = −ABC or I = −ABCD which is the only defining relation. The resulting 8-run design resolution is R$_{IV}$. The additional factor D is only used as a blocking variable to evaluate differences between the two blocks of 4 runs. If factor D appears significant during the analysis, the experimenter would investigate the cause of this difference.

	Factors			
Run	A	B	C	D = new factor
1	−1	−1	+1	−1
2	−1	+1	−1	−1
3	+1	−1	−1	−1
4	+1	+1	+1	−1
5	+1	+1	−1	+1
6	+1	−1	+1	+1
7	−1	+1	+1	+1
8	−1	−1	−1	+1

Table 3.13 Foldover Design Matrix

A foldover design is typically used if you initially attempted to model a process with a low resolution design (R_{III} or R_{IV}) and the confirmatory runs failed to validate the model due to suspected interaction aliasing problems. The foldover design will then provide a new augmented matrix with a higher resolution. For more information on foldover designs, see references [4] and [6].

3.5   Plackett-Burman Designs

These R_{III} designs were developed by Plackett and Burman in 1946 [8] for screening a large number of factors, i.e., reducing a large number of factors to a smaller set of important factors for subsequent experimentation. The disadvantage of using only fractional factorial designs for screening large numbers of factors is that the number of runs is a function of powers of 2, i.e., $n = 2^{k-q} = 4, 8, 16, 32, 64, 128 \ldots$. The **Plackett-Burman** (P−B) designs are based on Hadamard matrices which have more flexibility. The number of runs in a **Hadamard matrix** is a multiple of 4, i.e., $n = 4, 8, 12, 16, 20 \ldots$. Each column contains elements which are 1 or −1 and all (n−1) columns are balanced and pairwise orthogonal (see John [5]). Plackett and Burman use the Hadamard matrices for their designs in order to study up to (n−1) factors in n runs. When the number of runs in a P-B design is $n = 2^{k-q}$, the design is referred to as a geometric P-B design which is essentially the same as the 2-level fractional factorial discussed previously. If you built a fractional factorial design, multiplied the even ordered interactions by (−1) and then permuted the rows and columns, you would get a geometric P-B design. All other 2-level P-B designs, such as n = 12, 20, 24, 28, 36, 40, etc., are non-geometric and have a different confounding structure of the main effects and 2-way interactions. For geometric P-B designs of R_{III}, it can be shown that each 2-way interaction will be positively or negatively aliased with a main effect (see Table 3.14 and verify D = −AB). However, the non-geometric P-B designs contain main effects which are partially confounded (intercorrelated) with 2-way interactions (see Table 3.15). In fact, each 2-way interaction is partially confounded with each of the other main effects, i.e., the BC interaction of Table 3.15 is intercorrelated with the A, D, E, F, and G main effects. The simplest way to determine the amount of intercorrelation between any two effects is to use the correlation coefficient discussed in Chapter 4.

	Factors						
Run	A	B	C	D	E	F	G
1	+	−	−	+	−	+	+
2	+	+	−	−	+	−	+
3	+	+	+	−	−	+	−
4	−	+	+	+	−	−	+
5	+	−	+	+	+	−	−
6	−	+	−	+	+	+	−
7	−	−	+	−	+	+	+
8	−	−	−	−	−	−	−

Table 3.14 Plackett-Burman (P-B) 8 Run Design

	Factors										
Run	A	B	C	D	E	F	G	H	I	J	K
1	+	−	+	−	−	−	+	+	+	−	+
2	+	+	−	+	−	−	−	+	+	+	−
3	−	+	+	−	+	−	−	−	+	+	+
4	+	−	+	+	−	+	−	−	−	+	+
5	+	+	−	+	+	−	+	−	−	−	+
6	+	+	+	−	+	+	−	+	−	−	−
7	−	+	+	+	−	+	+	−	+	−	−
8	−	−	+	+	+	−	+	+	−	+	−
9	−	−	−	+	+	+	−	+	+	−	+
10	+	−	−	−	+	+	+	−	+	+	−
11	−	+	−	−	−	+	+	+	−	+	+
12	−	−	−	−	−	−	−	−	−	−	−

Table 3.15 Plackett-Burman (P-B) 12 Run Design

To build a Plackett-Burman design, take a single generating vector for a particular size n design and generate the remaining vectors. For example, if n = 8 combinations, the unique generating vector consists of the following (n−1) values (+ + + − + − −). The design matrix is completed as follows: (1) use the generating vector as column A, (2) build column B by making the last value in A the first value in B, then slide the rest of the A values below that value, (3) vector C is made by using the last value in B as the first value in C with the rest of B below it, (4) continue until all k columns are complete, and (5) add an nth row of all "−" values as shown in Table 3.14.

Generating columns for various P-B designs are as follows:

n = 8   (+ + + − + − −)
n = 12   (+ + − + + + − − − + −)
n = 16   (+ + + + − + − + + − − + − − −)
n = 20   (+ + − − + + + + − + − + − − − − + + −)
n = 24   (+ + + + + − + − + + − − + + − − + − + − − − −)

The obvious advantage of P-B designs is the limited number of runs to evaluate large numbers of factors. Since interactions are typically not evaluated for P-B designs, the effects for all the factors are placed in a Pareto chart and the obvious important factors can then be selected for more in-depth testing.

The disadvantages of P-B designs are tied to the assumptions required to evaluate up to (n−1) factors in n runs. The assumptions are that interactions (if they exist) are not strong enough to mask main effects and quadratic effects (if they exist) are closely related to the factor linear effects. For most processes and most types of factors evaluated over reasonable ranges, these assumptions are often fairly accurate.

As a final note, any R_{III} P-B designs can be folded over to achieve a design of resolution IV. For more information on P-B designs, see the article by Plackett and Burman [8]. (Also see Appendix M for Rules of Thumb on the use of screening designs.)

3.6   Non-Orthogonal 2-Level Designs

There are also non-orthogonal 2-level design types worthy of discussion: the random balanced, foldover Koshal, and D-optimal designs. The **random balanced** design is typically used as a supersaturated or saturated design and is formed by randomly generating the first n/2 runs for k factors. In order to balance the design, you fold it over, generating an equal number of "+" and "−" values in each column.   In this case, n can be determined independently of k, which can be advantageous.   The primary disadvantage is that the confounding structure for main effects and interactions is totally random and outside of the experimenter's control. However, a design can be generated and intercorrelations calculated (see Chapter 4) for all possible pairs of effects. Then, if any paired correlations are too high, they can be avoided or the design can be regenerated randomly until satisfactory levels of intercorrelation are achieved. The random balanced design is not highly recommended.

A balanced non-orthogonal design which is closely related to the one-at-a-time design is a **Foldover Koshal**. Basically, the Foldover Koshal is a one-at-a-time design for 2 levels which is folded over to achieve balance. Except for n=8, the design is far from orthogonal, which is why simulation studies indicate it is only about 60 to 70 % effective in discriminating significant main and 2-way interaction effects when the number of factors k is less than 20. Intercorrelations will remain low enough, provided k is less than or equal to 20.   Due to increased main effect correlations as k increases, the effectiveness deteriorates for larger k. There are three reasons for including this design in this book: (1) some consultants are using it to screen large numbers of variables even though better alternatives exist such as Plackett-Burman and fractional factorials, (2) if a one-at-a-time or Koshal design has already been run as shown in Table 3.16, then folding over the design results in the Foldover

Run	Factors			
	A	B	C	D
1	−	−	−	−
2	+	−	−	−
3	−	+	−	−
4	−	−	+	−
5	−	−	−	+

Table 3.16 One-at-a-Time Design (First Half of a Foldover Koshal)

Koshal of Table 3.17 provides the means for possibly salvaging a previously bad design, and (3) if you are conducting a failure analysis study and desire to swap components between a good (+) and bad (−) system, Shainin [10] has popularized a component swapping technique based on the foldover Koshal matrix.

Run	Factors			
	A	B	C	D
1	−	−	−	−
2	+	−	−	−
3	−	+	−	−
4	−	−	+	−
5	−	−	−	+
6	+	+	+	+
7	−	+	+	+
8	+	−	+	+
9	+	+	−	+
10	+	+	+	−

Table 3.17 A Foldover Koshal Design of 10 Runs

The last and most important non-orthogonal design to be discussed is the **D-optimal** design. If we were to call our design matrix X, then premultiplying X with its transpose is $X'X$. D-optimality refers to a design that maximizes the determinant of $X'X$ which is called the D value, i.e., $D = |X'X|$. Optimal values of D vary with n (the number of runs) and for some n values, D-optimal designs will be orthogonal. When n is other than a value associated with orthogonal designs, the columns of "+" and "−" values must be generated so as to maximize $|X'X|$. Sophisticated software packages such as RS/Discover, SAS, E-Chip, and Catalyst will do this for you, i.e., generate a D-optimal or close to D-optimal design for any desired number of runs, n. It should be noted that less than orthogonal designs ought to be avoided when possible. The lack of orthogonality will confound effects and complicate the analysis. However, in situations where it is too costly or impossible to generate a fractional design which is orthogonal, the D-optimal alternative is very helpful. Consider the following three examples where the D-optimal design is useful. In example one, assume we are interested in 4 factors at 2 levels with all 2-way interactions. Due to cost and time constraints, we need the number of runs to be 13 instead of 16. First of all, the odd number of runs will prevent the design from being balanced or orthogonal. The D-optimal design will generate 13 runs where the corresponding $|X'X|$ is maximized, thus, minimizing the amount of non-orthogonality.

As a second example, consider a design with 5 factors A, B, C, D, and E. Factor A is qualitative with 3 levels, B is qualitative with 4 levels, C is qualitative with 2 levels, and D and E are quantitative with 3 levels. Assume the $C \times D$ and $C \times E$ interactions are also of interest and the number of runs must be less than 25. In this case, standard designs will not be of much assistance. However, the D-optimal alternative will provide a design with $|X'X|$ close, if not equal, to its maximum for some specific n < 25.

The last example consists of 4 quantitative factors each at 3 levels where the (+, +, +, +) and (−, −, −, −) combinations are infeasible. Using the D-optimal option of the appropriate software package, you can generate a near orthogonal design excluding the infeasible points. Thus, the advantages of D-optimal designs include (i) ability to handle different types of factors with different numbers of levels, (ii) exclusion of infeasible points, and (iii) desired number of runs. Disadvantages include non-orthogonality and specific software needs. (See pages 5-44, 8-100, and 8-155 for some D-optimal design matrices contained in this text).

D-optimality is one of several related design optimality criteria which can be used to select the best design points from a list of candidate points. The other most common criteria are A-, G-, and V-optimality. All four types employ slightly different variance minimizing criteria. If we let $\underline{b}$ be the column matrix of regression coefficients (see multiple regression in Chapter 4) it can be shown that:

1) $\underline{b} = (X'X)^{-1} X'Y$
2) $\text{Var} (\underline{b}) = (X'X)^{-1} \sigma^2$
3) $\text{Var} (\hat{Y} (X)) = X'(X'X)^{-1}X \sigma^2$, where $\hat{Y} (X)$ is the predicted response at some point X.

A-optimality seeks to minimize the average variance of the elements of $\underline{b}$. G-optimality seeks to minimize the maximum prediction variance, max $[d = X'(X'X)^{-1} X\sigma^2]$, for a given set of design points. V-optimality minimizes the average value of d over a specified set of design points [11].

3.7   Fractional Factorials and Latin Squares for 3-Level Designs

Fractional factorials in designs of 3 levels, 3^{k-q} designs, are a bit more complicated. They are designs which include a mixture of corner points and mid-level points. Because of this combination of experimental points, these designs can be used to test all linear effects, all 2nd order (quadratic) effects, and limited linear and quadratic interactions. The 3-level fractional factorial and full factorial designs are the only designs discussed in this book which can evaluate quadratic interactions. An example of a quadratic interaction term is AB^2 (see more on quadratic interactions in Chapter 4). The drawbacks of the 3^{k-q} design are the number of runs required to estimate all desired effects, the limited number of interactions which can be evaluated, and the complexity of the confounding/aliasing patterns. Even if quadratic interactions and simple 2-way interactions are not estimated, n must still be large to estimate 2nd order terms for all factors. Since the aliasing pattern for a 3^{k-q} design is difficult to obtain, it will not be discussed in this book. Several software packages will do this for you, or you can use reference [7] or [1].

One method of formulating a 3^{k-1} fractional design or 1/3 fraction is through the use of **Latin Squares**. For example, given the Latin Square design in Table 3.18, a list of all combinations of the 3 factors results in a balanced orthogonal 3^{k-1} design shown in Table 3.19. When k = 4, a Latin cube is constructed to build a 3^{4-1} design and Latin hypercubes are the basis for designs with k > 4. As with the use of regular Latin Square designs, interactions are not typically available for evaluation.

		Factor B			
		−1	0	+1	
	−1	−1	0	+1	
Factor C	0	0	+1	−1	Factor A
	+1	+1	−1	0	(coded values are inside the square)

Table 3.18 Latin Square Used to Generate a Latin Square Design

The 3-level fractional factorial designs are very useful for qualitative factors which have few (if any) anticipated significant interactions. These designs are also used for 3-level screening designs. However, if all the factors are quantitative and you want to evaluate factor interactions, the 3 level fractional factorial design will be very inefficient (see the comparisons of 3-level design types in section 3.9).

	Factors		
Run	A	B	C
1	−	−	−
2	0	0	−
3	+	+	−
4	0	−	0
5	+	0	0
6	−	+	0
7	+	−	+
8	−	0	+
9	0	+	+

Table 3.19 Coded Matrix for Latin Square Design for 3 Factors, Each at 3 Levels

3.8 Box-Behnken Designs

An efficient and frequently used 3-level design for modeling quantitative factors with 3 levels is the **Box-Behnken** (B-B). Through the use of embedded 2^k designs while holding certain factors at their centerpoint, the B-B designs are much more efficient than 3^k full factorials and some types are easily blocked. In their 1960 article, Box and Behnken [2] provide tabled B-B designs for k up to 16 (excluding k = 8). When k = 4, the design, in shorthand notation, appears as shown in Table 3.20a. This design is divided into 3 orthogonal blocks of 9 runs where each ±1 refers to alternating columns of (− − + +) and (− + − +). The completed 27 run Box-Behnken for k = 4 is shown in Table 3.20b.

	Factors		
A	B	C	D
±1	±1	0	0
0	0	±1	±1
0	0	0	0
±1	0	0	±1
0	±1	±1	0
0	0	0	0
±1	0	±1	0
0	±1	0	±1
0	0	0	0

Table 3.20a Shorthand Notation for a Box-Behnken Design for 4 Factors

Run	A	B	C	D	
1	−	−	0	0	
2	−	+	0	0	
3	+	−	0	0	
4	+	+	0	0	
5	0	0	−	−	
6	0	0	−	+	
7	0	0	+	−	
8	0	0	+	+	
9	0	0	0	0	
10	−	0	0	−	
11	−	0	0	+	Note: Interaction and quadratic columns
12	+	0	0	−	can be developed by the appropriate
13	+	0	0	+	column value multiplications.
14	0	−	−	0	
15	0	−	+	0	The orthogonal blocks are separated by
16	0	+	−	0	dashed lines.
17	0	+	+	0	
18	0	0	0	0	
19	−	0	−	0	
20	−	0	+	0	
21	+	0	−	0	
22	+	0	+	0	
23	0	−	0	−	
24	0	−	0	+	
25	0	+	0	−	
26	0	+	0	+	
27	0	0	0	0	

Table 3.20b Complete Set of 27 Runs for a Box-Behnken Design Based on 4 Factors

The B-B design does not contain any corner points in the design space, which may or may not be of concern. In some experiments where corner points are infeasible, the B-B may be an attractive alternative. If only 3 factors were used, the design would appear graphically as shown in Figure 3.3. The B-B designs are nearly orthogonal resolution V designs that allow for estimating main effects, quadratic effects, and all linear 2-way interactions. The primary disadvantage of B-B designs is that the number of runs is large enough to estimate all factor 2nd order effects and all linear 2-way interactions, whether you want to or not. Typically, the B-B design will be less efficient with regard to the number of runs than the 3^{k-q} designs unless several 2-way interactions are of interest (see section 3.10 for a comparison). The slight non-orthogonality is not a disadvantage when the analysis is conducted by way of least square regression (see Chapter 4).

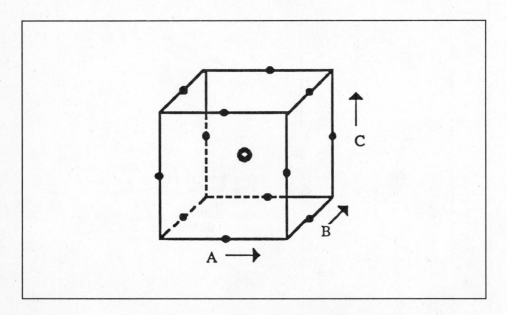

Figure 3.3 Geometric Representation of a Box-Behnken Design for 3 Factors

A summary of shorthand notation for Box-Behnken designs given k = 3, 4, and 5 is displayed in Table 3.21.

Box-Behnken Designs						
Number of factors (k)	Design Matrix				Number of Points	Blocking and Centerpoint Information
3		±1	±1	0	4	No orthogonal blocking 3 replicated center points
		±1	0	±1	4	
		0	±1	±1	4	
		0	0	0	3	
					n = 15	
4	±1	±1	0	0	4	3 blocks of 9 1 center point per block
	0	0	±1	±1	4	
	0	0	0	0	1	
	±1	0	0	±1	4	
	0	±1	±1	0	4	
	0	0	0	0	1	
	±1	0	±1	0	4	
	0	±1	0	±1	4	
	0	0	0	0	1	
					n = 27	
5	±1 ±1	0	0	0	4	2 blocks of 23 3 replicated center points per block
	0 0	±1	±1	0	4	
	0 ±1	0	0	±1	4	
	±1 0	±1	0	0	4	
	0 0	0	±1	±1	4	
	0 0	0	0	0	3	
	0 ±1	±1	0	0	4	
	±1 0	0	±1	0	4	
	0 0	±1	0	±1	4	
	±1 0	0	0	±1	4	
	0 ±1	0	±1	0	4	
	0 0	0	0	0	3	
					n = 46	

Table 3.21 Box-Behnken Designs for k = 3, 4, and 5

3.9   Box-Wilson (Central Composite) Designs

A very flexible and efficient 2nd order modeling design for quantitative factors is the Box-Wilson or **Central Composite Design** (CCD). For 3 factors (k = 3), CCD examples appear as shown in Tables 3.22a and b.

a) CCD k = 3, using 2^3					b) CCD k = 3, using 2^{3-1}				
	FACTOR					FACTOR			
Run	A	B	C		Run	A	B	C	
1	−	−	−		1	−	−	−	
2	−	−	+		2	−	+	+	F
3	−	+	−		3	+	−	+	
4	−	+	+	F	4	+	+	−	
5	+	−	−		5	0	0	0	
6	+	−	+		6	0	0	0	C
7	+	+	−		7	0	0	0	
8	+	+	+		8	α	0	0	
9	0	0	0		9	$-\alpha$	0	0	
10	0	0	0		10	0	α	0	
11	0	0	0	C	11	0	$-\alpha$	0	A
12	0	0	0		12	0	0	α	
13	0	0	0		13	0	0	$-\alpha$	
14	0	0	0						
15	α	0	0						
16	$-\alpha$	0	0		F =	factorial portion			
17	0	α	0	A	C =	centerpoint portion			
18	0	$-\alpha$	0		A =	axial portion			
19	0	0	α						
20	0	0	$-\alpha$						

Table 3.22 Central Composite Designs for 3 Factors a) Full Factorial 2-Level Portion   b) Fractional Factorial 2-Level Portion

Graphically, the design from Table 3.22a will appear as indicated in Figure 3.4.

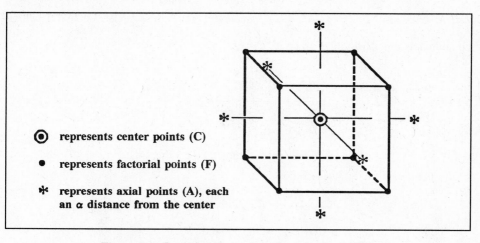

Figure 3.4 Graphical Representation of a k=3 CCD

If 2-way interactions are potentially important, you can use a R_{IV} or R_V design in the factorial or F portion of the design; whereas, if they are not you can save runs by using a R_{III}. The number of axial points will typically be 2k unless some factors don't lend themselves to quadratic effects. The value for α is usually set at $(n_F)^{1/4}$ where n_F is the number of runs in the factorial portion of the design. These values are chosen to produce **rotatability,** which simply implies that the predicted response is capable of being estimated with equal variance regardless of the direction from the center of the design space. Rotatability is a nice statistical property to have; however, it is not required for successful designs. You should notice that, in order to satisfy rotatability, α will take on values greater than 1 which means that ±1 no longer represents the factor minimum and maximum. An easy way to resolve the transformation to or from the design notation of 0, ±1, ±α is as follows:

(1) Set up a number line as shown below.

design values: $-\alpha$ -1 0 1 α

(2) Place the real factor value min and max (use 100 and 300 as an example) under $-\alpha$ and α, respectively. Place $\dfrac{\text{max} + \text{min}}{2}$ under 0. The values under -1 and 1 will be computed as the value $200 \pm$ some Δ.

design values:	$-\alpha$	-1	0	1	α
real values:	100	$(200-\Delta)$	200	$(200+\Delta)$	300

(3) To find the values for Δ, you need to know your α value (assume it is 1.5), and then form the following ratio:

$$\frac{300 - 200}{1.5} = \frac{\Delta}{1} \ , \quad \text{therefore,} \quad \Delta = 66.\overline{66}$$
$$\text{implying} \ \ 200 + \Delta = 266.\overline{66} \ \text{and} \ 200 - \Delta = 133.\overline{33}$$

The final result appears as follows:

design values:	$-\alpha$	-1	0	1	α
real values:	100	133.33	200	266.67	300

which indicates the real factor values for all 5 possible design levels. Sometimes it may be infeasible to run 5 levels of a factor, yet efficient 2nd order modeling is desired. In this case you can set $\alpha = 1$ for a Central Composite Faced (CCF) design. This design will not have rotatability and will suffer some loss of orthogonality for the quadratic terms; however, it has been used successfully in numerous case studies.

The number of center points, n_c , will vary, but basically, enough are needed to get a good estimate of pure experimental error and to maintain orthogonality. According to John [5], $n_c = 4\sqrt{n_F + 1} - 2k$ will produce orthogonality and rotatability for a CCD. For more information, see Box and Draper [3] or Myers [7]. Depending on which resolution you use in the 2^{k-q} portion of a CCD, it can be up to 47% more efficient in experimental runs than the B-B design [3]. In addition, the CCD is rotatable for $\alpha = (n_F)^{1/4}$ whereas the B-B is only rotatable for specific values of k.

The CCD has several advantages as shown below:

(i) It is the most effective and efficient 2nd order modeling design for factors which are quantitative.

(ii) It has flexibility in resolution selection, i.e., you can build the factorial portion for any resolution you might require.

(iii) Aliasing for R_{III} or R_{IV} designs is computed the same way you would for the 2-level fractional factorials.

(iv) CCD designs can be run sequentially to possibly save resources, i.e., run factorial and center point portions first and only add the axial portion if a quadratic relationship appears probable.

The primary disadvantage is that the CCD is best suited for quantitative factors. If only one qualitative factor exists, a CCD can still be modified to handle it, but when you have several qualitative factors you would need to use a fractional factorial or D-optimal design. The slight non-orthogonality of the CCD is not a disadvantage when the analysis is conducted by way of least squares regression (see Chapter 4).

Consider another example where the CCD might be your choice of design type. Assume you have 5 factors (all quantitative) and you wish to build a 2nd order model with all important 2-way linear interactions. Since interactions are a big concern, you require a R_{IV} or R_V design. Starting with the factorial (or 2 level) portion, you could:

(i) Let $n_F = 2^5 = 32$ runs, i.e., the full factorial which would allow for estimating all interactions.

(ii) Let $n_F = 2^{5-1} = 16$ runs where factor E = ABCD and I = ABCDE. This design is R_V and a big savings over the full factorial.

(iii) Let $n_F = 2^{5-2} = 8$ runs where D = AB, E = ABC, and I = ABD = ABCE = DCE. Since this design is less than R_{IV}, it appears that the best 2-level design for the factorial portion is a 2^{5-1}.

Next you need to determine the number of center points $n_c = 4\sqrt{n_F + 1} - 2k$ which results in $n_c = 4\sqrt{16 + 1} - 2(5) \doteq 7$. The number of axial points is $n_a = 2k = 10$ with a recommended α value of $\alpha = (n_F)^{1/4} = (16)^{1/4} = 2.0$. The complete CCD for this example is shown in Table 3.23.

Run	A	B	C	D	E = ABCD	
1	−	−	−	−	+	
2	−	−	−	+	−	
3	−	−	+	−	−	
4	−	−	+	+	+	
5	−	+	−	−	−	
6	−	+	−	+	+	
7	−	+	+	−	+	
8	−	+	+	+	−	
9	+	−	−	−	−	
10	+	−	−	+	+	
11	+	−	+	−	+	
12	+	−	+	+	−	
13	+	+	−	−	+	Note: All quadratic and 2-way
14	+	+	−	+	−	linear interaction columns can be
15	+	+	+	−	−	developed by appropriate column
16	+	+	+	+	+	value multiplications.
17	0	0	0	0	0	
18	0	0	0	0	0	
19	0	0	0	0	0	
20	0	0	0	0	0	
21	0	0	0	0	0	
22	0	0	0	0	0	
23	0	0	0	0	0	
24	α	0	0	0	0	
25	$-\alpha$	0	0	0	0	
26	0	α	0	0	0	
27	0	$-\alpha$	0	0	0	
28	0	0	α	0	0	
29	0	0	$-\alpha$	0	0	
30	0	0	0	α	0	
31	0	0	0	$-\alpha$	0	
32	0	0	0	0	α	
33	0	0	0	0	$-\alpha$	

Table 3.23 CCD for k = 5 with R_V

3.10 Comparison of 3-Level Designs

The information in Table 3.24 is provided as a comparison of the number of runs required for the 3^{k-q}, B-B, CCD, D-optimal (D-opt), and the full factorial given 3-level designs with various numbers of quantitative factors and 2-way interactions of interest. Numbers in parentheses indicate the number of replicated center points included in the design. The 3^{k-q} designs are similar to the Taguchi 3-level designs discussed in section 3.11.

Scenario	3^{k-q}	B-B	CCD**	D-opt***	Full Factorial
k = 3 with no interactions	9(0)	15(3)	13(3)*	7(0)	27(1)
k = 4 with 3 2-way linear interactions	27(0)	27(3)	20(4)*	11(0)	81(1)
k = 5 with 1 2-way linear interactions	27(0)	46(6)	20(2)*	12(0)	243(1)
k = 5 with all 2-way linear interactions	81(0)	46(6)	33(7)*	21(0)	243(1)
k = 7 with 7 2-way linear interactions	81(0)	62(6)	33(3)*	22(0)	2187(1)

$^{*}n_c = 4\sqrt{n_F + 1} - 2k$

**The total number of runs for the CCD includes 2k axial points. In many cases all the factors will not necessarily have a significant 2nd order term. Sequentially adding axial points in the CCD until there is no more evidence of another quadratic term will typically lead to less than 2k axial points.

***This is a minimum number of required runs; however most software packages will add 2 to these numbers.

Table 3.24 Comparison of Design Types for Quantitative Factor Scenarios

As you can see, the CCD is far more efficient than the 3^{k-q}, B-B, and Full Factorial in all but one of the situations presented. Keep in mind, though, the objective for each of the designs as shown in Table 3.25.

3.11 Taguchi Designs

The last set of designs to be discussed are those of Taguchi [9]. These designs are typically orthogonal with respect to main effects, but contain either aliased or confounded mains with 2-way interactions. They are essentially in an R_{III} category. The designs used

Design	Objective
3^{k-q}	• Used for qualitative or quantitative factors • Estimate all linear and quadratic effects and, when higher order resolutions are designed, you can also estimate linear and quadratic 2-way interactions. • Typically the 3^{k-q} is a good choice if you have qualitative factors with few or no interactions. This is not a good choice for factors that are quantitative.
B-B	• Used only for quantitative factors; however, for some k you can have one qualitative blocking factor. • Estimate all linear, quadratic, and 2-way linear interactions plus experimental error.
CCD	• Used primarily for quantitative factors; however, if you only have 1 qualitative factor, the CCD can still be useful. • Estimate all linear effects, selected quadratics, and selected 2-way linear interactions plus experimental error. • This is typically your best choice for quantitative factors each at 3 levels.
D-optimal	• Used for quantitative, qualitative, or a mix of both. • Can estimate any desired effect with reduced n. • This design will typically not be orthogonal, but it does minimize the non-orthogonality of the design.
Full Factorial	• Used for qualitative or quantitative factors. • Estimate all linear and quadratic effects plus all possible simple and higher order interactions.

Table 3.25 Summary of 3-Level Design Types

by Taguchi are not new, but simply a form of Plackett-Burman, Fractional Factorial, or Latin Square designs. The fact that they are tabled provides for easy use. The most frequently used Taguchi designs are shown in Table 3.26. Tables 3.27 through 3.33 provide more information on each of these design types. Notice that the 2-level Taguchi designs (L_4, L_8, L_{16}) are the same as the 2^{k-q} designs except that all even numbered interaction columns have been multiplied by (-1). See Appendix M for Rules of Thumb on the recommended use of these designs.

The most important contribution of Taguchi is not the orthogonal arrays which have been around for some time. Instead, it is his development of a loss function and robust designs which have made important contributions. These robust designs allow the investigation of factors which affect the response value and the variability or dispersion of the response. The result is a design which determines factor settings which optimize the response and minimize response variability. A more detailed discussion on Taguchi methods appears in Chapters 5 and 6.

Design	# of Levels	# of Factors for the Full Factorial	# of Factors to Maintain Resolution V	# of Factors for Screening
L_4	2	2	2	3
L_8	2	3	3	7
L_9	3	2	–	4
L_{12}	2	–	–	11
L_{16}	2	4	5	15
L_{18}	mixed	–	–	8
L_{27}	3	3	–	13

Table 3.26 Most Frequently Used Taguchi Designs

	L_4 Design		
Run #	1	2	3
1	−1	−1	−1
2	−1	+1	+1
3	+1	−1	+1
4	+1	+1	−1
Original factors and interactions used to generate the matrix	a	b	−ab
Strategy	# Factors	Which Columns?	
Full Factorial	2	1, 2	
Resolution V	2	1, 2	
Screening	3	1, 2, 3	

Table 3.27 L_4 Design [9] for 2-Level Factors

Run #	L₈ Design						
	1	2	3	4	5	6	7
1	−1	−1	−1	−1	−1	−1	−1
2	−1	−1	−1	+1	+1	+1	+1
3	−1	+1	+1	−1	−1	+1	+1
4	−1	+1	+1	+1	+1	−1	−1
5	+1	−1	+1	−1	+1	−1	+1
6	+1	−1	+1	+1	−1	+1	−1
7	+1	+1	−1	−1	+1	+1	−1
8	+1	+1	−1	+1	−1	−1	+1
Original factors and interactions used to generate the matrix	a	b	−ab	c	−ac	−bc	abc

Strategy	# Factors	Which Columns?
Full Factorial	3	1, 2, 4
Resolution V	3	1, 2, 4
Screening	4-7	1, 2, 4, 7, ...

Table 3.28 L₈ Design [9] for 2-Level Factors

	L_9 **Design**			
Run #	1	2	3	4
1	−1	−1	−1	−1
2	−1	0	0	0
3	−1	+1	+1	+1
4	0	−1	0	+1
5	0	0	+1	−1
6	0	+1	−1	0
7	+1	−1	+1	0
8	+1	0	−1	+1
9	+1	+1	0	−1
Original factors and interactions used to generate the matrix	a	b	−ab	ab^2
Strategy	**# Factors**		**Which Columns?**	
Full Factorial	2		1, 2	
Screening	3-4		1, 2, 3, ...	

Table 3.29 L_9 Design [9] for 3-Level Factors

					L_{12} **Design**						
Run #	1	2	3	4	5	6	7	8	9	10	11
1	−1	−1	−1	−1	−1	−1	−1	−1	−1	−1	−1
2	−1	−1	−1	−1	−1	+1	+1	+1	+1	+1	+1
3	−1	−1	+1	+1	+1	−1	−1	−1	+1	+1	+1
4	−1	+1	−1	+1	+1	−1	+1	+1	−1	−1	+1
5	−1	+1	+1	−1	+1	+1	−1	+1	−1	+1	−1
6	−1	+1	+1	+1	−1	+1	+1	−1	+1	−1	−1
7	+1	−1	+1	+1	−1	−1	+1	+1	−1	+1	−1
8	+1	−1	+1	−1	+1	+1	+1	−1	−1	−1	+1
9	+1	−1	−1	+1	+1	+1	−1	+1	+1	−1	−1
10	+1	+1	+1	−1	−1	−1	−1	+1	+1	−1	+1
11	+1	+1	−1	+1	−1	+1	−1	−1	−1	+1	+1
12	+1	+1	−1	−1	+1	−1	+1	−1	+1	+1	−1

Strategy	# Factors	Which Columns?
Screening	8-11	any

Table 3.30 L_{12} Design [9] for 2-Level Factors
(This is the same matrix as the P-B, 12 run design; however,
the columns and rows have been permuted.)

							L_{16} Design								
Run #	1	2	3	4	5	6	7	8	9	10	11	12	13	14	15
1	−1	−1	−1	−1	−1	−1	−1	−1	−1	−1	−1	−1	−1	−1	−1
2	−1	−1	−1	−1	−1	−1	−1	+1	+1	+1	+1	+1	+1	+1	+1
3	−1	−1	−1	+1	+1	+1	+1	−1	−1	−1	−1	+1	+1	+1	+1
4	−1	−1	−1	+1	+1	+1	+1	+1	+1	+1	+1	−1	−1	−1	−1
5	−1	+1	+1	−1	−1	+1	+1	−1	−1	+1	+1	−1	−1	+1	+1
6	−1	+1	+1	−1	−1	+1	+1	+1	+1	−1	−1	+1	+1	−1	−1
7	−1	+1	+1	+1	+1	−1	−1	−1	−1	+1	+1	+1	+1	−1	−1
8	−1	+1	+1	+1	+1	−1	−1	+1	+1	−1	−1	−1	−1	+1	+1
9	+1	−1	+1	−1	+1	−1	+1	−1	+1	−1	+1	−1	+1	−1	+1
10	+1	−1	+1	−1	+1	−1	+1	+1	−1	+1	−1	+1	−1	+1	−1
11	+1	−1	+1	+1	−1	+1	−1	−1	+1	−1	+1	+1	−1	+1	−1
12	+1	−1	+1	+1	−1	+1	−1	+1	−1	+1	−1	−1	+1	−1	+1
13	+1	+1	−1	−1	+1	+1	−1	−1	+1	+1	−1	−1	+1	+1	−1
14	+1	+1	−1	−1	+1	+1	−1	+1	−1	−1	+1	+1	−1	−1	+1
15	+1	+1	−1	+1	−1	−1	+1	−1	+1	+1	−1	+1	−1	−1	+1
16	+1	+1	−1	+1	−1	−1	+1	+1	−1	−1	+1	−1	+1	+1	−1
	a	b	−ab	c	−ac	−bc	abc	d	−ad	−bd	abd	−cd	acd	bcd	−abcd

Strategy	# Factors	Which Column?
Full Factorial	4	1, 2, 4, 8
Resolution V	5	1, 2, 4, 8, 15
Screening	6-15	1, 2, 4, 8, 14, 7, 11, 13, 15, 12, 10, ...

Table 3.31 L_{16} Design [9] for 2-Level Factors

Run #	1	2	3	4	5	6	7	8
				L$_{18}$ Design				
1	−1	−1	−1	−1	−1	−1	−1	−1
2	−1	−1	0	0	0	0	0	0
3	−1	−1	+1	+1	+1	+1	+1	+1
4	−1	0	−1	−1	0	0	+1	+1
5	−1	0	0	0	+1	+1	−1	−1
6	−1	0	+1	+1	−1	−1	0	0
7	−1	+1	−1	0	−1	+1	0	+1
8	−1	+1	0	+1	0	−1	+1	−1
9	−1	+1	+1	−1	+1	0	−1	0
10	+1	−1	−1	+1	+1	0	0	−1
11	+1	−1	0	−1	−1	+1	+1	0
12	+1	−1	+1	0	0	−1	−1	+1
13	+1	0	−1	0	+1	−1	+1	0
14	+1	0	0	+1	−1	0	−1	+1
15	+1	0	+1	−1	0	+1	0	−1
16	+1	+1	−1	+1	0	+1	−1	0
17	+1	+1	0	−1	+1	−1	0	+1
18	+1	+1	+1	0	−1	0	+1	−1

Strategy	# Factors	Which Column?
Screening	8 (max)	2-level factor in column 1
		3-level factors in columns 2-8

Table 3.32 L$_{18}$ Design [9] for (1) 2-Level Factor and (7) 3-Level Factors

								L$_{27}$ Design					
Run #	1	2	3	4	5	6	7	8	9	10	11	12	13
1	−1	−1	−1	−1	−1	−1	−1	−1	−1	−1	−1	−1	−1
2	−1	−1	−1	−1	0	0	0	0	0	0	0	0	0
3	−1	−1	−1	−1	+1	+1	+1	+1	+1	+1	+1	+1	+1
4	−1	0	0	0	−1	−1	−1	0	0	0	+1	+1	+1
5	−1	0	0	0	0	0	0	+1	+1	+1	−1	−1	−1
6	−1	0	0	0	+1	+1	+1	−1	−1	−1	0	0	0
7	−1	+1	+1	+1	−1	−1	−1	+1	+1	+1	0	0	0
8	−1	+1	+1	+1	0	0	0	−1	−1	−1	+1	+1	+1
9	−1	+1	+1	+1	+1	+1	+1	0	0	0	−1	−1	−1
10	0	−1	0	+1	−1	0	+1	−1	0	+1	−1	0	+1
11	0	−1	0	+1	0	+1	−1	0	+1	−1	0	+1	−1
12	0	−1	0	+1	+1	−1	0	+1	−1	0	+1	−1	0
13	0	0	+1	−1	−1	0	+1	0	+1	−1	+1	−1	0
14	0	0	+1	−1	0	+1	−1	+1	−1	0	−1	0	+1
15	0	0	+1	−1	+1	−1	0	−1	0	+1	0	+1	−1
16	0	+1	−1	0	−1	0	+1	+1	−1	0	0	+1	−1
17	0	+1	−1	0	0	+1	−1	−1	0	+1	+1	−1	0
18	0	+1	−1	0	+1	−1	0	0	+1	−1	−1	0	+1
19	+1	−1	+1	0	−1	+1	0	−1	+1	0	−1	+1	0
20	+1	−1	+1	0	0	−1	+1	0	−1	+1	0	−1	+1
21	+1	−1	+1	0	+1	0	−1	+1	0	−1	+1	0	−1
22	+1	0	−1	+1	−1	+1	0	0	−1	+1	+1	0	−1
23	+1	0	−1	+1	0	−1	+1	+1	0	−1	−1	+1	0
24	+1	0	−1	+1	+1	0	−1	−1	+1	0	0	−1	+1
25	+1	+1	0	−1	−1	+1	0	0	0	−1	0	−1	+1
26	+1	+1	0	−1	0	−1	+1	+1	+1	0	+1	0	−1
27	+1	+1	0	−1	+1	0	−1	−1	−1	+1	−1	+1	0
	a	b	−ab	ab^2	c	−ac	ac^2	−bc	abc	ab^{2}c^2	bc^2	−ab^{2}c	−abc^2

Strategy	# Factors	Which Columns?
Full Factorial	3	1, 2, 5
Screening	8-13	1, 2, 5, ...

Table 3.33 L$_{27}$ Design [9] for 3-Level Factors

3.12 Design Selection Scenarios

The following scenarios have been written to give you a feel for the process of design selection. Each time you run an experiment a similar process should take place.

Scenario 1: You have 11 quantitative factors and you wish to i) separate the important factors from the unimportant factors, and ii) build a model relating the important factors with the response.

Screening Phase Options:
(A)   Use a P-B 12 run design and Pareto the effects to select the important factors.
(B)   Use a 2^{11-7} fractional factorial design to separate the important factors while also assessing a few potentially strong interactions.
(C)   Use an L_{27} design to screen important factors from a linear and a quadratic perspective.
(D)   Use a D-optimal design built for studying all linear effects, desired quadratics, and desired interactions.

Modeling Options: (Assume 4 factors were screened out in the previous phase − each factor is quantitative. In addition, quadratics and 2-way linear interactions are possible.)
(A)   Use an L_{27} design which provides most of the interaction information plus all linear and all quadratic information.
(B)   Use a B-B design for 4 factors with n = 27 (see Table 3.21). This design will be R_V, allowing you to estimate all linear, all quadratic, and all 2-way linear interactions.
(C)   Use a CCD design for (i) R_{IV} or (ii) R_V.
   (i)   Factorial portion has $n_F = 2^{4-1} = 8$ runs (R_{IV}).
         $n_c = 4\sqrt{8 + 1} - 2(4) = 4$ runs
         $n_a = 2(k) = 8$ runs
         $\alpha = (n_F)^{1/4} = (8)^{1/4} = 1.68$ or use $\alpha = 1$ for a CCF
         n(total) = 20 runs

        (ii)   factorial portion has $n_F = 2^4 = 16$ runs (full factorial)

$$n_c = 4\sqrt{16 + 1} - 2(4) = 9 \text{ runs}$$

$$n_a = 2(k) = 8 \text{ runs}$$

$$\alpha = (n_F)^{1/4} = (16)^{1/4} = 2.0 \text{ or use}$$

$$\alpha = 1 \text{ for a CCF design.}$$

$$n(\text{total}) = 33$$

(D)   Use a D-optimal design where $n_{min} = 4 + 4 + 6 + 1 = 15$ is derived from <u>4</u>

linear terms, <u>4</u> quadratics, $\binom{4}{2} = \dfrac{4!}{2!\ 2!} = \underline{6}$ 2-way linear interactions,

and <u>1</u> constant. Most software packages will automatically add two more tests when building the D-optimal design.

The actual design(s) to address scenario 1 will depend on several key issues such as:

        (i)   time and dollars available

        (ii)   time and cost per run

        (iii)   amount of prior knowledge of the process

        (iv)   software support for generating and analyzing designs

        (v)   experience level of the experimenter

As in most cases, there is no perfect choice of designs. All of the choices have advantages and disadvantages – some choices are better than others. You eventually have to choose which strategy best suits your needs.

Scenario 2: You have 1 qualitative factor with 2 levels and 5 quantitative factors with potential quadratics and 2-way interactions. Your objective is to model the response with respect to the input factors. Consider the following approaches.

(A)   (i)   Screen the factors with an L_{18} to identify important factors based on linear and quadratic effects.

        (ii)   Model important factors with a CCD. If the qualitative factor is found important from step i) then you could run 2 modeling experiments for each level of the qualitative factor. Another approach would be to leave the qualitative factor in the CCD and run half the center points at its $(-)$ and the other half at its $(+)$. The axial points can all be run at either the $(-)$ or $(+)$

setting for the qualitative factor. Some orthogonality is lost in the approach, but usually not enough to be overly concerned.

(B) (i) Screen the factors with a L_8 design.

(ii) Model the important factors with CCD described in (A) (ii) or use a D-optimal design.

(C) (i) Model the 6 factors using a D-optimal design where n_{min} = $\underline{1}$ (for the qualitative factor) + $\underline{5}$ (for the quantitative linear effects) + $\underline{5}$ (for the quantitative quadratic effects) + $\underline{15}$ (for all 2-way linear interactions) + $\underline{1}$ (for estimating the constant). Therefore, n_{min} = 27.

As in scenario 1, we cannot say that you should always take a particular approach. Obviously the choice you make depends on many concerns. The important point in this section is for you to realize that any experimental design can be approached in different ways. It is always best to brainstorm all the possibilities and then select the best choice for you and for meeting your objective. It is also important to use the Rules of Thumb from Appendix M to determine the sample size per combination and thus the total number of resources required for each design option. The option which provides the most information for the least number of resources is usually the best one.

Chapter 3 Problem Set

1. Describe the similarities among the following design types: (i) a Taguchi L_8 design, (ii) a fractional factorial 2^{7-4}, and (iii) a fractional factorial 2^{4-1}.

2. What are the similarities/differences between fractional factorial, geometric Plackett-Burman, non-geometric Plackett-Burman, and Taguchi designs?

3. Define the following terms:
    - Orthogonal designs

    - Resolution of a design

    - Aliased

    - Confounded

    - Generator

    - Defining word

    - Defining relationship

4. Discuss when you would use the following designs:

    - 3-level fractional factorial (Taguchi)

    - Box-Behnken

    - Box-Wilson (CCD)

    - D-optimal

5.   How can you use sequential experimentation to increase resolution from III to IV?

6.   Given a 2^{7-2} design where F=ABCD and G=BCDE, determine the defining relationship.

7.   For the example on page 3-10, let E=ABCD. How does this affect the defining relationship?

8.   Why should you use balanced and orthogonal designs?

9.   What is a saturated design?

10.   Given the generating line for a n=16 Plackett-Burman design (see page 3-17), write the entire design matrix. Compare this design with a 16 run fractional factorial and a L_{16}.

11.   What 3 level designs can you use for qualitative factors? Quantitative factors? Mixtures of both?

12.   If Taguchi designs are not new, what has been his major contribution to experimental design?

13.   A chemical engineer with a major oil refinery wanted to try a number of different additives to boost the octane level of the unleaded fuel produced at the refinery. For the sake of secrecy his additives were uniquely coded: A, B, C, D, E. All additives were varied from their lowest to their highest. The engineer knew from prior experience that the following additives would interact with each other: AB, AE, BD, DE, CE, and AD. To make matters worse, his boss told him that if his experiment did not produce positive results, he could forget about any further raises. The

engineer ran the following screening design:
Note: D=AC, E=BC

Runs	A	B	C	D = AC	E = BC
1	+1	+1	+1	+1	+1
2	+1	+1	−1	−1	−1
3	+1	−1	+1	+1	−1
4	+1	−1	−1	−1	+1
5	−1	+1	+1	−1	+1
6	−1	+1	−1	+1	−1
7	−1	−1	+1	−1	−1
8	−1	−1	−1	+1	+1

(a)   What is the defining relation?

(b)   Are any important interactions aliased with main effects? If you answer "yes", list the alias problems you found?

(c)   Generate a foldover of the design generated for this problem.

(d)   What is the defining relation for the foldover design?

(e)   Would a foldover design salvage this experiment?

(f)   Should the engineer forget about his future raises?

14.   Given the following factor conditions:

(a)   3 quantitative factors (3 levels for each factor)

(b)   3 quantitative factors (3 levels for each factor)
      1 qualitative factors (3 levels for each factor)

    (c)   3 quantitative factors (3 levels for each factor)
               2 qualitative factors (3 levels for each factor)
               2 qualitative factors (4 levels for each factor)

For each scenario, determine the minimum number of runs for a D-optimal design (Hint: you will have to count up the total degrees of freedom of the effects to be estimated).

15.   A scrubber is used to remove particles from a wafer surface in semiconductor processing. An engineer would like to determine the most important variables that optimize the scrubber's ability to remove particles. There are 8 factors that can be manipulated. The engineer does not want to spend too much time doing this experiment. Given the following data:

Factors	Levels	Units
A: Pre-Rinse Speed (PRS)	500 – 1500	RPM
B: Pre-Rinse Time (PRT)	10 – 25	RPM
C: Scrub Speed (SRS)	500 – 1500	RPM
D: Scrub Time (SRT)	10 – 50	Sec
E: Post-Rinse Speed (PSRS)	500 – 1500	RPM
F: Post-Rinse Time (PSRT)	10 – 25	Sec
G: Dry Speed (DRS)	4000 – 5000	RPM
H: Dry Time (DRT)	25 – 50	Sec

Note: The interactions of interest are AB, CD, EF, and GH

(a)   Generate the most effective screening experiment.

(b)   What is the defining relationship?

(c)   Are any of your important interactions aliased with any main effects?

(d)   What is the resolution of your design?

16.   A researcher in immunology wanted to determine what effect various concentrations of nitrogen dioxide would have on the ability of the lungs to fight off airborne infections such as pneumonia. Her factors were as follows:

Factors	Absolute Low and High
A: Nitrogen Dioxide Concentration	.5% to 2%
B: Environmental Humidity	10% to 80%
C: Environmental Temperature	50°F to 90°F

She decides to create a Central Composite Design.

(a)   What is the α distance for this experiment?

(b)   What are the design and real values for each factor?

(c)   How many center points should this design have in order to be rotatable with uniform precision?

(d)   How many total runs are needed for this experiment?

17.   In the following design matrix, determine the degree of confounding between factors (i.e., factor intercorrelations).

Runs	A	B	C
1	+1	+1	+1
2	+1	+1	−1
3	−1	+1	+1
4	−1	−1	−1
5	+1	−1	+1
6	+1	−1	−1
7	−1	+1	+1
8	−1	−1	−1

Hint: To find the correlation between A and B, regress A on B and look at R^2.

18.   If an engineer was running a 8 factor screening experiment, how many possible 2 factor interactions are there?

Chapter 3 Bibliography

1.   Barker, Thomas B. (1985), *Quality By Experimental Design*, Marcel Dekker Inc.

2.   Box, George E.P. and Behnken, D.W. (1960), "Some New Three Level Designs for the Study of Quantitative Variables", *Technometrics*, 2, pp 455-475.

3.   Box, George E.P. and Draper, Norman R. (1987), *Empirical Model-Building and Response Surfaces*, John Wiley and Sons, Inc.

4.   Box, George E.P.; Hunter, William G.; and Hunter, J. Stuart (1978), *Statistics for Experimenters*, John Wiley and Sons, Inc.

5.   John, Pere W.M. (1971), *Statistical Design and Analysis of Experiments*, MacMillan Company.

6.   Montgomery, Douglas C. (1984), *Design and Analysis of Experiments*, John Wiley and Sons, 2nd Edition.

7.   Myers, Raymond H. (1976), *Response Surface Methodology*, Virginia Polytechnic Institute.

8.   Plackett,R.L. and Burman, J.P. (1946), "The Design of Optimum Multifactorial Experiments", *Biometricka*, 33, pp 305-325.

9.   Taguchi, G. and Konishi, S. (1987), *Taguchi Methods: Orthogonal Arrays and Linear Graphs*, American Supplier Institute, Inc.

10.   Bhote, Keki R. (1988), *World Class Quality*, AMA Membership Publications Division.

11.   Cornell, John A. (1990), *Experiments with Mixtures*, 2nd Edition, John Wiley and Sons, New York.

Chapter 4

Statistical Techniques

4.1   Introduction

In order to fully comprehend subsequent discussions on experimental design, the reader should have a working knowledge of inferential statistical techniques. This chapter covers the broad topics of Analysis of Variance (ANOVA), Simple Linear Regression (SLR), and Multiple Linear Regression (MLR) in enough detail to provide the required level of expertise for subsequent use. In no way does this chapter exhaust these three topics.

4.2   Analysis of Variance

In Chapter 2 we discussed a non-statistical approach for determining which factor effects and/or interactions should go into the prediction equation and which ones should be left out. In this section we provide a statistical approach to assist in making that decision.

As an example, a company wants to investigate ways to improve a process measured by the output variable y. It is decided to test the process at two different temperatures ($70°$ and $90°$) to determine if there is a temperature effect. In order to infer that temperature did or did not cause a difference in response, random assignments of experimental units to each temperature are made. The randomization procedure results in spreading all possible uncontrolled variable effects evenly over the two treatments. Thus, if there does exist a significant difference in group means, it can be concluded that temperature did in fact cause that difference, and not some uncontrolled factor.

To conduct the randomization procedure, you could simply flip a coin where a head implies $70°$ and a tail is $90°$. Say the first coin flip resulted in a head; then run #1 would be made with the temperature at 70°. The process continues until a predetermined number of runs are made for each temperature. Random numbers, such as those found in Appendix A, are typically used when the number of experimental conditions (runs) are greater than two.

Sample data used to demonstrate ANOVA when there is one input factor is presented in Table 4.1.

Run #	Factor Temperature (A)	Replicated Response Values				$\bar{y}_r$	S^2_r	df_r*
		y_1	y_2	y_3	y_4			
1	70° (−)	2.2	2.8	3.2	3.6	2.95	.357	3
2	90° (+)	4.6	5.0	5.4	5.8	5.20	.267	3
Average @ (−) 2.95								
Average @ (+) 5.20			grand mean = 4.075					
Δ 2.25								

Table 4.1 Sample Data for 1-Factor ANOVA

The average response and variance for each run, r, in the design matrix are found by using Equations 4.1 and 4.2.

$$\bar{y}_r = \sum_1^{n_r} \frac{y_i}{n_r} \quad \text{where } n_r = \text{number of response values per run}$$

Equation 4.1 Average Response for each Run in the Design Matrix

$$S^2_r = \sum_1^{n_r} \frac{(y_i - \bar{y}_r)^2}{(n_r - 1)}$$

Equation 4.2 Sample Variance for each Run in Design Matrix

* The df_r column, which refers to degrees of freedom, is $(n_r - 1)$ for each run.

Using run #1 as an example,

$$\bar{y}_1 = \frac{2.2 + 2.8 + 3.2 + 3.6}{4} = 2.95$$

$$S_1^2 = \frac{(2.2 - 2.95)^2 + (2.8 - 2.95)^2 + (3.2 - 2.95)^2 + (3.6 - 2.95)^2}{3} = .357$$

The Δ in Table 4.1 represents the difference in response average for temperature at the $(+)$ minus the response average when temperature is at the $(-)$ (i.e., $\bar{y}_{(+)} - \bar{y}_{(-)}$). It is obvious from the data that $\bar{y}_{(+)} > \bar{y}_{(-)}$, but is this difference large enough, based on sample variations, to conclude that the population means, $\mu_{(+)}$ and $\mu_{(-)}$, are different? If we conclude a significant difference in $\mu_{(+)}$ and $\mu_{(-)}$, then temperature would be included in the prediction equation (i.e., $\hat{y} = \bar{y} + (\Delta/2)A$, where A is in coded values of -1 through $+1$). However, if $\mu_{(+)}$ and $\mu_{(-)}$ are not found to be significantly different, then our one factor experiment would have a simple prediction equation, $\hat{y} = \bar{y}$. To complete the statistical test requires some knowledge of hypothesis tests.

A statistical hypothesis is a statement or claim about some unrealized true state of nature. The actual hypotheses to be tested consist of two complementary statements about the true state of nature. For our example, the hypotheses would appear as follows:

H_0: Mean response at low temperature is equal to the mean response at high temperature.

H_1: Mean response at low temperature differs from the mean response at high temperature.

These two complementary statements are defined as the null hypothesis (H_0) and the alternative hypothesis (H_1). If we conclude H_0, temperature does not go into the prediction equation; whereas, concluding H_1 implies temperature does go into the prediction equation. Since the true state of nature is seldom (if ever) known with 100% certainty, the two statements serve as strawmen to be tested based on sample data results. Probability and statistics are combined to infer from the sample data results to the entire population (the true state of nature) with a measurable amount of uncertainty.

An analogy of hypothesis testing can be drawn from our legal system where an accused on trial is presupposed to be innocent unless the prosecution presents overwhelming evidence to convict him. In this example, the hypotheses to be tested are stated as:

H_0:   Defendant is Innocent

H_1:   Defendant is Guilty

Regardless of the jury's conclusion, they are never really sure of the true state of nature. Concluding "H_0: Defendant is Innocent" does not mean that the defendant is in fact innocent. An H_0 conclusion simply means that the evidence was not overwhelming enough to justify a conviction. On the other hand, concluding H_1 does not prove guilt; rather, it implies that the evidence is so overwhelming that the jury can have a high level of confidence in a guilty verdict.

Since verdicts are arrived at with less than 100% certainty, either conclusion has some probability of error. Consider Table 4.2

| | | True State of Nature | |
		H_0	H_1
Conclusion Drawn	H_0	Conclusion is Correct	Conclusion results in a Type II error
	H_1	Conclusion results in a Type I error	Conclusion is Correct

Table 4.2 Type I or II Error Occurs if Conclusion Is Not Correct

The probability of committing a Type I error is defined as α $(0 \leq \alpha \leq 1)^*$ and the probability of committing a Type II error is β $(0 \leq \beta \leq 1)$.

In the courtroom example, α, the probability of convicting an innocent person, is of critical concern. To minimize the risk of such an erroneous conclusion, our court system

---

* This is not the same α described in the Central Composite Designs.

requires overwhelming evidence to conclude H_1. Although minimizing α has its advantages, it should be obvious that requiring overwhelming evidence to conclude H_1 will in turn increase β, the probability of a Type II error. To resolve this dilemma, statistical hypothesis tests are designed such that

(1) The most critical decision error is a Type I error.

(2) α is set at a minimum level, usually .10, .05, .01, or .001, depending on the degree of criticality of such an error.

(3) Based on (1) and (2) above, the hypothesis statement to be tested with $(1 - \alpha)100\%$ confidence is placed in H_1.

(4) The nature of most statistical hypothesis tests requires that the equality condition be placed in H_0.

(5) To minimize β while holding α constant requires increased sample sizes.

In our example, α represents the risk of placing temperature in the prediction equation when it is really an unimportant factor. When optimizing any prediction equation, the optimal settings for a factor represented in the equation will depend on the size and sign of its coefficient ($\Delta/2$ for 2-level designs). Therefore, placing an unimportant factor in the prediction equation (committing a Type I error) might limit our ability to also minimize cost. To minimize this type of error, α is usually set at .05 for industrial experiments.

The term "analysis of variance" indicates we will be testing the hypotheses

$$H_0: \quad \mu_{(+)} = \mu_{(-)}$$
$$H_1: \quad \mu_{(+)} \neq \mu_{(-)}$$

with an analysis based on estimates of variance. Since variation exists within each run of replicated response values and also between each low and high average response, we can obtain two estimates of the overall population variability, σ^2 (also referred to as experimental variance).

The variance from each run could be used to estimate σ^2; however, a better estimate will be obtained by pooling each of the run S_r^2 values. This pooled estimate of run variances is referred to as the mean square error (MSE) and is found as shown in equation 4.3.

$$MSE = \frac{\sum (df_r)(S_r^2)}{\sum (df_r)} = \frac{\sum (n_r - 1)(S_r^2)}{\sum (n_r - 1)}$$

where each sum, Σ, is over all design matrix runs.

Equation 4.3 Estimating σ^2 with MSE

Assuming H_0 is true, $\bar{y}_{(+)}$ and $\bar{y}_{(-)}$ are both estimates of μ and vary about μ with variance $\sigma_{\bar{y}}^2 = \sigma^2 / n'$ where n' is the number of data values averaged to obtain a $\bar{y}$. In terms of the population variance, $\sigma^2 = (n')(\sigma_{\bar{y}}^2)$. Thus, another estimate for σ^2 called mean square between (MSB) can be found using equation 4.4.

$$MSB = (n')(S_{\bar{y}}^2)$$

which for 2-level designs results in a more simple form,

$$MSB = \frac{N}{4}(\Delta^2)$$

where N is the total number of response values obtained in the experimental matrix.

Equation 4.4 Estimating σ^2 with MSB

When H_0 is true, both MSE and MSB are independent estimates of σ^2 and thus their observed ratio, $F_0 = \frac{MSB}{MSE}$, has an F distribution. In addition, H_0, being true, results in an F_0 value which is usually close to 1.0. If H_0 is not true, there should not be any discernable change in MSE; however, MSB should increase due to an expected increased difference in $\bar{y}_{(+)}$ and $\bar{y}_{(-)}$, causing $S_{\bar{y}}^2$ to increase. Therefore, values of $\frac{MSB}{MSE}$ larger than 1.0 are evaluated using the F distribution to assess their probability (p) of being that large due to chance alone or because H_1 is true. Low probabilities (i.e., less than .05) typically indicate that the ratio is large not due to chance when H_0 is true, but rather because H_0 is not true. This leads to the conclusion that there is a significant factor effect. The test is formalized by the following steps:

Step (1)   Define hypotheses. H_0: $\mu_{(+)} = \mu_{(-)}$
H_1: $\mu_{(+)} \neq \mu_{(-)}$

Step (2)   Determine α.

Step (3)   Compute MSB, MSE, and $F_0 = \dfrac{MSB}{MSE}$. The degrees of freedom for MSB, df_B, is the number of levels minus 1, i.e., $df_B = 1$ for 2-level designs. The degrees of freedom for MSE, $df_E = \Sigma(n_r - 1)$. These degrees of freedom are used to determine a critical F value, F_C , which is compared with F_0.

Step (4)   In an F table, look up $F(1 - \alpha, df_B, df_E) = F_C$. Using Appendix F, $a = 1 - \alpha$.

Step (5)   Compare the ratio $F_0 = \dfrac{MSB}{MSE}$ to F_C from (4).

If $F_0 \leq F_C$, fail to reject H_0.

If $F_0 > F_C$, reject H_0 with $(1 - \alpha)100\%$ confidence of being correct.

(See Appendix M for Rules of Thumb on significant F values from a designed experiment.)

For our example, the five steps appear as follows:

Step (1) H_0: $\mu_{(+)} = \mu_{(-)}$
H_1: $\mu_{(+)} \neq \mu_{(-)}$

Step (2)   select $\alpha = .05$

Step (3) $MSB = \dfrac{N}{4}(\Delta^2) = \dfrac{8}{4}(2.25)^2 = 10.125$

$$MSE = \frac{\Sigma (n_r - 1)(S_r^2)}{\Sigma (n_r - 1)} = \frac{(4 - 1)(.357) + (4 - 1)(.267)}{(4 - 1) + (4 - 1)} = .312$$

$$F_0 = \frac{10.125}{.312} = 32.452$$

Step (4) $F_C = F(.95, 1, 6) = 5.99$

Step (5)   Since $F_0 > F_C$, reject H_0 and conclude H_1 with at least 95% confidence. Also notice that $F(.99, 1, 6) = 13.75$ implies we could be as high as 99% confident in concluding H_1. Thus, the prediction equation generated from our experiment will be $\hat{y} = 4.075 + 1.125A$, where A is the coded temperature value.

For a second one-factor design example, assume you are working on a new composite material and you desire to test the effect of two different processing temperature settings on the strength of the material. Based on engineering knowledge, the autoclave temperature settings to be tested are 200°F and 300°F. Samples of 9 composite materials are taken at each setting. The sample results are shown in Table 4.3.

Run #	Factor Temp(A)	Strength Measurements										
		y_1	y_2	y_3	y_4	y_5	y_6	y_7	y_8	y_9	$\bar{y}$	S^2
1	200°(−)	2.8	3.6	6.1	4.2	5.2	4.0	6.3	5.5	4.5	4.6889	1.3761
2	300°(+)	7.0	4.1	5.7	6.4	7.3	4.7	6.6	5.9	5.1	5.8667	1.1575

Avg (−)	4.6889	Grand Average = 5.2778
Ave (+)	5.8667	
Δ	1.1778	

Table 4.3 Composite Material Data (Low = 200°, High = 300°)

The hypotheses to be tested are:

H_0: $\mu_{(+)} = \mu_{(-)}$

H_1: $\mu_{(+)} \neq \mu_{(-)}$

These hypotheses are portrayed graphically in Figure 4.1.

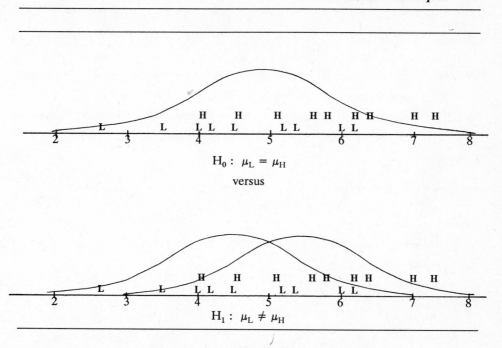

$$H_0 : \mu_L = \mu_H$$

versus

$$H_1 : \mu_L \neq \mu_H$$

Figure 4.1 Illustration of a 1 Factor (2 Level) Hypotheses Test
(where low and high setting responses are identified with
an L and H respectively)

The graphics in Figure 4.1 seem to indicate that H_1 is very likely; however, without conducting the statistical test, it is difficult to determine if the evidence is overwhelming. The actual hypothesis test produces the following.

Step (1) H_0: $\mu_{(+)} = \mu_{(-)}$

             H_1: $\mu_{(+)} \neq \mu_{(-)}$

Step (2) $\alpha = .05$

Step (3) $\text{MSB} = \dfrac{N}{4}(\Delta^2) = \dfrac{18}{4}(1.1778)^2 = 6.2425$

$$\text{MSE} = \frac{\sum (n_r - 1)(S_r^2)}{\sum (n_r - 1)} = \frac{8(1.3761) + 8(1.1575)}{8 + 8} = 1.2668$$

$$F_0 = \frac{MSB}{MSE} = \frac{6.2425}{1.2668} = 4.9278$$

Step (4) $F_C = F(.95, 1, 16) = 4.49$ (Since Appendix F is missing, $v_2 = 16$, you would find the most conservative $F_C = F(.95, 1, 15) = 4.54$.)

Step (5)   Since $F_0 > F_C$, we reject H_0. (See Appendix M for Rules of Thumb on significant F values from a designed experiment.)

When factors in the experiment have only 2 levels, we can also complete the hypotheses test using the t distribution. A t test for this same problem would be:

Step (1) H_0: $\mu_{(+)} = \mu_{(-)}$

H_1: $\mu_{(+)} \neq \mu_{(-)}$

Step (2) $\alpha = .05$

Step (3) $t_0 = \dfrac{\bar{y}_{(+)} - \bar{y}_{(-)}}{S_p \sqrt{\dfrac{1}{n_{(+)}} + \dfrac{1}{n_{(-)}}}}$ where $S_p = \sqrt{MSE}$

$$t_0 = \frac{5.8667 - 4.6889}{\sqrt{1.2668}\sqrt{\dfrac{1}{9} + \dfrac{1}{9}}} = 2.2199$$

Step (4) $t_C = t\left(1 - \dfrac{\alpha}{2}, \sum (n_r - 1)\right)$

$$= t\,(.975, 16) = 2.120$$

Step (5)   Since $|t_0| > t_C$, we reject H_0 just as we did using the F test. Note that for 2-level designs, these tests are equivalent. The resulting prediction equation is $\hat{y} = 5.2778 + \dfrac{1.1778}{2}$ (Temp), where Temp is in coded form.

Note: A quick and easy way to accomplish the same procedure as an F or t test can be found in Appendix M. The End Count method described in Appendix M is almost as accurate (statistically) as the F or t test with much less complexity. Also see the case study on page 8-209.

4.3   2 Factor ANOVA

Assume that in the first example problem of section 4.2 there exists another variable of interest, pressure. A one-factor-at-a-time experimental approach implies that we collect data on two temperatures as shown in section 4.2 and then collect more data on a desired number of pressure levels. In each case, we would hold the other variable at some constant level. As previously discussed in Chapter 1, this approach is grossly inefficient and it is possible to determine that neither temperature nor pressure has an effect when in fact the contrary may be true. The missing effect when conducting one-at-a-time designs is the interaction of two effects when analyzed together. To estimate whether or not a **main effect** and/or an **interaction effect** exists in our example, a **factorial design** must be used. Assuming that we are interested in only two levels of pressure, 100 and 200 psi, Table 4.4 displays the different combinations possible for temperature and pressure.

2 Factor Combinations of Temperature and Pressure

Run	Temp	Pressure
1	70	100
2	70	200
3	90	100
4	90	200

Table 4.4 Combinations for 2 Factors, Temperature and Pressure

If the average observed values in runs 1, 2, 3, and 4 are 10, 30, 30, and 50, respectively, they would appear graphically as shown in Figure 4.2.

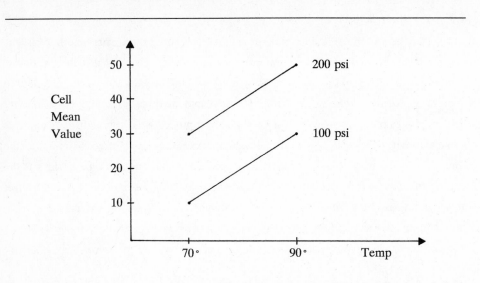

Figure 4.2 Graphical Display of 2 Factors that Do Not Interact

The graph in Figure 4.2 indicates a potential temperature effect (average response over pressure levels increases for temperature changes from 70° to 90°), a potential pressure effect (average response for 200 psi is consistently above that for 100 psi), and there appears to be no 2-way interaction because the change in response from 70° to 90° for 100 psi is equal to the change at 200 psi (i.e., the lines are approximately parallel). An example of run average values with a potential 2-way interaction would be 20, 50, 40, and 30 for runs 1, 2, 3, and 4, respectively. Graphically, these values appear as shown in Figure 4.3. The non-parallel nature of these lines (whether they cross or not) is an indicator of an interaction. The non-horizontal nature of each pressure line over the two temperatures indicates a potential positive temperature effect when pressure is 100 psi and a potential negative temperature effect when pressure is 200 psi. Notice that if you average the response over levels of pressure, there will appear to be no temperature effect; thus, the importance of the interaction term for understanding the process.

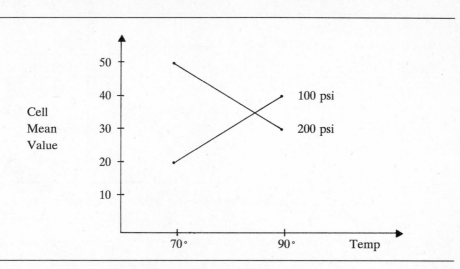

Figure 4.3 Graphic of a 2-Factor Interaction

For an example of a 2 factor (2 level) ANOVA, see the data presented in Table 4.5.

Run	Temp A	Press B	Response y_1	Response y_2	$\bar{y}_r$	S_r^2
1	70	100	2.2	2.8	2.5	.18
2	70	200	3.2	3.6	3.4	.08
3	90	100	5.0	4.6	4.8	.08
4	90	200	5.8	5.4	5.6	.08

Table 4.5 Data for a 2 Factor (2 Level) ANOVA Example

Replacing the low factor settings with a (-1) and the high factor settings with a $(+1)$, the above design matrix with the response values is shown in Table 4.6.

Run	A	B	A × B	Response y_1	Response y_2	$\bar{y}_r$	S_r^2	df_r
1	−1	−1	+1	2.2	2.8	2.5	.18	1
2	−1	+1	−1	3.2	3.6	3.4	.08	1
3	+1	−1	−1	5.0	4.6	4.8	.08	1
4	+1	+1	+1	5.8	5.4	5.6	.08	1

Table 4.6 Coded Matrix and Response Values for 2-Factor (2 Levels) Example

Notice that the A × B column is generated by multiplying the (+1) and (−1) values in columns A and B for each row. This new column represents the 2-way linear interaction between A and B.

To analyze the significance of the A, B, and A × B effects, you need only complete a table similar to the one shown in Table 4.7.

	A	B	A × B
(−1) Level Response Avg	2.95	3.65	4.10
(+1) Level Response Avg	5.20	4.50	4.05
(+1) Avg − (−1) Avg = Δ	2.25	.85	−.05
MSB_{effect}	10.125	1.445	.005
F_{effect}	96.428	13.762	.0476
Note: MSE = .105			

Table 4.7 Analysis Table for 2-Factor (2 Levels) ANOVA Example

To complete Table 4.7, you will need the following hypotheses tests for each column in the matrix:

Test I: Testing for a significant A effect.

Step (1) H_0: $\mu_{A(+)} = \mu_{A(-)}$

 H_1: $\mu_{A(+)} \neq \mu_{A(-)}$

Step (2) $\alpha = .05$

Step (3) $MSB_A = \dfrac{N}{4}(\Delta_A)^2 = \dfrac{8}{4}(2.25)^2 = 10.125$

$$MSE = \frac{\sum (n_r - 1)(S_r^2)}{\sum (n_r - 1)} = \frac{\sum (df_r)(S_r^2)}{\sum (df_r)}$$

$$= \frac{1(.18) + 1(.08) + 1(.08) + 1(.08)}{1 + 1 + 1 + 1} = \frac{.42}{4} = .105$$

The observed F_0 for the A effect is $F_0 = \dfrac{MSB_A}{MSE} = \dfrac{10.125}{.105} = 96.428$

Step (4) $F_C = F(.95, 1, \Sigma(n_r - 1)) = F(.95, 1, 4) = 7.71$

Step (5)   Since $F_0 > F_C$, conclude H_1, i.e., factor A is significant and it belongs in the prediction equation. (See Appendix M for Rules of Thumb on significant F values from a designed experiment).

Test II: Testing for a significant B effect.

Step (1) H_0: $\mu_{B(+)} = \mu_{B(-)}$

 H_1: $\mu_{B(+)} \neq \mu_{B(-)}$

Step (2) $\alpha = .05$

Step (3) $MSB_B = \dfrac{8}{4}(.85)^2 = 1.445$

 $MSE = .105$ (same as before)

$$F_0 = \frac{1.445}{.105} = 13.762$$

Step (4) $F_C = F(.95, 1, 4) = 7.71$ (same as before)

Step (5)   Since $F_0 > F_C$, conclude H_1, i.e., factor B is significant and it belongs in the prediction equation.

Test III: Testing for a significant A × B interaction effect.

Step (1) H_0: $\mu_{AB(+)} = \mu_{AB(-)}$

            H_1: $\mu_{AB(+)} \neq \mu_{AB(-)}$

Step (2) $\alpha = .05$

Step (3) $MSB_{AB} = \dfrac{8}{4}(-.05)^2 = .005$

            $MSE = .105$ (same as before)

            $F_0 = \dfrac{.005}{.105} = .0476$

Step (4) $F_C = F(.95, 1, 4) = 7.71$ (same as before)
Step (5)   Since $F_0 < F_C$, conclude H_0, i.e., the A × B interaction effect is not significant and it does not belong in the prediction equation.

Now that all the statistical tests are complete, you should review the statistical conclusions to see if they make practical and engineering sense. Assuming the statistical conclusions match the non-statistical conclusions, the resulting prediction equation will be as follows:

$$\hat{y} = 4.075 + 1.125A + .425B$$

Using the ANOVA approach just demonstrated, it now becomes a fairly easy task to perform ANOVA for any number of factors provided they all have 2 levels. For example, consider the following 3-factor ANOVA using a Taguchi L_8 matrix.

Run	A	B	−AB	C	−AC	−BC	ABC	y_1 y_2 y_3	$\bar{y}_r$	S_r^2	df_r
1	−	−	−	−	−	−	−	5 7 6	6.0	1.00	2
2	−	−	−	+	+	+	+	20 21 23	21.33	2.33	2
3	−	+	+	−	−	+	+	6 4 5	5.0	1.00	2
4	−	+	+	+	+	−	−	23 19 18	20	7.00	2
5	+	−	+	−	+	−	+	21 21 20	20.67	.33	2
6	+	−	+	+	−	+	−	4 5 6	5.0	1.00	2
7	+	+	−	−	+	+	−	24 21 19	21.33	6.33	2
8	+	+	−	+	−	−	+	5 8 7	6.67	2.33	2

	A	B	−AB	C	−AC	−BC	ABC	
Avg (−)	13.08	13.25	13.83	13.25	5.66	13.33	13.08	$\bar{y} = 13.25$
Avg (+)	13.42	13.25	12.67	13.25	20.83	13.16	13.42	
Δ=Avg(+) −Avg(−)	.34	0.0	−1.16	0.0	15.17	−.17	.34	$F_C=F.95,1,16)$ = 4.49
MSB_{effect}	.6939	0.0	8.0736	0.0	1380.77	.1734	.6936	
F_{effect}	.260	0.0	3.029	0.0	518.11	.065	.260	

$$MSE = \frac{2(1) + 2(2.33) + \ldots + 2(2.33)}{2 + 2 + \ldots + 2} = \frac{42.64}{16} = 2.665$$

The reason for using the negative of the two-way interaction columns is to comply with the Taguchi design matrix in Chapter 3.

Table 4.8 Taguchi L_8 Matrix and ANOVA Results for a 3-Factor Design

Since most of the effects are not significant, the only significant terms for the prediction equation are the grand mean and the A × C interaction. Using the rule of hierarchy, our prediction equation becomes:

$$\hat{y} = 13.25 + .17A + 0 \times C + 7.585(-A \times C) = 13.25 + .17A - 7.585(A \times C)$$

In order to use this equation to predict the response for various factor settings, you must ensure the levels for A and C are in the coded form, i.e., somewhere between −1 and +1.

If your objective is to minimize or maximize the response, you can determine the optimum settings by using the prediction equation. For example, consider minimizing the average predicted response, $\hat{y}$. Since the coefficient for A is positive, setting A at (-1) will move the $\hat{y}$ in the minimum direction. The negative coefficient of A $\times$ C and the previously determined negative setting for A will dictate that factor C be set to (-1). The resulting predicted average response will be:

$$\hat{y} = 13.25 + .17(-1) - 7.585(-1)(-1) = 5.495$$

If your objective is to hit a target value for $\hat{y}$, say 16, then the factor settings are determined as follows:

(1)   Set $\hat{y} = 16$ as shown below.

$$16 = 13.25 + .17A - 7.585 \ (A \times C)$$

(2)   Since we have one equation with two variables, select a coded setting for either A or C based on economics. For example, suppose A is twice as expensive to set at $(+1)$ than at (-1). Therefore, you would want to set A at (-1) to minimize cost.

(3)   Solve the equation for C.

$$16 = 13.25 + .17(-1) - 7.585(-1)(C)$$

$$C = \frac{2.92}{7.585} = .385$$

(4)   The actual values for A and C are easily decoded using the following equation from Chapter 3.

$$f_j = \frac{x_j d_j}{2} + \bar{f}$$

where f_j is the actual value for factor j

$\bar{f}_j$ is the actual average for factor j

d_j is the range for factor j actual values (i.e., $f_{max} - f_{min}$)

x_j is the coded value for factor j

Equation 4.5 Transformation Equation for Coded to Actual Factor Values

As an example, assume the actual low and high values for factor C are 100 and 400. Converting the coded C = .385, the actual factor C value becomes:

$$f = \frac{.385(300)}{2} + 250 = 307.75$$

Graphically, the linear interpolation appears as shown in Figure 4.4.

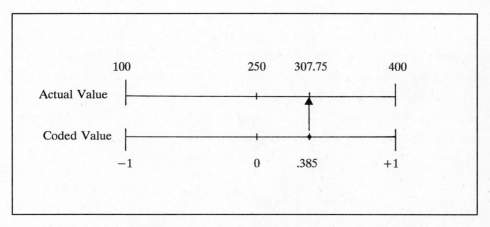

Figure 4.4 Linear Interpolation from Coded to Actual Factor Value

4.4   Introduction to Simple Linear Regression (SLR)

Consider the scenario of measuring the yield or response of a certain process at four levels of temperature: 70°, 80°, 90°, and 100°. Our objective is to develop a model which will allow us to estimate the response at levels other than those mentioned above and be able to place prediction intervals about these estimates. We also need a measure of model effectiveness. The technique to be discussed next, *Simple Linear Regression (SLR)*, will assist us in accomplishing these objectives. Assume we have collected three responses or observations from each of the four temperature settings. The data is shown in Table 4.9.

	Temperature		
70°	80°	90°	100°
2.3	2.5	3.0	3.3
2.6	2.9	3.1	3.5
2.1	2.4	2.8	3.0

Table 4.9 Three Yields at Each of Four Temperatures

A graph of the data is given in Figure 4.5. The line represents an "eyeball fit" or free-hand regression line. The closeness of all the observations to the line indicates the accuracy of the predicted values of y for any given temperature.

The objective in placing the line is to attempt to minimize the distance that observations are from the line. Using the slope-intercept formula [$E(y) = \beta_0 + \beta_1 x$], we can estimate "$\beta_0$" graphically as .1 by looking at Figure 4.5. The slope, β_1, is estimated by measuring the change in y, (Δy), for some specific change in x, (Δx), i.e.,

$$\frac{\Delta y}{\Delta x} = \frac{2.7 - 2.3}{80 - 70} = 0.04.$$

Thus our "eyeballed" regression line takes the form of y = .1 + .04x, where x is now in actual values of temperature. Since all observations do not lie exactly on the line, there is obviously some error in our straight line estimate. To incorporate this *error* in the formula for predicting y given any x value, we use $y = \beta_0 + \beta_1 x + \epsilon$, where ϵ represents the error term which is assumed to be distributed normally about 0; the ϵ's have equal variability for all levels of x; and the error values are assumed to be independent.

The relationship or model $y = \beta_0 + \beta_1 x + \epsilon$ is applicable for the population data (i.e., the set of *all* possible x and y values). The true regression line imbedded in this model is represented by $E(y) = \beta_0 + \beta_1 x$. Unfortunately, β_0 and β_1 are population parameters which are not known. Thus, the true regression line is not known. Data used in experimentation and process control are almost always sample data (a subset of population data); therefore, we use $\hat{y} = b_0 + b_1 x$ to approximate the true regression line; i.e., $\hat{y}$, b_0,

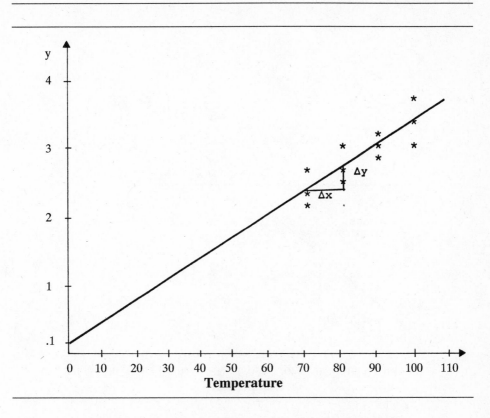

Figure 4.5 Free-Hand Regression Line Through SLR Data

and b_1 estimate E(y), β_0, and β_1, respectively. In addition, $e_i = y_i - \hat{y}_i$ is called the ith residual which estimates ϵ_i. These terms are displayed graphically in Figure 4.6. The subscript " i " as used here is associated with the " i^{th} " observation.

From Figure 4.6, it can also be seen that each observation's deviation from $\bar{y}$ can be partitioned such that

$$(y_i - \bar{y}) = (y_i - \hat{y}_i) + (\hat{y}_i - \bar{y}).$$

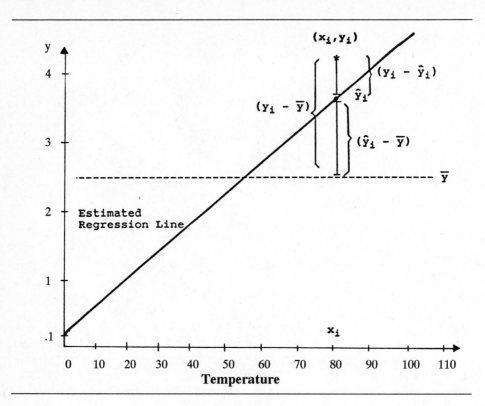

Figure 4.6 Partitioning of $(y_i - \overline{y})$

Squaring both sides and then summing both sides over all observations (i.e., over i) yields, after some algebra, the following equation:

$$\sum_{i=1}^{n} (y_i - \overline{y})^2 = \sum_{i=1}^{n} (y_i - \hat{y}_i)^2 + \sum_{i=1}^{n} (\hat{y}_i - \overline{y})^2$$

where n is the number of paired data points and

$$SST = \sum_{i=1}^{n} (y_i - \overline{y})^2$$

is the numerator of what we have already seen to be the variance of y. SST is called the sum of squares total (or total sum of squares). The two terms partitioning SST are called the sum of squares due to regression, SSR, and the sum of squares due to error, SSE. SSR and SSE are given by

$$SSR = \sum_{i=1}^{n} (\hat{y}_i - \bar{y})^2 \quad \text{and} \quad SSE = \sum_{i=1}^{n} (y_i - \hat{y}_i)^2$$

respectively. Thus, in abbreviated notation,

$$SST = SSE + SSR$$

If we don't use x to predict y, then the best prediction for some future y is $\bar{y}$. The variance of observations about $\bar{y}$ is

$$S_y^2 = \sum_{i=1}^{n} \frac{(y_i - \bar{y})^2}{n - 1} \ .$$

When information on x is used to predict y, then $\hat{y} = b_0 + b_1x$ is the prediction and the variance of observations about $\hat{y}$ is given by

$$\sum_{i=1}^{n} \frac{(y_i - \hat{y}_i)^2}{n - 2} = \frac{SSE}{n - 2} = MSE.$$

MSE is referred to as the mean square error or the error variance. The objective of SLR is to develop a regression line such that MSE is much smaller than s_y^2. The size of the error variance determines how good a fit the regression line will be and it plays a key role in the prediction intervals to be developed later in this chapter.

One mathematical method for finding b_0 and b_1 is the *method of least squares*. The name is derived from the property of minimizing the sum of squared deviations from the line, i.e., minimize Σe_i^2. The appropriate formulas are presented in Equation 4.6.

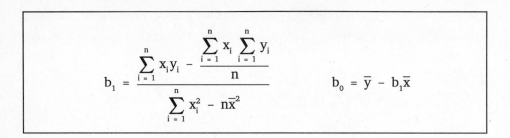

$$b_1 = \frac{\sum\limits_{i=1}^{n} x_i y_i - \dfrac{\sum\limits_{i=1}^{n} x_i \sum\limits_{i=1}^{n} y_i}{n}}{\sum\limits_{i=1}^{n} x_i^2 - n\bar{x}^2} \qquad b_0 = \bar{y} - b_1\bar{x}$$

Equation 4.6 Least Squares Estimates for β_1 and β_0 in SLR

For the example data, the calculations are shown in Table 4.10. As you can see, the estimates from the free-hand regression, namely $b_1 = .04$ and $b_0 = .1$, are close to those of least squares for this example. However, when the data are more spread out and more variables are added, the "eyeball fit" becomes an impossible task.

n	x	y	x^2	xy	y^2
1	70	2.3	4900	161	5.29
2	70	2.6	4900	182	6.76
3	70	2.1	4900	147	4.41
4	80	2.5	6400	200	6.25
5	80	2.9	6400	232	8.41
6	80	2.4	6400	192	5.76
7	90	3.0	8100	270	9.00
8	90	3.1	8100	279	9.61
9	90	2.8	8100	252	7.84
10	100	3.3	10000	330	10.89
11	100	3.5	10000	350	12.25
12	100	3.0	10000	300	9.00
	1020	33.5	88200	2895	95.47

$$\bar{x} = 85 \qquad\qquad \bar{y} = 2.79$$

$$b_1 = \frac{\sum xy - \dfrac{\sum x \sum y}{n}}{\sum x^2 - n\bar{x}^2} = \frac{2895 - \dfrac{1020\,(33.5)}{12}}{88200 - 12\,(85)^2} = \frac{47.5}{1500} = 0.032$$

$$b_0 = \bar{y} - b_1\bar{x} = 2.79 - 0.032(85) = 0.07$$

$\hat{y} = .07 + .032x$ The $\hat{y}$ indicates that an average value of y is a function of x derived from sample data.

Table 4.10 Simple Linear Regression Calculations

4.5   Strength of the Linear Relationship and ANOVA for SLR

A measure of strength of the linear relationship of y with temperature is the correlation coefficient,

$$R = \frac{\sum xy - \frac{\sum x \sum y}{n}}{\sqrt{\left(\sum x^2 - n\bar{x}^2\right)\left(\sum y^2 - n\bar{y}^2\right)}}$$

Equation 4.7 Correlation Coefficient

For our example,

$$R = \frac{2895 - \frac{1020(33.5)}{12}}{\sqrt{[88200 - 12(85)^2]\,[95.47 - 12(2.792)^2]}} = \frac{47.5}{53.763} = .883$$

The value of R is limited to the interval $[-1, 1]$ where -1 indicates perfect negative correlation, 1 indicates a perfect positive correlation and 0 implies no linear relationship between y and x. For SLR, the sign of R will be the same as that for b_1.

The proportion of variability in y which is explained by y's relationship with x is measured by R^2; to understand the meaning of R^2, recall that SST = SSR + SSE. If we divide both sides of this equation by SST, we obtain

$$1 = \frac{SSR}{SST} + \frac{SSE}{SST}$$

Since SST represents the sum of squares total, the ratio SSE/SST represents the proportion of total variability remaining about the regression line and SSR/SST is the proportion of the total variability that has been explained (or removed) by the regression line. Using algebra, it can be shown that

$$\frac{SSR}{SST} = \frac{\sum_{i=1}^{n} (\hat{y}_i - \bar{y})^2}{\sum_{i=1}^{n} (y_i - \bar{y})^2}$$

is equivalent to R^2 and thus

$$\frac{SSE}{SST} = 1 - R^2$$

If there is no linear relationship between x and y, then SSE = SST and thus $R^2 = 0$ (see Figure 4.7a). If all the observations fall on the $\hat{y}$ (or regression) line, then SSE = 0 which implies that $R^2 = 1$ (see Figure 4.7b).

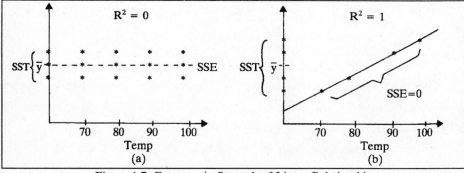

Figure 4.7 Extremes in Strength of Linear Relationship

In the example data, $R^2 = .780$ indicates that about 78.0% of the variability in y is explained through y's linear relationship with x. The strength of the linear relationship between the two variables is directly related to the amount of variability in y that can be accounted for by the variable x. In this example, the strength of the linear relationship between the two variables is considered to be fairly strong.

Recall that the variance of y without regard to x is

$$S_y^2 = \frac{SST}{n-1} = \frac{\sum y^2 - n\bar{y}^2}{n-1} = \frac{95.47 - 12(2.792)^2}{11} = .1752$$

We can now solve for SST in the above relationship by computing

$$SST = S_y^2 \cdot (n-1) = .1752(11) = 1.927.$$

Since we know SST = 1.927 and R^2 = .78, we can solve for SSR in the relationship R^2 = SSR/SST to get

$$SSR = R^2 \cdot SST = .78(1.972) = 1.503$$

The last of the three sums of squares, SSE, can be obtained from the previous two by way of the relationship SSE = SST − SSR. We have

$$SSE = SST - SSR = 1.927 - 1.503 = .424.$$

It is customary to formalize the partitioning of the total sums of squares in what is called an Analysis of Variance (ANOVA) table. Table 4.11 gives the format of an SLR ANOVA table.

Source	SS	df	MS	F_0
Regression	$SSR = R^2 \cdot SST$	1	SSR/1	MSR/MSE
Error	$SSE = (1 - R^2) \cdot SST$	n − 2	SSE/(n − 2)	
Total	$SST = s_y^2 \cdot (n - 1)$	n − 1		

Table 4.11 Format for SLR Analysis of Variance (ANOVA) Table

The degrees of freedom (df) for the error term in a regression model are equal to n minus the number of parameters estimated. In SLR, b_0 and b_1 are estimates (obtained from sample data) of the two population parameters β_0 and β_1, respectively. Therefore, df_E = n − 2. The degrees of freedom for regression in SLR are the number of parameters estimated minus 1, which is 2 − 1 = 1. The F_0 in Table 4.11 is compared with F_C = $F(1 - \alpha; df_R, df_E)$ from Appendix F. If $F_0 \leq F_C$ the model, with all the terms included for this test, is insignificant, i.e., the best model is $\hat{y} = \bar{y}$. If $F_0 > F_C$, the model represented in the test is significant.

Since R^2 is computed from sample data, it is only an estimate of the strength of the population linear relationship between y and x. We would like to test the value of R^2 to insure that it is significantly different from zero, i.e., we will test the hypotheses:

$$H_0: \ R^2 = 0$$
$$H_1: \ R^2 \neq 0$$

For SLR the hypothesis test for R^2 is equivalent to testing the slope parameter, i.e.,

$$H_0: \ \beta_1 = 0$$
$$H_1: \ \beta_1 \neq 0$$

In either case, concluding H_0 implies that the sample data did not provide overwhelming evidence to indicate a significant linear relationship between y and x. Concluding H_1 indicates the presence of a significant linear relationship with $(1 - P)100\%$ confidence. The symbol P refers to the minimum value that α (the risk of a Type I error) can have in order to still conclude H_1. The MSE is an estimate of the variance of y at any level of x. Table 4.12 displays the example data in the format of Table 4.11. Note that the F-test indicates the model, $\hat{y} = b_0 + b_1 x$, is significant in representing y's relationship with x.

Source	SS	df	MS	F_0
Regression	1.503	1	1.503	35.448*
Error	.424	10	.0424	
Total	1.927	11		
	* From Appendix F, F(.999; 1, 10) = 21.0 implies P < .001. Thus, we can conclude H_1: $R^2 \neq 0$ or H_1: $\beta_1 \neq 0$ with at least 99.9% confidence.			

Table 4.12 SLR ANOVA Table for Example Data

4.6   Prediction Using SLR

To obtain a point estimate for y, given any x value (x_h), we simply insert the value for x_h in $\hat{y} = b_o + b_1 x_h$ (Note: whereas the subscript "i" is associated with "observed" values, the subscript "h" is used to denote *any* possible or hypothetical level of x). Point estimates, however, do not include any reference to precision. In order to indicate how precise our estimate is, a **prediction interval** is formed in which the researcher is $(1 - \alpha)100\%$ confident that the true value of y will occur. Since the data is just one sample, it is reasonable to take another sample and compute different values of b_0 and b_1. This implies that there exists sample variability in the slope and intercept of the regression line, in addition to observation variability about the regression line itself. The slope and intercept variability will cause larger deviations in the predicted response ($\hat{y}$) as the desired level of x (x_h) moves further away from the center of the domain of x ($\bar{x}$). For these reasons, the prediction limits for the response at any given value of x (labeled x_h) will appear as shown below:

$$\hat{y}\,|_{x_h} \; \pm \; t\left(\frac{\alpha}{2}, n - 2\right) \sqrt{MSE\left(1 + \frac{1}{n} + \frac{(x_h - \bar{x})^2}{\sum x^2 - n\bar{x}^2}\right)}^{\,*}$$

While MSE represents the variability of observations about the regression line, the quantity

$$MSE\left[\frac{1}{n} + \frac{(x_h - \bar{x})^2}{\sum x^2 - n\bar{x}^2}\right]$$

represents the variability in the average response due to the variability in slope, b_1, and intercept, b_0.

For our temperature example, the prediction intervals (y_L , y_U) based on 95% confidence are displayed in Table 4.13 for all x values previously used to develop the model. For example, if the process temperature is set at 80°, we would be 95% confidence that all future y values will occur between 2.15 and 3.11.

---

* When actually performing the calculations, $n\bar{x}^2$ is more prone to numerical error than its equivalent, $(\Sigma x_i)^2 / n$.

x	$\hat{y}$	y_L	y_U	Δ	
70	2.31	1.80	2.82	1.02	t(.975, 10) = 2.228
* 80	2.63	2.15	3.11	.96	MSE = .0424
90	2.95	2.47	3.43	.96	$\Sigma x^2 - n\bar{x}^2 = 1500$
100	3.27	2.76	3.78	1.02	

* Using $x_h = 80$, $\bar{x} = 85$

$\hat{y}\,|\,x_h = .07 + .032(80) = 2.63$

$$y_L = 2.63 - 2.228 \sqrt{.0424\left(1 + \frac{1}{12} + \frac{(80-85)^2}{1500}\right)} = 2.15$$

$$y_U = 2.63 + 2.28 \sqrt{.0424\left(1 + \frac{1}{12} + \frac{(80-85)^2}{1500}\right)} = 3.11$$

Table 4.13 Prediction Intervals for y Based on 95% Confidence

Predicting values of y for $x_h < x_{min}$ or $x_h > x_{max}$ is called **extrapolation,** which assumes the same relationship between y and x outside the region of the sampled x as it is within the sampled region. Notice that the 95% prediction interval would be wider outside the interval [70, 100] than it will be inside. This lack of accurate predictability and uncertainty of the relationship between y and x outside the region of sampled x is why extrapolation is not recommended. If we need to predict outside the sampled region for x, it is best to expand the sampled region. If extrapolation is used, confirmation data should be collected as soon as possible to validate the prediction.

When prediction intervals are too wide for a particular application, there are ways to reduce these intervals. The researcher can:

(1)   decrease the confidence level $(1 - \alpha)$; then $t(1 - \alpha/2, n - 2)$ will get smaller

(2)   increase the sample size; then t and 1/n become smaller

(3)   increase $[\Sigma x^2 - n\bar{x}^2]$

It is not recommended that one alter the confidence level to reduce interval width. In fact, the subject of experimental design discussed in this text addresses only items (2) and (3) as the major areas of interest. For non-experimental data, the primary method of reducing prediction interval widths is increasing sample size. However, in conducting industrial experiments, the sample size is desired to be a minimum, thus putting even more emphasis on the remaining option, i.e., maximizing $[\Sigma x^2 - n\overline{x}^2]$. This is equivalent to maximizing the variance of x, which is accomplished by placing n/2 observations at each of x_{min} and x_{max}. The problem with sampling at only two levels of x is that we are restricted to linear estimates. For this reason, center points can be added to 2-level designs to allow for curvature examination and if necessary axial points are added to complete a CCD as described in Chapter 3. The idea of maximizing $[\Sigma x^2 - n\overline{x}^2]$ has another effect in that

$$S^2(b_1) = \frac{MSE}{\sum x^2 - n\overline{x}^2}$$

can be minimized.* Therefore, the variance of the slope is minimized if $[\Sigma x^2 - n\overline{x}^2]$ is maximized, creating more power in the F test for H_0: $R^2 = 0$ or H_0: $\beta_1 = 0$. This implies that if we minimize $S^2(b_1)$, we are more likely to find a real linear relationship if it in fact exists. Once multiple regression is covered, the reader will also see that maximizing $[\Sigma x^2 - n\overline{x}^2]$ is related to the D-optimal experimental designs discussed in Chapter 3.

4.7   Polynomial Regression

Suppose that the example temperature data included 3 more observations taken at x = 50°, which when added to Table 4.9 produces the example data shown in Table 4.14. Graphically, the data are presented in Figure 4.8. This graphic indicates the danger in extrapolation because the function takes on a different shape outside the original sampled space of x.

---

* Sample variance will be represented interchangeably by both S^2, s^2, and $\hat{\sigma}^2$ throughout this book.

		Temperature		
50°	70°	80°	90°	100°
3.3	2.3	2.5	3.0	3.3
2.8	2.6	2.9	3.1	3.5
2.9	2.1	2.4	2.8	3.0

Table 4.14 Example Data for Polynomial Regression

As you can see, the linear dashed line is no longer a good fit. In fact, if we try SLR on the new data set, the result is $b_1 = 0$ which implies $R^2 = 0$.

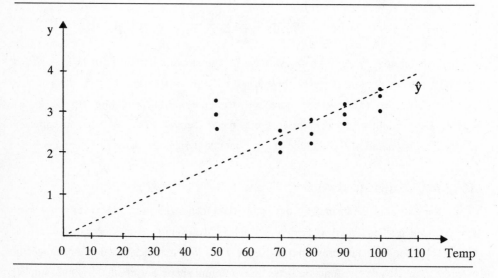

Figure 4.8 Graph of Example Data for Polynomial Regression

However, the more appropriate thing to do is to try a quadratic term. Staying with SLR, the new model becomes $\hat{y} = b_0 + b_{11}x_1^2$, where the subscript "11" is not read "eleven" but rather "one, one" to indicate that b_{11} is the coefficient of x_1x_1 or x_1^2. Table 4.15 shows the computations for SLR.

		Computations for $\hat{y} = b_0 + b_{11}x_1^2$		
(x_1^2)	y	$(x_1^2)^2$	y^2	$(x_1^2)y$
2500	3.3			
2500	2.8			
2500	2.9			
4900	2.3			
4900	2.6			
4900	2.1			
6400	2.5			
6400	2.9			
6400	2.4			
8100	3.0			
8100	3.1			
8100	2.8			
10000	3.3			
10000	3.5			
10000	3.0			
95700	42.5	710490000	122.61	276810

$$\hat{y} = 2.472 + 0.000056(x_1^2) \qquad R^2 = .146$$

Table 4.15 SLR Using a Quadratic Term

The low R^2 indicates that something is wrong because we would expect a better fit. What happened? Creating x^2 as a new variable and using it in the SLR model above resulted in fitting the points with a parabola whose vertex (lowest point) is on the y-axis. Graphically, we can see that the vertex should be far from $x = 0$.

To properly conduct polynomial regression in the manner in which the problem was stated, we must include both x_1 and x_1^2 in the model. That is, the correct model should be $\hat{y} = b_0 + b_1x_1 + b_{11}x_1^2$ which supports the use of the previously discussed hierarchy rule. This second order polynomial regression model is similar to a multiple regression problem. The computer output shown in Table 4.16 indicates that this model is a fairly good fit, i.e., the probability of an incorrect model is very small ($P = .001$).

Summary of Computer Output for Polynomial Regression

Model: $\hat{y} = 7.960 - .15374x_1 + .00108x_1^2$ $R^2 = .673$ $P = .001$

Parameter	Estimate	P-value
b_0	7.96000	.0001
b_1	−0.15374	.001
b_{11}	0.00108	.001

Table 4.16 Polynomial Regression on Example Data

Differential calculus can be used to find the x values which produce a minimum for y. That is, setting dy/dx =0 and solving for x, we have

$$\frac{dy}{dx} = -.15374 + .00216x_1 = 0$$

and

$$x_1 = \frac{.15374}{.00216} = 71.176$$

Thus, y is a minimum at $x_1 = 71.176°$. The graph of $\hat{y} = 7.96 - .15374x_1 + .00108x_1^2$ in Figure 4.9 shows that the minimum of y does indeed occur when $x_1 = 71.176$.

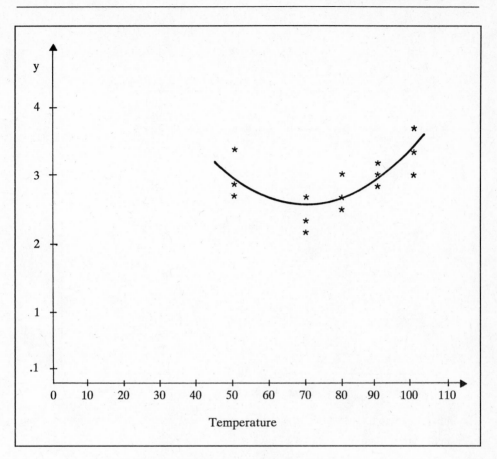

Figure 4.9 Graph of Polynomial Regression

4.8 Multiple Regression*

If a variable for pressure was also included in our model, the multivariable model would be referred to as a **Multiple Regression Model**. A new set of example data appears in Table 4.17.

---

* This material is slightly more complex than the previous sections. The reader should realize that evaluating 3-level designs will require the use of multiple regression and the interpretation of software output.

	MLR Data	
y	x_1 = Temp	x_2 = Pressure
3.1	50	100
3.3	50	100
2.6	50	200
2.4	50	200
2.5	70	100
2.6	70	100
3.0	70	200
3.3	70	200
2.4	80	100
2.3	80	100
3.2	80	200
3.5	80	200
2.8	90	100
2.6	90	100
3.1	90	200
3.0	90	200
3.2	100	100
3.4	100	100
2.5	100	200
2.4	100	200

Table 4.17 Example Data for Multiple Linear Regression (MLR)

The initial model to be tested is $y = \beta_0 + \beta_1 x_1 + \beta_2 x_2 + \epsilon$, which can be represented in an equivalent matrix notation as shown below.

$$\underline{Y} = X \underline{\beta} + \underline{\epsilon} \qquad \text{where}$$

$$\underline{Y} = \begin{bmatrix} y_1 \\ y_2 \\ y_3 \\ . \\ . \\ . \\ y_n \end{bmatrix} \qquad X = \begin{bmatrix} 1 & x_{11} & x_{12} \\ 1 & x_{21} & x_{22} \\ 1 & x_{31} & x_{32} \\ . & . & . \\ . & . & . \\ . & . & . \\ 1 & x_{n1} & x_{n2} \end{bmatrix} \qquad \underline{\beta} = \begin{bmatrix} \beta_0 \\ \beta_1 \\ \beta_2 \end{bmatrix} \qquad \underline{\epsilon} = \begin{bmatrix} \epsilon_1 \\ \epsilon_2 \\ \epsilon_3 \\ . \\ . \\ . \\ \epsilon_n \end{bmatrix}$$

As in SLR, $\underline{b}$ is used to estimate $\underline{\beta}$ and $\underline{e}$ is an estimate of $\underline{\epsilon}$. Suppose n is the number of observations and p is the number of parameters to be estimated. In general, $\underline{Y}$ is an (n × 1) vector of the response values; X is an (n × p) matrix whose first column is all 1's (to estimate β_0) and whose other columns are the levels of the k = p − 1 independent variables; $\underline{b}$ is a (p × 1) vector of p regression coefficients; and $\underline{e}$ is an (n × 1) vector of random error values (one for each of the n observations). For our example, data, $\underline{Y}$, X, $\underline{b}$, and $\underline{e}$ appear as

$$\underline{Y} = \begin{bmatrix} 3.1 \\ 3.3 \\ 2.6 \\ 2.4 \\ 2.5 \\ \cdot \\ \cdot \\ \cdot \\ 2.4 \end{bmatrix} \qquad X = \begin{bmatrix} 1 & 50 & 100 \\ 1 & 50 & 100 \\ 1 & 50 & 200 \\ 1 & 50 & 200 \\ 1 & 70 & 100 \\ \cdot & \cdot & \cdot \\ \cdot & \cdot & \cdot \\ \cdot & \cdot & \cdot \\ 1 & 100 & 200 \end{bmatrix}$$

$$\underline{b} = \begin{bmatrix} b_0 \\ b_1 \\ b_2 \end{bmatrix} \qquad \underline{e} = \begin{bmatrix} e_1 \\ e_2 \\ e_3 \\ \cdot \\ \cdot \\ \cdot \\ e_n \end{bmatrix}$$

Values of $\underline{b}$ and $\underline{e}$ are determined from the least squares multiple regression which has the objective of minimizing [$\underline{e}' \underline{e}$]. Note: $\underline{e}'$ denotes the transpose of $\underline{e}$ and z^{-1} is the inverse of z. To find $\underline{b}$ such that $\underline{Y} = X\underline{b}$ we use $\underline{b} = (X'X)^{-1}X'\underline{Y}$, and then after computing $\underline{Y} = X\underline{b}$ we can compute $\underline{e} = \underline{Y} - \underline{Y}$. The variance-covariance matrix of the regression coefficients is the (p × p) matrix given by

$$\text{cov}(\underline{b}) \;=\; \hat{\sigma}^2 \cdot (X' \, X)^{-1} \;=\; \begin{bmatrix} \sigma_{b_0}^{\,2} & \sigma_{b_0 b_1} & \sigma_{b_0 b_2} \\ & \sigma_{b_1}^{\,2} & \sigma_{b_1 b_2} \\ & & \sigma_{b_2}^{\,2} \end{bmatrix}$$

$\sigma_{b_0}^2$ is the variance of b_0 and $\sigma_{b_1 b_2}$ denotes the covariance of b_1 and b_2. The estimate for σ^2 is based on the variance of the residuals which have $n - p$ degrees of freedom. It is given by

$$\text{MSE} \;=\; \frac{\displaystyle\sum_{i=1}^{n} (y_i - \hat{y}_i)^2}{n - p}.$$

$\text{MSE} = \text{SSE}/(n - p)$ is referred to as the regression error variance. For our MLR example data in Table 4.17, the computer output is displayed in Table 4.18.

Dependent Variable: Y	N: 20	Multiple R: .108		Squared Multiple R: .012	
Adjusted Squared Multiple R: .000			Standard Error of Estimate: 0.410		
Variable	Coefficient	Std Error	Std Coef	Tolerance	T	P(2 Tail)**
Constant	2.695	0.506	0.000	1.0000000	5.322	0.000
x_1	0.001	0.005	0.026	1.0000000	0.108	0.915
x_2	0.001	0.002	0.105	1.0000000	0.437	0.668

Analysis of Variance

Source	Sum-of-Squares	DF	Mean-Square	F-Ratio	P
Regression	0.034	2	0.017	0.101	0.904
Residual*	2.854	17	0.168		

Table 4.18 Computer Output for MLR Model $\hat{y} = b_0 + b_1 x_1 + b_2 x_2$

---

* The *MYSTAT Software* which accompanies [3] refers to the error as Residual.
**See Appendix M for Rules of Thumb on significant P(2 Tail) values.

Since $R^2 = .012$ with $P = .904$, the model is inadequate for predicting the response. In order to calculate the model's lack of fit, we partition SSE into pure error and lack of fit error. That is, $SSE = SSE_{pure} + SSE_{LF}$. The pure error, SSE_{pure}, is acquired from repeated observations of y at the same levels of x. Since the example data has 2 observations of y for the 10 different combinations of x_1 and x_2, the pure error is estimated by

$$SSE_{pure} = \sum_{ij = 1}^{10} (n_{ij} - 1) \, s_{ij}^2 = 0.19$$

$MSE_{pure} = SSE_{pure}/df_{pure} = .19/10 = .019^*$, where $df_{pure} = \Sigma(n_{ij} - 1)$. The sum of squares for lack of fit, SSE_{LF}, is calculated using $SSE_{LF} = SSE - SSE_{pure}$. For our example, $SSE_{LF} = 2.854 - .19 = 2.664$ and the degrees of freedom for testing lack of fit are $df_{LF} = df_E - df_{pure}$ where $df_E = n - p$ and p is the number of parameters estimated. Therefore, the example data has a mean square for lack of fit of

$$MSE_{LF} = \frac{SSE_{LF}}{df_{LF}} = \frac{2.664}{17 - 10} = 0.3806.$$

The F test for lack of fit is $F_0 = MSE_{LF}/MSE_{pure} = .3806/.019 = 20.032$. Comparing this to $F(1 - \alpha, df_{LF}, df_{pure}) = F(.95, 7, 10) = 3.14$, we see that the lack of fit is significant. Thus, it is necessary to investigate a better model using interactions and quadratic effects. If we test a 2nd order model, it will appear as

$$\hat{y} = b_0 + b_1x_1 + b_2x_2 + b_{12}x_1x_2 + b_{11}x_1^2$$

Notice that we do not estimate a quadratic effect for x_2 because it has only 2 levels. The computer output is shown is Table 4.19.

---

* If the design does not contain any replicated points at designated levels of x, replications can be made at the center point (in our example, this would be 75° and 150 psi) and the SSE_{pure} would then be calculated for this one combination of independent variables. If using replicated center points, then $df_{pure} = n_c - 1$, where n_c is the number of center points.

Computer Output for $\hat{y} = b_0 + b_1x_1 + b_2x_2 + b_{12}x_1x_2 + b_{11}x_1^2$

Dependent Variable: Y   N: 20   Multiple R: .115   Squared Multiple R: .013

Adjusted Squared Multiple R: .000   Standard Error of Estimate: .0436

Variable	Coefficient	STD Error	STD Coef	Tolerance	T	P(2 Tail)
Constant	2.973	2.371	0.000	1.0000000	1.254	0.229
x_1	-0.004	0.057	-0.203	.0100250	-0.079	0.938
x_2	0.000	0.009	-0.054	.0463950	-0.046	0.964
$x_1 x_2$	0.000	0.000	0.194	.0327289	0.137	0.893
$x_1 x_2$	0.000	0.000	0.124	.0110192	0.051	0.960

Analysis of Variance

Source	Sum-of-Squares	DF	Mean-Square	F-Ratio	P
Regression	0.038	4	0.010	0.50	0.995
Residual	2.850	15	0.190		

Table 4.19 Computer Output for 2nd Order Regression Model

Testing this model for lack of fit, we have $SSE_{LF} = SSE - SSE_{pure} = 2.850 - .19 = 2.66$

(about the same as before). The F test for lack of fit is $F_0 = \dfrac{2.66/(15-10)}{0.19/10} = 28.0$ which

is still significant at the .05 level. What's wrong? The problem is that 2nd order models can also have quadratic interaction terms. Therefore, the new model to be tested will include a quadratic interaction of x_1^2 with variable x_2:

$$\hat{y} = b_0 + b_1x_1 + b_2x_2 + b_{12}x_1x_2 + b_{(11)2}x_1^2x_2 + b_{11}x_1^2$$

The results are shown in Table 4.20. It should be of concern that adding the $x_1^2x_2$ term increased R^2 to .901 which is significant, but the coefficient for $x_1^2x_2$ is 0 (the problem here is due to the use of only 3 decimal place accuracy. Formatting the output to 4 or more places will produce non-zero coefficients). There is, however, another more insidious problem occurring in this analysis.

Computer Output for $\hat{y} = b_0 + b_1 x_1 + b_2 x_2 + b_{12} x_1 x_2 + b_{(11)2} x_1^2 x_2 + b_{11} x_1^2$

Dependent Variable: Y   N: 20   Multiple R: .949   Squared Multiple R: .901

Adjusted Squared Multiple R: .866   Standard Error of Estimate: .0143

Variable	Coefficient	Std Error	Std Coef	Tolerance	T	P(2 Tail)
Constant	23.815	2.015	0.000	1.0000000	11.818	0.000
x_1	-0.596	0.056	-26.984	.0011019	-10.656	0.000
x_2	-0.139	0.013	-18.337	.0025125	-10.934	0.000
$x_1 x_2$	0.004	0.000	49.539	.0003606	11.192	0.000
$x_1 x_1$	0.004	0.000	27.053	.0006840	10.683	0.000
$x_1 x_1 x_2$	0.000	0.000	-36.030	.0006840	-11.210	0.000

Analysis of Variance

Source	Sum-of-Squares	DF	Mean-Square	F-Ratio	P
Regression	2.602	5	0.520	25.504	0.000
Residual	0.286	14	0.020		

Table 4.20 Computer Output for Quadratic Interaction Model

The tolerance column represents $1 - R_j^2$, where R_j^2 is the squared correlation coefficient obtained by regressing predictor variable j on all other predictor variables. Tolerance values other than 1.0 indicate that a phenomenon referred to as **multicollinearity** exists. Tolerance values close to zero imply strong multicollinearity conditions which can grossly affect the accuracy of the estimated coefficients. Since the vectors of actual values for x_1, x_2, and $x_1^2 x_2$ are intercorrelated, the predictor variable matrix is not orthogonal and, therefore, the estimates of the β_i's can be very ambiguous. In fact, sometimes a variable with a known positive effect on the response will have a negative coefficient due to high multicollinearity. In addition, multicollinearity increases the off-diagonal values of $(X'X)^{-1}$ causing inflated variance of the coefficients which is estimated by $MSE(X'X)^{-1}$ This can result in significant F-ratios without significant regression weights. When including terms for a 2nd order model, multicollinearity is difficult to avoid unless a transformation of the predictor values is used. A commonly used transformation of non-experimental data is "centering the data." That is, we center each predictor column by transforming each x_{ij} into $(x_{ij} - \bar{x}_j)$. When possible orthogonal coding should be used as discussed in Chapter 3. A final note is that the MLR prediction limits for a future observation at some independent level $\underline{x}_h{}' = (x_{1h}, x_{2h}, ..., x_{kh})$ can be obtained by using

$$\hat{y}\Big|_{\underline{x}_h} \pm t\left(1 - \frac{\alpha}{2}, n - p\right) \sqrt{MSE\left(1 + x_h{}' \left(X' \ X\right)^{-1}\underline{x}_h\right)}$$

4.9 Example Demonstrating the Need to Use Coded Data When Examining Second Order Models

Table 4.21 contains data from a 3 factor Box-Behnken design where the run order has been randomized. Building a 2nd order model on the actual data results in the multiple regression output displayed in Table 4.22.

Run	Factors			Coded Factors			$\bar{Y}^*$
	A	B	C	x_1	x_2	x_3	
1	75	60	71	0	−	+	1332
2	70	60	68	−	−	0	1370
3	75	60	65	0	−	−	1386
4	80	60	68	+	−	0	1377
5	75	65	68	0	0	0	1387
6	80	65	71	+	0	+	1365
7	80	65	65	+	0	−	1392
8	75	65	68	0	0	0	1411
9	70	65	71	−	0	+	1372
10	70	65	65	−	0	−	1417
11	75	65	68	0	0	0	1397
12	75	70	71	0	+	+	1394
13	80	70	68	+	+	0	1405
14	75	70	65	0	+	−	1410
15	70	70	68	−	+	0	1399

Table 4.21 Data from a 3 Factor Box-Behnken Design

Using the average response in a 2nd order model results in the following computer output.

---

* $\bar{Y}$ is the average of five data points.

Dependent Variable: Y	N: 15	Multiple R: .951		Squared Multiple R: .904		
Adj Squared Multiple R: .731		Std Error of Estimate: 11.397		F = 5.232	P = .042	
Variable	Coefficient	Std Error	Std Coef	Tolerance	T	P(2 Tail)
Constant	−934.713	4599.512	0.000	1.0000000	−0.203	0.847
A	−6.475	46.409	−1.113	0.0003	−0.140	0.894
B	4.375	43.719	0.752	0.0003	0.100	0.924
C	74.602	97.241	7.696	0.0002	0.767	0.478
A^2	−0.092	0.237	−2.365	0.0051	−0.386	0.715
B^2	−0.332	0.237	−7.416	0.0007	−1.398	0.221
C^2	−1.060	0.659	−14.876	0.0002	−1.609	0.169
A•B	−0.010	0.228	−0.171	0.0013	−0.044	0.967
A•C	0.300	0.380	4.207	0.0007	0.790	0.465
B•C	0.633	0.380	8.539	0.0007	1.667	0.156

Table 4.22 Computer Output for Row Data in Table 4.21

The low tolerance values in Table 4.22 indicate high multicollinearity which is due to modeling the actual data. High multicollinearity inflates the standard error estimates which results in large P(2 Tail) values. The following simple example illustrates the problem which can result from using actual data in non-linear models.

Actual Data			Coded Data	
A	A^2	vs	x	x^2
1	1		−1	1
1	1		−1	1
2	4		0	0
2	4		0	0
3	9		1	1
3	9		1	1
$R^2 = 0.979$			$R^2 = 0$	

Using the coded data in Table 4.21 to build a 2nd order model, you will obtain output as shown in Table 4.23.

Variable	Coefficient	Std Error	Std Coef	Tolerance	T	P(2 Tail)
Constant	1398.333	6.708	0.000	1.0000	208.464	0.000
x_1	−2.375	4.108	−0.081	1.0000	−0.578	0.588
x_2	18.250	4.108	0.625	1.0000	4.443	0.007
x_3	−17.375	4.108	−0.595	1.0000	−4.230	0.008
x_1^2	−2.667	6.046	−0.062	0.9890	−0.441	0.678
x_2^2	−7.917	6.046	−0.185	0.9890	−1.309	0.247
x_3^2	−9.167	6.046	−0.215	0.9890	−1.516	0.190
$x_1 \cdot x_2$	−0.250	5.809	−0.006	1.0000	−0.043	0.967
$x_1 \cdot x_3$	4.500	5.809	0.109	1.0000	0.775	0.474
$x_2 \cdot x_3$	10.250	5.809	0.248	1.0000	1.764	0.138

Dependent Variable: Y   N: 15   Multiple R: .949   Squared Multiple R: .901

Adj Squared Multiple R: .723   Std Error of Estimate: 11.618   F = 5.052   P = .045

Table 4.23 Computer Output for Coded Data in Table 4.21

The high tolerance values in Table 4.23 indicate there is little or no multicollinearity. Because of this, the P(2 Tail) values are useful in finding which effects are significant. You can interpret the P(2 Tail) values as follows: $(1 - P)100\%$ is the confidence of that term belonging in the model. Using the coded data, the parsimonious model is: $Y = f(x_2, x_3)$, which results in the following coded value prediction equation:

$$\hat{Y} = 1398.333 + 18.25(x_2) - 17.375(x_3)$$

Predicting Y for $(x_1 = 1, x_2 = 0, x_3 = 0)$ results in $Y = 1387.600$.
The prediction interval is

$$\hat{Y}\,|_{\underline{x}_h} \pm t\left(1 - \frac{\alpha}{2}, n - p\right) \hat{\sigma}\sqrt{1 + \underline{x}_h{}' \left(X'\ X\right)^{-1}\underline{x}_h}$$

where $\underline{x}_h$ = coordinates of desired point $(x_{1_h}, x_{2_h}, x_{3_h})$.

The matrix algebra used to generate this prediction interval is presented as follows.

$$X = \begin{bmatrix} 1 & x_{11} & x_{21} & x_{31} \\ 1 & x_{12} & x_{22} & x_{32} \\ \cdot & \cdot & \cdot & \cdot \\ \cdot & \cdot & \cdot & \cdot \\ \cdot & \cdot & \cdot & \cdot \\ 1 & x_{1n} & x_{2n} & x_{3n} \end{bmatrix} \qquad X'X = \begin{bmatrix} 15 & 0 & 0 & 0 \\ 0 & 8 & 0 & 0 \\ 0 & 0 & 8 & 0 \\ 0 & 0 & 0 & 8 \end{bmatrix} \qquad (X'X)^{-1} = \begin{bmatrix} \frac{1}{15} & 0 & 0 & 0 \\ 0 & \frac{1}{8} & 0 & 0 \\ 0 & 0 & \frac{1}{8} & 0 \\ 0 & 0 & 0 & \frac{1}{8} \end{bmatrix}$$

Therefore, a 95% prediction interval at $(1, 1, 0, 0)$ is

$$1398.333 \pm 2.179(11.618) \sqrt{1 + (1, 1, 0, 0) \begin{bmatrix} \frac{1}{15} & 0 & 0 & 0 \\ 0 & \frac{1}{8} & 0 & 0 \\ 0 & 0 & \frac{1}{8} & 0 \\ 0 & 0 & 0 & \frac{1}{8} \end{bmatrix} \begin{bmatrix} 1 \\ 1 \\ 0 \\ 0 \end{bmatrix}}$$

$= 1398.333 \pm 2.179(11.618) \sqrt{1.19167}$

$= 1398.333 \pm 26.146$

$= [1372.19, 1424.48]$

4.10   Crossvalidation

When conducting SLR or MLR, researchers typically use R^2 as a measure of the strength of the model. It should be noted that R^2 is dependent on the model, the sample, and the sample size. It is possible to continue to add variables up to $k = n-1$ to produce an R^2 which is monotonically increasing up to 1.0. This results in an overfit condition which inaccurately predicts the strength of the model. To compensate for models with large numbers of terms compared to the number of observations, some software packages such as MYSTAT will calculate an adjusted R^2 value. The formula for adjusted R^2 is shown below.

$$\text{Adj } R^2 = 1 - \left(\frac{n-1}{n-p}\right)(1-R^2)$$

where n = total number of observations

p = total number of terms in the model (including the constant)

A more accurate measure of model strength is to use half of the sample data to build the model and the other half for crossvalidation. Another strategy is to build the model on sample data and obtain a crossvalidated R^2 from confirmation experiments. To get the crossvalidated estimate of R^2, calculate $e_i = y_i - \hat{y}_i$, where the $\hat{y}_i$ are based on the estimates obtained from the original data and the y_i are the observed y values in the crossvalidation sample data. As before, we then compute

$$R^2 = 1 - \frac{\sum e_i^2}{SST_2}$$

where SST_2 is the total sum of squares for the data in the crossvalidation sample. It is also possible to correlate the y_i from the original data with the $\hat{y}_i$ model based on the second data set.

Since splitting a sample into two parts requires a large amount of data, a more efficient measure for estimating model strength is the Relative Press (Rel PRESS) constant. This measure of the strength of a model is calculated through the following steps:

(1) Compute the first of a series of regression model coefficients by withholding (or setting aside) the first observation (i).

(2) Use the model obtained in (1) to predict the observation (i) not used in building the model, namely $\hat{y}_{i,-i}$.

(3) Calculate the press residual $(e_{i,-i}) = (y_i - \hat{y}_{i,-i})$ for observation (i) not in the model.

(4) Repeat steps (1) through (3) n times, deleting a different observation each time.

(5) Calculate the Prediction Sum of Squares, referred to as

$$\text{PRESS} = \sum_{1}^{n} (e_{i,-1})^2 \ .$$

(6) Calculate Relative Press as follows:

$$\text{Rel PRESS} = 1 - \frac{\text{PRESS}}{\text{SST}} \, ,$$

where SST is the Total Sum of Squares for all n actual values of the response y. For a more efficient approach to calculating PRESS and more information on Rel PRESS see [1].

One way to reduce overfit is to ensure that for any regression model, n is sufficiently large compared to the number of parameters estimated. A good rule of thumb for the final prediction model is to have n be at least 10 times more than the number of parameters in the model.

4.11 Notes of Caution

When using SLR or MLR on "happenstance or non-experimental data", one should be cautious of high correlations, for they may be significant only because a spurious condition unknown to the analyst exists. This problem can be averted by designing proper experiments instead of analyzing non-experimental data. Furthermore, there are times when a researcher has properly designed an experiment, but obtains unanticipated results due to improper levels for independent variables. For example, consider the relationship between y and x in Figure 4.10. If the levels of x are a and d, you will find a significant linear relationship between y and x. However, if the range on x is b to c, this relationship may not be discovered.

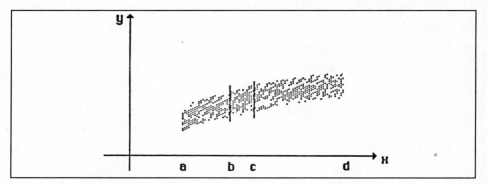

Figure 4.10 Effect of Varying Levels of an Independent Variable

A less likely, but certainly possible, scenario appears in Figure 4.11. Due to the quadratic relationship, if the sampled levels of x are a and b for a SLR, a non-significant relationship will be found even though a significant quadratic relationship exists. To find a significant linear relationship, the sampled region of x must be [a, c] or [c, b]. Obviously, using a quardratic (or 2nd order) model over [a, b], one could detect a significant quadratic relationship. Thus, experimenters should carefully plan not only which variables to put in the model, but also which levels are necessary to detect anticipated relationships.

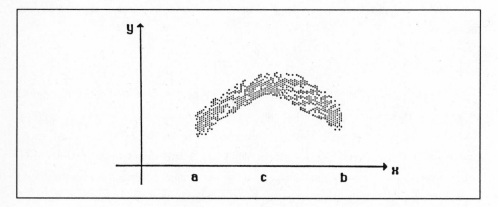

Figure 4.11 Quadratic Relationship on the Interval [a, b]

Finally, it will be briefly stated that outliers in the data can significantly alter regression results. Before any regression is conducted or any other data analysis takes place, one should explore the data distribution for each variable, identify outliers, ascertain their accuracy, and take the appropriate action. One can also use sophisticated techniques to identify outliers in a multi-variable setting through the use of residual analysis [1].

4.12 ANOVA for Factors with More Than 2 Levels (Optional)

Consider the following one-way ANOVA example where factor A has 5 levels (L).

Factor A

	A_1	A_2	A_3	A_4	A_5
$\bar{y}_j$	8	6	7	12	10
s_j^2	4.888	7.333	12.222	9.333	10.667
n_j	10	10	10	10	10

(1) H_0: $\mu_1 = \mu_2 = \mu_3 = \mu_4 = \mu_5$

   H_1: not all μ_i are equal

(2) $\alpha = .01$

(3) $\bar{y} = \dfrac{\sum n_j \bar{y}_j}{\sum n_j} = \dfrac{10\ (8 + 6 + 7 + 12 + 10)}{50} = 8.6$

$$MSB = \frac{\sum_1^L n_j(\bar{y}_j - \bar{y})^2}{L - 1} = \frac{10(-.6)^2 + 10(-2.6)^2 + 10(-1.6)^2 + 10(3.4)^2 + 10(1.4)^2}{5 - 1} = 58$$

$$MSE = \frac{\sum_1^L (n_j - 1)S_j^2}{\sum_1^L (n_j - 1)} = \frac{9(4.88) + 9(7.333) + 9(12.222) + 9(9.333) + 9(10.667)}{9 + 9 + 9 + 9 + 9}$$

$$= \frac{400}{45} = 8.887$$

(4) $F(.99, 4.45) = 3.83$

(5) Since $F_0 > 3.83$, reject H_0 with 99% confidence.

In this case the conclusion is H_1: μ_i are not all equal. The next question is "Which levels are different and which are not?" Using a regular t test or F test on each possible pair of treatments, i.e., H_0: $\mu_i = \mu_j$ for all $i \leq j$, would result in an increase in the overall α known as the experiment-wise α, α_{EW}. The α_{EW} is also referred to as the probability of committing at least one type one error after conducting all the individual pairwise tests. A procedure which allows for testing all pairwise levels at a specified α_{EW} is the Tukey Post Hoc comparison test. Use of the test is only valid when the factor overall F_0 is significant. The test is fairly simple and requires the use of the studentized ranges provided in Appendix H. The Tukey test will be illustrated using the previous example.

Since factor A is significant, it is desired to determine which levels differ and which do not. The procedure is as follows:

(a) Use the Tukey comparison test and compute the critical Tukey

distance $\bar{d} = q_T \sqrt{\dfrac{MSE}{n_i}}$ where q_T is found in Appendix H and all n_j are assumed equal.

(b) In Appendix H, p is the number of factor levels. Thus, for our example $p = 5$. Since $df_E = 45$ and given $\alpha = .05$, then $q_T = 3.171$. Therefore,

$$d_T = 3.171 \sqrt{\frac{8.887}{10}} = 2.989.$$

(c) Compute all factor level mean pairwise differences and conclude those differences greater than $\bar{d}_T$ as significant at the chosen α level (see the chart shown below).

Conclude μ_4 is significantly different than μ_2, μ_3, and μ_1. Also conclude μ_5 is significantly different than μ_2 and μ_3; fail to reject H_0 for all other pairwise comparisons. What does this mean to the experiment? If your objective is to maximize the response, then you are 95% confident that material 4 produces a larger response than materials 2, 3, and 1. In addition, material 5 produces a larger response than 2 or 3. Therefore, the best material type is either 4 or 5. Unless 5 is substantially cheaper than 4, the most conservative decision is to select type 4 because you failed to reject μ_5 different from μ_1. However, there exists probabilities of error, α (rejecting H_0 incorrectly) and β (failing to reject H_0 when you should). Small sample sizes contribute to large β values. Our sample size of ten for each factor level may not provide enough overwhelming information to reject H_0: $\mu_5 = \mu_1$ when we should.

4.13 Determining Sample Size (Optional)

Often the question is asked, "How large should my sample be?" The answer to this question depends on the following:

(1)   The desired level for α.

(2)   The desired level for $1 - \beta$ (β is the probability of concluding H_0 given that H_1 is true; therefore, $1 - \beta$ is the probability of concluding H_1 correctly. The term $1 - \beta$ is also referred to as the power of the test).

(3)   Estimated amount of experimental error, $\hat{\sigma}^2$.

(4)   A specific alternative hypothesis deemed critical to detect with $(1 - \beta)100\%$ confidence.

(5)   Whether or not we're estimating $\hat{y}$ models or $\hat{s}$ models or both, items (1) through (4) are easily used for sample size determination. When modeling variation, $\hat{s}$, we should as a rule of thumb have 16 degrees of freedom for any average standard deviation used in building the

model. (Chapter 5 provides more details. Also see Appendix M for simple Rules of Thumb.))

The appropriate equation containing terms (1) through (4) is:

$$\Phi^2 = n \left[\frac{\displaystyle\sum_{1}^{k} (\mu_i - \mu)^2}{k} \right] \Big/ \hat{\sigma}^2$$

where $\hat{\sigma}^2$, α, n', and degrees of freedom are used to enter the power charts in Appendix I. A step-by-step procedure for determining the approximate n' is as follows:

(1)   Determine k, the number of factor levels you intend to investigate.

(2)   Determine the alternative hypothesis of interest either with specific values of μ_i or with the anticipated range, R, of the μ_i values and their suspected variability (low, medium, or high) over that range.

Low variability implies that one μ_i is at the maximum, one μ_i is at the minimum and the other $(K - 2)\mu_i$ values occur at a point in between.

Medium variability implies that the μ_i are approximately normally distributed,

i.e., $\displaystyle\sum_{1}^{k} \frac{(\mu_i - \mu)^2}{k} \doteq \frac{R^2}{16}$.

High Variability implies that half of the μ_1 are at the maximum and the other half are at the minimum.

(3)   Estimate $\hat{\sigma}^2$ via

    (a)   historical data

    (b)   pilot study

(c) $\dfrac{R_i^2}{16}$ where R_i is the anticipated range of raw data for some factor level i. This method assumes $Y_{ij} \sim N(\mu_i, \sigma_i^2)$ for each level i.

(4) Select an estimated value of a factor level sample size, n'.

(5) Compute Φ.

(6) Enter the appropriate chart from Appendix I where v_1 is the degrees of freedom for the factor effect and v_2 is the degrees of freedom for error.

(7) If $(1 - \beta)$ is larger than desired, lower n' otherwise raise n'.

(8) Continue until step 7 converges on the desired $(1 - \beta)$.
An example follows:

Assume $\hat{\sigma}^2 = 20$, k = 4, $\sigma = .05$ and the desired power is 0.8 for detecting

$\mu_1 = 16, \mu_2 = 21, \mu_3 = 23, \mu_4 = 20.$

$$\Phi^2 \;=\; \frac{\dfrac{n'\left[(16-20)^2 + (21-20)^2 + (23-20)^2 + (20-20)^2\right]}{4}}{20} = .325n'$$

Thus $\Phi \;=\; .57\sqrt{n'}$

Using the table with $v_1 = 3$, you could iterate n' as follows:

n'	V_2	Φ	1 - ß	Comments
5	16	1.28	.45	Too small, increase n'
7	24	1.51	.64	Too small, increase n'
9	32	1.72	.78	Slightlly less than .8
10	36	1.80	.83*	Use n = 10 for each level

4.14 Comparison of Confidence Interval, t-Test, ANOVA and Regression Approaches for Analyzing a Factor with 2 Levels (Optional)

Given some factor A measured at 2 levels, A1 and A2, this section will compare four approaches which could be used to test the following hypotheses: H0: $\mu_1 = \mu_2$ versus H1: $\mu_1 \neq \mu_2$.

The following sample data will be used to make the comparison.

SAMPLE DATA			
	A_1	A_2	
	4	14	
	10	8	
	2	16	
	12	6	
Mean ($\bar{A}_j$)	7	11	Grand Mean, $\bar{Y}$ = 9
Variance (S_j^2)	22.667	22.667	
Sample size (n_j)	4	4	

Table 4.24 Sample Data

For the sake of this comparison, α will be set to .05.

(1) Confidence Interval Approach

$(1 - \alpha)100\%$ confidence interval for $\mu_2 - \mu_1$ is:

$$\left(\overline{A}_2 - \overline{A}_1\right) \pm t\left(1 - \frac{\alpha}{2}, n_1 + n_2 - 2\right)\sqrt{MSE\left(\frac{1}{n_1} + \frac{1}{n_2}\right)}$$

where $MSE = \dfrac{\sum(n_j - 1)s_j^2}{\sum(n_j - 1)}$. The sample data results are:

$$\left[(11 - 7) \pm t\,(.975, 6)\sqrt{22.667\left(\frac{1}{4} + \frac{1}{4}\right)}\right]$$

$$= [(4) \pm (2.447)(3.366)]$$
$$= [-4.236,\ 12.236]$$

Since the 95% confidence interval for $\mu_2 - \mu_1$ contains zero, your conclusion will be to fail to reject H_0.

(2) t-Test Approach

The 5 step method for the t-test is:

(1) H_0: $\mu_1 = \mu_2$
      H_1: $\mu_1 \neq \mu_2$

(2) $\alpha = .05$

(3) $t_0 = \dfrac{\overline{A}_2 - \overline{A}_1}{\sqrt{MSE\left(\dfrac{1}{n_1} + \dfrac{1}{n_2}\right)}} = \dfrac{4}{\sqrt{22.667\left(\dfrac{1}{4} + \dfrac{1}{4}\right)}} = 1.188$

(4) $\quad t_{CRITICAL} = t \left(1 - \dfrac{\alpha}{2}, n_1 + n_2 - 2 \right) = t \ (.975, \ 6) = 2.447$

(5) $\quad$ Since $[\,|t_0| \ = \ 1.188\,]$ is less than $[t_{CRITICAL} \ = \ 2.447]$ your conclusion will be to fail to reject H_0.

(3) $\quad$ **Analysis of Variance (ANOVA) Approach**

The 5 step method for ANOVA is:

(1) $\quad H_0: \quad \mu_1 = \mu_2$

$\qquad\quad H_1: \quad \mu_1 \neq \mu_2$

(2) $\quad \alpha = .05$

(3) $\quad F_0 \ = \ \dfrac{MSB}{MSE} = \dfrac{32}{22.667} = 1.412$

ANOVA TABLE				
SOURCE	SS	df	MS	F
BETWEEN	32	1	32	1.412
ERROR	136	6	22.667	
TOTAL	168	7		

Table 4.25 ANOVA for Sample Data

(4) $\quad F_{CRITICAL} = F(.95, \ 1, \ 6) = 5.99$

(5) $\quad$ Since $[F_0 = 1.412]$ is less than $[F_{CRITICAL} = 5.99]$, your conclusion will be to fail to reject H_0.

Before completing a regression approach to this problem, it is worth reviewing the following:

- All previous conclusions were the same.

- $t_0^2 = F_0$

- $\left[t \left(1 - \dfrac{\alpha}{2}, 6 \right) \right]^2 = F(1 - \alpha, 1, 6)$

(4) Regression Approach

To solve a two group problem by way of regression, you must form a group membership column. Although there are several ways to code this column, the most common would be using effect coding as shown below.

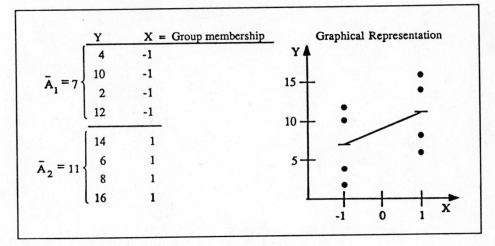

The regression equation is built based on the slope intercept equation, i.e.,

$$\hat{y} = \text{intercept} + \text{slope } X = b_0 + b_1 X$$

$$9 + \frac{11 - 7}{2}X$$

$$= 9 + 2X$$

Using the formula for R^2 in section 4.5, R^2 is calculated to be .1904.

The 5 step hypotheses test would be:

(1a) H_0: $R^2 = 0$ or   (1b) H_0: $B_1 = 0$

  H_1: $R^2 \neq 0$ H_1: $B_1 \neq 0$

In either case, H_0 implies no group differences, i.e., $\mu_1 = \mu_2$.

(2) $\alpha = .05$

(3a) To test H_0: $R^2 = 0$

$$F_0 = \frac{\dfrac{R^2}{1}}{\dfrac{(1 - R^2)}{(N - 2)}} = \frac{.1904}{\dfrac{(1 - .1904)}{(8 - 2)}} = 1.412$$

(4a) $F_{CRITICAL} = F(.95, 1, 6) = 5.99$

(5a) Since $F_0 < F_{CRITICAL}$, you fail to reject H_0.

(3b) To test H_0: $B_1 = 0$

$$t_0 = \frac{b_1}{\sqrt{\dfrac{MSE}{\left(\sum x^2 - n\bar{x}^2\right)}}} = \frac{2}{\sqrt{\dfrac{22.667}{8}}} = 1.188$$

(4b) $t_{CRITICAL} = t(.975, 6) = 2.447$

(5b) Since $|t_0| < t_{CRITICAL}$, you fail to reject H_0.

The conclusion is that all 4 methods produced identical results.* In addition, the t_0 values were all the same and equal to $\sqrt{F_0}$. Likewise, the $t_{CRITICAL}$ values were all the same and equal to $\sqrt{F_{CRITICAL}}$.

* Note: A similar conclusion would be made using the end count technique described on page 8-209 and in Appendix M.

Chapter 4 Problem Set

1. Given 2 different soft bake methods for a photoresist process, determine if the mean critical dimension in microns for the feature of interest is the same for each process. (Conduct this test using Confidence Intervals, t-test, and ANOVA.)

Setup #	Convection Bake	Hot Plate Bake
1	4.1	3.8
2	3.9	3.5
3	4.3	4.1
4	4.1	4.1
5	4.1	3.8

    All data is in microns, and the exposure was constant in all cases.

2. A precision bearing manufacturer who wants to determine a minimum sample size to check the accuracy of two groups of bearings manufactured 2 months apart. The manufacturer wants to insure that he can detect a specific H_1: $\mu_1 - \mu_2 = 10$ microns with 90% confidence given: $\sigma^2 = 81$ and $\alpha = 0.05$.

    What is the minimum sample size?

3. (a) Given the following etch rate data as a function of DC bias voltage, determine which levels of DC bias voltage are significantly different.

Level
A1 = -220 volts
A2 = -240 volts
A3 = -260 volts
A4 = -280 volts
A5 = -300 volts

Etch Rate Data: Angstroms/Min.

A1	A2	A3	A4	A5
1280	1285	1301	1345	1401
1277	1296	1310	1350	1390
1275	1291	1298	1340	1395
1282	1287	1316	1356	1402
1272	1294	1300	1351	1391
1289	1281	1285	1360	1405
1271	1285	1305	1347	1406

$\alpha = 0.01$

(b) If the objective is to maximize the etch rate, which level(s) should be selected?

4. A manufacturer of a proprietary plating additive wants to understand the reaction kinetics of the additive. The manufacturer assumes that the following reaction represents his process:

Concentration of Additive C_A > Rate of Reaction $-r_A$.

Given this Data:

n	1	2	3	4	5
C_A	5.6	9.2	15.6	22.5	26.2
$-r_A$	1.5	3.2	7.3	13.9	22

(a) Calculate the slope and intercept of the linear model between C_A and $-r_A$ where $-r_A$ is the dependent variable.

(b) Find the correlation coefficient, R, and also the proportion of variability explained by the regression.

(c) Graph C_A versus $-r_A$. Is a linear relationship a good assumption?

(d) The manufacturer now thinks the relationship is:

$$-rA = k \, C_A{}^N , \; k = \text{rate constant}$$
$$N = \text{order of reaction}$$

Taking the natural log of this equation:
$$\ln(-r_A) = \ln(k) + N \cdot \ln(C_A)$$

Which is a linear model? Take the natural log of the raw data and repeat (a) and (b) to find the order of the reaction, N.

5.   A graduate student was asked to determine if a relationship existed between sulfuric acid concentration (wt. %) and temperature (°C), and their effect on a 1st order rate constant (k) with respect to the hydrolysis of cellulose.

(a) Using the data he derived, generate a ANOVA table.

(b) Is there an interaction between wt. % H_2SO_4 and temperature?

(c) Is there a significant effect by temperature of wt. % H_2SO_4?

T(°C)	wt. % H_2SO_4	k_1	k_2
170	0.5	0.0196	0.0201
190	0.5	0.0376	0.0355
210	0.5	0.8710	0.9000
170	1.0	0.0808	0.0799
190	1.0	0.1660	0.1500
210	1.0	0.7420	0.7210
170	2.0	0.0736	0.0691
190	2.0	0.1630	0.1700
210	2.0	0.4480	0.4350

6.   Given the following data:

Run	A	B	−AB	C	−AC	D	E	Y_1	Y_2	$\bar{Y}$	S
1	−1	−1	−1	−1	−1	−1	−1	66	62	64	2.83
2	−1	−1	−1	+1	+1	+1	+1	68	63	65.5	3.54
3	−1	+1	+1	−1	−1	+1	+1	88	80	84	5.66
4	−1	+1	+1	+1	+1	−1	−1	63	65	64	1.41
5	+1	−1	+1	−1	+1	−1	+1	73	71	72	1.41
6	+1	−1	+1	+1	−1	+1	−1	37	42	39.5	3.54
7	+1	+1	−1	−1	+1	+1	−1	38	39	38.5	0.71
8	+1	+1	−1	+1	−1	−1	+1	57	48	52.5	6.36

a) Complete the following table:

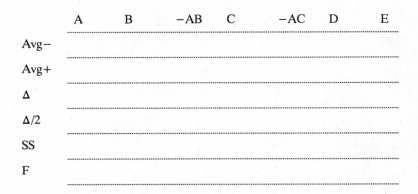

	A	B	−AB	C	−AC	D	E
Avg−							
Avg+							
Δ							
Δ/2							
SS							
F							

b) Complete a standard ANOVA table.

c) Determine the prediction equation: Y

d) What are the optimal settings to minimize Y ?

e) What is the prediction response for the settings in (c) above for Y ?

7.   If only one variable is being tested at 2 levels, is there any difference between using ANOVA or a t-test?

8. What is the effect of using pairwise t-test if multiple μ_i are found not to be equal? Why is it bad to increase α_{EW}?

9. What does R^2 measure in regression modeling?

10. $\left[t\left(1 - \dfrac{\alpha}{2}, n\right) \right]^2 = F(\ \underline{\quad},\ \underline{\quad},\ \underline{\quad})$

11. What is an example of a blocking variable? How can an experimenter deal with a blocking variable?

12. What is a prediction interval and how would you interpret it?

Chapter 4 Bibliography

1. Myers, Raymond H. (1986), *Classical and Modern Regression with Applications*, Duxbury Press, Boston, Mass.

2. Walpole, R. E. and Myers, R. H. (1985), *Probability and Statistics for Engineers and Scientists*, MacMillian, Inc., New York, NY.

3. Kiemele, M. J. and Schmidt, S. R. (1991), *Basic Statistics: Tools for Continuous Improvement*, 2nd Edition, Air Academy Press, Colorado Springs, CO.

Chapter 5

Design Evaluation

5.1 Introduction

After collecting good data through a well designed experiment, the analysis phase is used to turn the data into information for making sound decisions. The purpose of this chapter is to expand upon the simple analysis techniques presented in Chapter 2 and incorporate the statistical methods discussed in Chapter 4.

The type of analysis technique you use will depend primarily on your experimental objective and the capability of your software.* In this chapter, we will discuss analysis techniques for the following design objectives: screening, modelling (also referred to as characterization), variance reduction, and robustness. Response surface methods and optimization are presented in Chapter 7.

If your objective is factor screening, i.e., reducing a large number of factors to a manageable subset of important factors, then resolution III designs will typically be used such as those presented in Chapter 3. The analysis will most likely be a graphical approach such as plotting the averages for each effect, a Pareto diagram of the absolute value of the half effects ($|\Delta/2|$), or a normal probability plot. Analysis of Variance (ANOVA) and/or regression could also be utilized to get a statistical representation of important factors.

Modelling or characterizing a process will usually result in a resolution IV or V design (depending on your prior knowledge of interaction effects). To determine which terms

---

* Most of the techniques discussed in this chapter can be performed on the accompanying software *Q-Edge* or on *MYSTAT* which accompanies *Basic Statistics: Tools for Continuous Improvement*, Air Academy Press.

appear in the final model will require a Pareto of $|\Delta/2|$ values and/or regression as a minimum. Other techniques such as ANOVA and normal probability plots might also be used to provide additional insight into the data.

Selecting the analysis technique(s) for variance reduction and robustness is similar to that for modelling. The primary difference is that we are now modelling the variation through $\hat{s}$ or $\ln \hat{s}$ equations instead of modelling the average with the $\hat{y}$ equation.

5.2   Analyzing a 2-Level Design

Assume that you have collected the data for an 8-run Taguchi design matrix of 7 factors, each at 2 levels, as shown in Table 5.1. Since this is a resolution III design, chances are you are using it as a screening design; however, we will use this data to discuss several analysis techniques.

				Factors				Response
Run	A	B	C	D	E	F	G	y
1	−	−	−	−	−	−	−	18
2	−	−	−	+	+	+	+	20
3	−	+	+	−	−	+	+	12
4	−	+	+	+	+	−	−	10
5	+	−	+	−	+	−	+	12
6	+	−	+	+	−	+	−	14
7	+	+	−	−	+	+	−	19
8	+	+	−	+	−	−	+	21

Table 5.1 L_8 Design Matrix for 7 Factors at 2 Levels

If this design is intended as a trouble-shooting exercise to find the best settings for the factors, the simplest analysis technique is to look at the response values for the experimental runs and select the response that best satisfies the objective (i.e., minimum, maximum, or closest to a target). This method, referred to as **pick the winner** has obvious advantages: quick, easy, insensitive to interaction aliasing, and already contains a confirmation result. Despite these advantages, there also exist several disadvantages in that factor importance is unknown, no characterization model is built, factor settings will not necessarily minimize cost, and several possible combinations have been ignored. For example, if a small response is desired, then the "pick the winner" factor settings from Table 5.1 correspond to the highs (+) and the lows (−) of run number 4. However, the 7 factors at 2 levels produce a total of 128 different combinations and we've only considered 8. To look at the remaining 120 possibilities, we can analyze the marginal effects through average effect plots, Pareto diagrams of half effects, normal probability plots, ANOVA, and/or regression. Some of these procedures will not provide tests of significance of the data in Table 5.1 unless there is some replication. This problem will be discussed as we examine each of these methods.

To analyze the data by way of **average effect plots**, consider Figure 5.1. Assuming we want to minimize the response, the average effect plots in Figure 5.1 indicate the best settings would be $A_- B_+ C_+ D_- E_+ F_- G_-$. The predicted average response based on these factor settings is computed as follows:

$$\hat{y} = \bar{y} + \frac{\Delta_A}{2}A + \frac{\Delta_B}{2}B + \ldots + \frac{\Delta_G}{2}G$$

$$= 15.75 + \frac{1.5}{2}A_- - \frac{.5}{2}B_+ - \frac{7.5}{2}C_+ + \frac{1}{2}D_- - \frac{1}{2}E_+ + \frac{1}{2}F_- + \frac{1}{2}G_-$$

$$= 15.75 - .75 - .25 - 3.75 - .5 - .5 - .5 - .5 = 9$$

This is an improvement over run #4; however, this estimate could be overly optimistic if some of the included effects occurred only due to chance and in reality are not significantly different from zero.

One way to assess the significance of effects is to look at a **Pareto diagram** of the absolute value of all half effect values found in the equation for $\hat{y}$ (see Figure 5.2). It appears that the only factor which stands out beyond what might be considered the noise level is factor C. The remaining effects could be considered to be different from zero due only to experimental error. This being the case, the best factor settings would consist of C_+ and the remaining factors set at levels which are based on: (1) economics, (2)

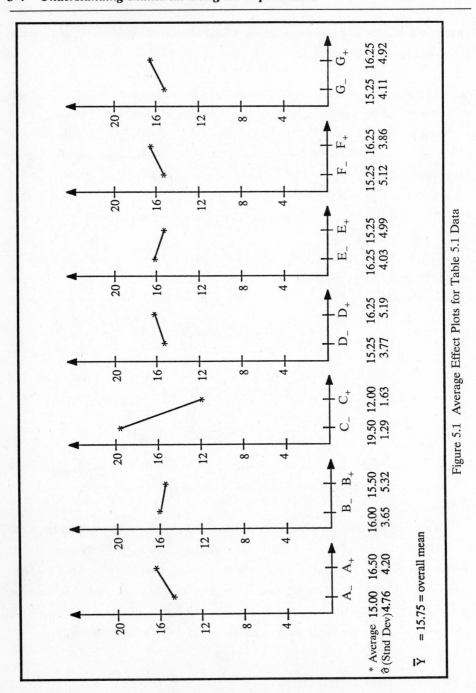

Figure 5.1 Average Effect Plots for Table 5.1 Data

convenience, (3) status quo, and/or (4) providing less variability in the response. The new predicted average response would be:

$$\hat{y} = 15.75 - 3.75C_+ = 12$$

which is slightly different from run #4. Since only factor C appears to be significant, production costs may be reduced through more flexibility in selecting levels for the remaining factors.

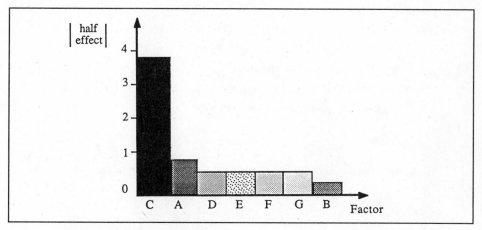

Figure 5.2 Pareto Diagram of Table 5.1 Half Effects

Not all Pareto diagrams provide as clear a picture of effect significance as Figure 5.2. Another approach would be the **normal probability plot (NPP)**. This method is especially useful when designs are saturated and there are no degrees of freedom for error. Since the design in Table 5.1 has n = 8 runs and k = 7 factors, there are no degrees of freedom for estimating error and, therefore, ANOVA and regression results with significance tests are not possible without some sort of modification. In these situations, the NPP or the Pareto diagram are the recommended techniques. To complete a NPP:

(1)   For each factor and interaction, calculate the difference in average response for high and low levels (i.e., calculate the column effects).

e.g., $\bar{A}_+ - \bar{A}_-$, $\bar{B}_+ - \bar{B}_-$, . . . , $\bar{G}_+ - \bar{G}_-$, $\overline{A \times B}_+ - \overline{A \times B}_-$, etc.

(2)   Assign a rank, i, to the differences in (1) starting from lowest to highest (i = 1, 2, . . . , M). M = the number of effects, i.e., the number of columns in the design matrix.

(3)   Convert each effect rank to a percentile using $p = \dfrac{100\,(i - 0.5)}{M}$.

(4)   Scale the horizontal axis of NPP paper to satisfy the values found in (1).

(5)   On NPP paper, plot p versus values found in (1).

(6)   Draw a straight line through the majority of points which appear to line up (discounting any obvious "outliers"). When establishing the line, give priority to effects close to zero.

(7)   Any point which deviates considerably from the straight line represents a potentially significant effect.

Using the data set in Table 5.1, steps (1) through (7) result in the following:

	Factor	A	B	C	D	E	F	G
(1)	Effect	1.5	−.5	−7.5	1.0	−1.0	1.0	1.0
(2)*	Rank	7	3	1	5	2	5	5
(3)	p	92.9	35.7	7.1	64.3	21.4	64.3	64.3

(4) (5) (6)   See Figure 5.3

(7)   The only factor which appears significant is C. Thus, if our objective for the design in Table 5.1 is factor screening, the Pareto diagram and the NPP both indicate that only factor C is important.

---

* When two or more effects have equal values, their ranks become the average of what would otherwise be assigned. For example, the ranks for D, F, and G would be 4, 5, and 6, but since all effect values are the same, each receives an average rank of 5.0.

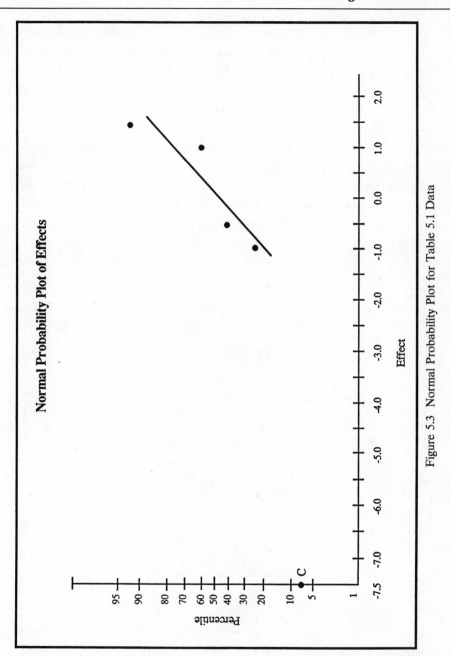

Figure 5.3 Normal Probability Plot for Table 5.1 Data

Graphical analyses provide a visual understanding of the relative importance of each effect, but do not provide a quantitative measure of confidence in our conclusion. To estimate this confidence, statistical methods such as **ANOVA** or **regression** must be used. To use ANOVA or regression, there must exist some degree of freedom for error. This is best accomplished by replicating the entire design or, more efficiently, replicating at one point. If runs are costly, replicating at one point is more economical, but the question is, at which point? To answer this question, you must first determine if there is any need to address non-linear effects. If so, you should replicate at the center point (0) and, thus, be able to graphically test for curvature and estimate error. If non-linear effects are not reasonable to test, then replicate at the level chosen for the best response. This replication serves as a way to estimate error and also provides confirmation runs.

To demonstrate the use of ANOVA and regression, we will use the data from Table 5.1 with an additional replicate for each of the 8 runs. The new data set appears as shown in Table 5.2

Factors								Response		Average	Variance	
Run	A	B	C	D	E	F	G	y_1	y_2	$\bar{y}_r$	s^2_r	df_r
1	−	−	−	−	−	−	−	18	23	20.5	12.5	1
2	−	−	−	+	+	+	+	20	18	19.0	2.0	1
3	−	+	+	−	−	+	+	12	13	12.5	0.5	1
4	−	+	+	+	+	−	−	10	14	12.0	8.0	1
5	+	−	+	−	+	−	+	12	11	11.5	0.5	1
6	+	−	+	+	−	+	−	14	9	11.5	12.5	1
7	+	+	−	−	+	+	−	19	25	22.0	18.0	1
8	+	+	−	+	−	−	+	21	20	20.5	0.5	1

	A	B	C	D	E	F	G	
Avg −	16.0	15.625	20.5	16.625	16.25	16.125	16.5	$\bar{y} = 16.1873$
Avg +	16.375	16.75	11.875	15.75	16.125	16.25	15.875	
Δ	.375	1.125	−8.625	−.875	−.125	.125	−.625	
MSB_{effect}	.563	5.063	297.563	3.063	.063	.063	1.563	$F_c = F(.95,1,8)$
F_{effect}	<1	<1	43.68	<1	<1	<1	<1	= 5.32

$$MSE = \frac{\sum (df_r)s^2_r}{\sum (df_r)} = \frac{1(12.5) + 1(2.0) + 1(.5) + \ldots + 1(18) + 1(.5)}{1 + 1 + 1 + 1 + 1 + 1 + 1 + 1} = 6.813$$

Table 5.2 Analysis of the Data from Table 5.1 with an Additional Replicate

Recall from Chapter 4 that 2-level designs will result in $MSB_{effect} = \frac{N}{4}(\Delta)^2$ where N is the total number of response values (for our example in Table 5.2, N = 16). Using the MSB_{effect} formula, the reader should verify the MSB_{effect} values in Table 5.2. The critical F value in Table 5.2 was obtained from Appendix F using $F_C = F(1 - \alpha, 1, \Sigma(df_r))$.

For readers familiar with a standard ANOVA table display, the results in Table 5.2 can be displayed as shown in Table 5.3. To interpret the P column, use (1 − P)100% as the confidence in that term being important. The ANOVA results indicate that factor C is the only significant factor. The reader should realize that the preciseness of the P value is based on the normality of the response. Because response data is not always normally distributed [11], we strongly recommend that ANOVA only be used to supplement the previous approaches of Pareto charts, NPP, size of the half effects, and prior knowledge.

ANOVA TABLE

Source	SS*	df	MS	F**	P***
A	.563	1	.5625	<1	.781
B	5.063	1	5.0625	<1	.414
C	297.563	1	297.5625	43.68	.000
D	3.063	1	3.0625	<1	.521
E	.063	1	.0625	<1	.926
F	.063	1	.0625	<1	.926
G	1.563	1	1.5625	<1	.645
Error	54.5	8	6.8125	–	
Total	362.438	15			

Table 5.3 Standard ANOVA Table for Table 5.2 Data Results

If we entered the 16 coded combinations from Table 5.2 into a regression package and regressed the response value against factors A through G, the output would appear as shown in Table 5.4.

---

* For orthogonal 2-level designs, $SS_A = N \bullet b_A^2$ where N is the total number of runs (in this case N = 16) and b_A is the regression coefficient for factor A obtained when you regress the N response values on the design matrix of (+)'s and (−)'s. SS_A is also equal to $MSB_A \bullet df_A$. Other SS values are found in a similar fashion.

** See Appendix M for Rules of Thumb on significant F values.

***See Appendix M for Rules of Thumb on significant P values.

Dependent Variable: Y	N: 16	Multiple R: .922		Squared Multiple R: .850		

Adjusted Squared Multiple R: .718   Standard Error of Estimate: 2.610

Variable	Coefficient	Std Error	Std Coef	Tolerance	T	P(2 Tail)*
Constant	16.188	0.653	0.000	1.0000000	24.808	0.000
A	0.188	0.653	0.039	1.0000000	0.287	0.781
B	0.563	0.653	0.118	1.0000000	0.862	0.414
C	−4.313	0.653	−0.906	1.0000000	−6.609	0.000
D	−0.438	0.653	−0.092	1.0000000	−0.670	0.521
E	−0.063	0.653	−0.013	1.0000000	−0.096	0.926
F	0.063	0.653	0.013	1.0000000	0.096	0.926
G	−0.313	0.653	−0.066	1.0000000	−0.479	0.645

Analysis of Variance

Source	Sum-of-Squares	df	Mean-Square	F-Ratio	P
Regression	307.938	7	43.991	6.457	0.009
Residual	54.500	8	6.813		

Table 5.4 Regression Output for Table 5.2 Example Data

For an orthogonal 2-level design, the coefficients for each factor are equal to

[response average at "+1" − response average at "−1"] ÷ 2,

thus, regression analysis is very similar to previous methods. Reviewing Tables 5.3 and 5.4 indicates that both regression and ANOVA find only factor C to be significant. In fact the ANOVA and regression results are identical (compare the P columns for each effect). The added benefit of regression is that we obtain the coefficients for the prediction equation (see the variable and coefficient columns in Table 5.4).

The prediction equation derived from Table 5.4 is:

$$\hat{y} = 16.188 - 4.313C$$

Since our objective is to minimize the response, setting C at the high level (+1 orthogonal coded value) results in a predicted average response of 11.875. Assuming we had some degree of confidence in this model, the next step would be to make as many confirmation runs as possible (4 to 20) and determine the confirmation run average $\overline{CR}$. To test if the

---

* See Appendix M on Rules of Thumb for significant P(2 tail) values.

confirmation runs produced anticipated results you could conduct a t-test where

H_0: μ = 11.875 and H_1: $\mu \neq$ 11.875 using $t_0 = \dfrac{\overline{CR} - 11.875}{S_{CR} / \sqrt{n_{CR}}}$ and $t_c \left(1 - \dfrac{\alpha}{2}, n_{CR} - 1\right)$

(see Chapter 4 for steps required for a t-test). If you fail to reject H_0, you have confirmed your results. However, if you reject H_0, i.e., conclude H_1: $\mu \neq$ 11.875, you must conduct an organized investigation of what may have caused the difference between the confirmation run average and the predicted average response. The following is a partial list of what could have caused the difference.

(1)   Factors not included in the experiment varying in an uncontrolled and non-random fashion.

(2)   Poor experiment discipline (i.e., settings were not made correctly, response values read incorrectly, no randomization, etc.)

(3)   Interactions aliased with main effects.

(4)   Model inadequacy: (i) a second order model is appropriate, but only a first order model is estimated due to a two-level type design; (ii) too few terms in the model; (iii) too many terms in the model

(5)   Something changed between the initial experiment and the confirmation runs (i.e., different type of raw material, different ambient conditions, different machines, etc.).

(6)   Excessive variation in the data coupled with an inadequate number of response values per experimental run.

At this point you would want to review items (1) through (6) above and try to rule out as many as possible. Unfortunately, if you cannot rule out items (1) or (2), you may not recover at all without redoing the entire experiment. If you suspect item (4) as the primary cause, you'd simply play with the prediction equation or add runs to allow for non-linear modelling (see CCD in Chapter 3). Provided all factors were set at the extremes, i.e., their (+) or (−) setting, you can rule out non-linearity as a possible cause of not confirming. Investigating item (5) requires some group brainstorming; whereas recovery from item (6) necessitates more replicates. In our example, seven factors were placed in an eight-run design which results in a resolution of III. Recall from Chapter 3 that resolution III implies that main effects are aliased with two-way interactions. Thus, there is a strong possibility that any problem with confirmation runs in our example is associated with a strong two-way interaction, item (3). The next section will discuss the importance of including two-way interactions in the design matrix.

5.3 Importance of Modelling Interactions

How important is it to model interactions? This question can be addressed by way of a simple example. Consider the design below.

Run	A	B	A•B
1	+	+	+
2	+	−	−
3	−	+	−
4	−	−	+

Assume the researcher de-emphasizes interactions and he is also interested in evaluating factor C. The 4-run design above could be used to test the A, B, and C main effects by aliasing A•B with factor C. Now, consider the following results:

Run	A	B	C	Y
1	+	+	+	9
2	+	−	−	6
3	−	+	−	6
4	−	−	+	9

The average effect plots on the following page indicate no A or B effect, but a substantial C effect.

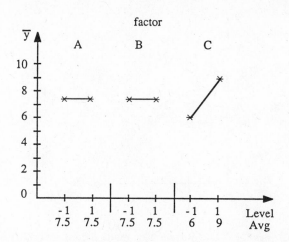

If "larger is better" then the researcher would suggest setting C at the high level and A and B settings to be based on economic considerations. For the sake of discussion, let's set A low and B high. Thus, the recommended settings are $A_{(-)}$ $B_{(+)}$ $C_{(+)}$ with a predicted average response of 9. We now make 5 confirmation runs with the following results.

Run	y
1	6
2	4
3	5
4	5
5	7
$\overline{CR}$	5.4
S_{CR}	1.14

H_0: $\mu = 9$ (confirmed)
H_1: $\mu \neq 9$ (did not confirm)
$\alpha = .05$

$$t_0 = \frac{5.4 - 9}{1.14 / \sqrt{5}} = -7.06$$

$$t_C = t(.975, 4) = 2.776$$

Since $|t_0| > t_C$, we would reject H_0, i.e., conclude H_1. What has gone wrong? The problem in this hypothetical example is that we originally assumed the $A \cdot B$ effect was not applicable and aliased factor C with $A \cdot B$. In our analysis, the A and B effects appeared unimportant indicating that where we set them will not affect the response. However, if $A \cdot B$ is important and C is not, then a true representation of A and B will appear as shown in the following graphic.

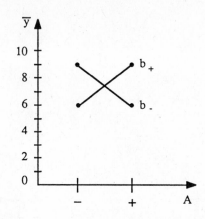

In this case, the settings for A and B cannot be made independently. For example, if A is set at (+), then y is maximized by setting B_+. However, if A is set at (−), then y is maximized by setting B_-. The results for this combination A_-B_+ in our confirmation runs is now predicted to be about 6.0 which is close to the 5.4 obtained.

The point to be made is that one should not ignore factor interactions just to simplify the design and to minimize the number of experimental runs. Even though it is true in most applications that strong interactions occur infrequently, the researcher should carefully consider whether or not to analyze interactions (see the Anodize case study from Boeing in Chapter 8 which further demonstrates the importance of identifying interactions). If you choose not to evaluate interactions, at least be aware of ways to recover from problems similar to the one presented above, i.e., if you use an R_{III} design which provides confusing results, you can fold it over to get an R_{IV}, separating 2-way interactions from the main effects. See Chapter 3 for more information on a foldover design.

5.4 Identifying Dispersion Factors (Factors which shift the Response Variability)*

Thus far, this analysis chapter has only discussed the identification of factors that shift the average response (i.e., location effects). Recent publicity of the Taguchi approach to parameter design for robust products and processes has emphasized the need to also identify

---

* Much of this section is taken from the Schmidt, S. R. and Boudot, J. R. paper presented at the 1989 Rocky Mountain Quality Conference and the 1989 ORSA/TIMS Annual Conference [8].

which factors contribute to changes in variability of the response. The use of Taguchi's loss function described in Chapters 1 and 6 provides the motivation for experiments to focus on target values instead of engineering tolerances or specifications. The average loss for any product is based upon the product variability and product average deviation from a designed target. This being the case, a set of objectives for a designed experiment should be as follows:

(1)   Identify factors which shift the average response (location effects) and then select those factor settings which minimize the difference between the average response and the desired target value (see Figure 5.4a).

(2)   Identify factors which contribute to changes in the response variability (dispersion effects) and then select the factor settings which minimize response variability (see Figure 5.4b).

(3)   Identify factors which shift the average response and the response variability and then select the factor settings which best increase the quality of the product or process (see Figure 5.4c).

(4)   Identify factors which have no effect on the response and then set these factors at levels that result in lower costs and/or faster through-put (see Figure 5.4d).

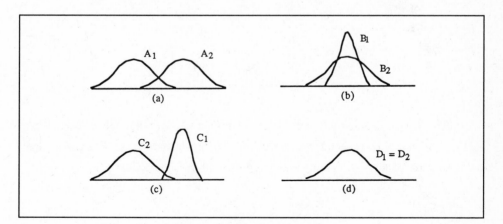

Figure 5.4 Types of Factors

The bottom line is that the engineer needs to know which factors or knobs he/she can use to adjust the response average, which ones will reduce the variance, and which ones do not make a difference with regard to the average or the variance.

To accomplish the four previously stated objectives, competing strategies have emerged. Since methods for finding location effects have already been discussed, the remainder of this section will concentrate on the following three commonly used methods for identifying dispersion effects: (1) signal-to-noise ratios, (2) s or ln s, and (3) residual analysis for replicated or unreplicated designs.

Taguchi's three signal-to-noise ratios are derived from loss functions associated with the following experimental goals: (1) maximize the response, (2) minimize the response, and (3) adjust the response to a specified target or nominal value. The resulting signal-to-noise ratios are shown below.

(1) $S/N_L = -10 \, \text{Log}_{10} \, \dfrac{1}{n_r} \sum_1^{n_r} \left(\dfrac{1}{y_i^2} \right)$ for maximizing the response.

(2) $S/N_S = -10 \, \text{Log}_{10} \, \dfrac{1}{n_r} \sum_1^{n_r} \left(y_i^2 \right)$ for minimizing the response.

(3) $S/N_N = 10 \, \text{Log}_{10} \, \dfrac{1}{n_r} \left(\dfrac{S_m - V_e}{V_e} \right)$ for a target response*

where $S_m = n_r \bar{y}^2$ and $V_e = \dfrac{\sum_1^{n_r} y_i^2 - \dfrac{1}{n} \left(\sum_1^{n_r} y_i \right)^2}{n - 1}$.

The ln s method is based on the logarithm of the standard deviation associated with responses for each experimental condition or run. To demonstrate the use of signal-to-noise ratios and ln s, consider the simple example presented in Table 5.5.

---

* (S/N_N is also proportional to $10 \, \text{Log}_{10} \left(\dfrac{\bar{y}^2}{s^2} \right)$.)

	Factors						
Run	A	B	C	y_1	y_2	$\bar{y}_r$	s_r
1	1	1	2	20	30	25	7.071
2	2	1	1	9	11	10	1.414
3	1	2	1	26	24	25	1.414
4	2	2	2	15	5	10	7.071

Table 5.5 Example Used to Demonstrate S/N Ratio and Ln s Analysis

If you look at the row changes in $\bar{y}$ and s and how these changes are correlated with certain design columns, you should be able to see that the A factor adjusts the mean response, the C factor adjusts the variability of the response and the B factor has no effect at all. A graphic display of each factor's effect on y appears in Figure 5.5.

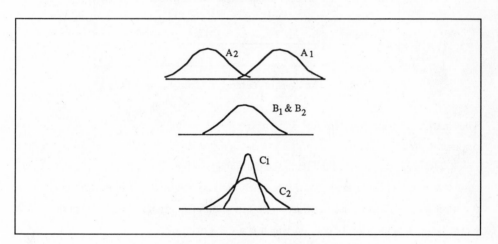

Figure 5.5 Graphical Representation of Factor Effects in Table 5.5

If you find the S/N_L values for each row in Table 5.5 and then calculate marginal averages of S/N_L for each factor, you should obtain results similar to those in Table 5.6. Using the analysis results in Table 5.6, the S/N_L strategy identified the correct location effect, but not the dispersion effect.

	Factor		
	A	B	C
Avg 2	18.20	22.24	21.98
Avg 1	27.69	23.65	23.90
Δ	−9.49	−1.41	−1.92

Table 5.6 Analysis of Table 5.5 Data Using S/N_L

Had we analyzed the data using S/N_S the marginal averages are displayed in Table 5.7.

	Factor		
	A	B	C
Avg 2	−20.51	−24.47	−24.55
Avg 1	−28.05	−24.09	−24.00
Δ	7.54	−.38	−.55

Table 5.7 Analysis of Table 5.5 Data Using S/N_S

The marginal mean analysis for S/N_S also identified the location effect, but not the dispersion effect.

Using S/N_N to analyze the data, the marginal averages are shown in Table 5.8. In this case, the marginal mean analysis for S/N_N identified both A and C as important factors, but not until the subsequent use of 10 $Log_{10}(S_m)$ in another marginal table can you determine which factor is a location effect and which is a dispersion effect.

	Factors		
	A	B	C
Avg 2	10.00	13.98	6.99
Avg 1	17.96	13.98	20.97
Δ	−7.96	0	−13.98

Table 5.8 Analysis of Table 5.5 Data Using S/N_N

The marginal mean analysis using ln s is shown in Table 5.9.

	Factor		
	A	B	C
Avg 2	1.15	1.15	1.95
Avg 1	1.15	1.15	.35
Δ	0	0	1.60

Table 5.9 Analysis of Table 5.5 Data Using Ln s

The analysis with ln s clearly indicates that factor C is a dispersion factor. Thus, given the type of data presented, the use of ln s appears to be an easier and more effective method for locating dispersion effects. Using ln s as the preferred approach for finding dispersion effects can also be verified by referring back to the example data in Table 5.2. To analyze the variability effects of factors A through G, you can perform an analysis similar to the regression used for identifying location effects; however, the s^2 column of Table 5.2 is used as the response. (As previously demonstrated, you could also use s, ln s, or ln s^2 as the response. The best models will be built on ln s and s because they are less skewed than s^2.) Using an orthogonally coded design matrix (i.e., low = -1 and high = $+1$) and s^2 as the response, the regression results appear in Table 5.10.

Dependent Variable: s^2 N: 8 Multiple R: 1.000 Squared Multiple R: 1.000						
Adjusted Squared Multiple R: 1.000 Standard Error of Estimate: 0.000						
Variable	Coefficient	Std Error	Std Coef	Tolerance	T	P(2 Tail)
Constant	6.813	0.000	-	1.0000000	-	-
A	1.063	0.000	-	1.0000000	-	-
B	−0.063	0.000	-	1.0000000	-	-
C	−1.438	0.000	-	1.0000000	-	-
D	−1.063	0.000	-	1.0000000	-	-
E	0.313	0.000	-	1.0000000	-	-
F	1.438	0.000	-	1.0000000	-	-
G	−5.938	0.000	-	1.0000000	-	-

Table 5.10 Using Table 5.2 Data to Model the Variance, s^2

The R^2 value of 1.0 and the 0 values for standard error are due to a perfect overfit (i.e., the number of effects are one less than the number of runs). This model is obviously inappropriate for significance tests, but because of the orthogonal design, the coefficients can be used to estimate effects. The coefficients for each factor are computed from average variance at the factor high (+) minus the average variance at the low (−), i.e., these coefficient values are equivalent to factor half effects using s^2 as the response $((\bar{s}_2^2 - \bar{s}_1^2)$ ÷ 2). For factor A, the average run variance for levels (−) and (+) are 5.75 and 7.875 respectively. Thus, $(\bar{s}_{A2}^2 - \bar{s}_{A1}^2)$ ÷ 2 is identical to the regression coefficient. You can now use a Pareto diagram or NPP to determine which factors (if any) have substantial dispersion effects. Based on the coefficients in Table 5.10, there appears to only be a G dispersion effect. The negative sign indicates that there is lower dispersion at the G = +1 setting.

If you prefer to use ln s as the dependent variable, the results appear in Table 5.11. The same conclusion is made either way; however, the distribution of ln s is less skewed than that of s^2. Since ln s analysis is less sensitive to occasional extreme values, it is the preferred response.

Dependent Variable: ln s N: 8 Multiple R: 1.000 Squared Multiple R: 1.000						
Adjusted Squared Multiple R: 1.000 Standard Error of Estimate: 0.000						
Variable	Coefficient*	Std Error	Std Coef	Tolerance	T	P(2 Tail)
Constant	0.540	0.000	-	1.0000000	-	-
A	−0.036	0.000	-	1.0000000	-	-
B	−0.092	0.000	-	1.0000000	-	-
C	−0.137	0.000	-	1.0000000	-	-
D	−0.036	0.000	-	1.0000000	-	-
E	0.082	0.000	-	1.0000000	-	-
F	0.137	0.000	-	1.0000000	-	-
G	−0.713	0.000	-	1.0000000	-	-

Table 5.11 Using Table 5.2 Data to Model Ln s

---

* See Appendix M for Rules of Thumb on significant terms for s and ln s models.

In conclusion, the complete analysis of the example data in Table 5.2 indicates that you set factor C at the high (+1) for minimizing the response average and factor G at the high (+1) for minimizing the response variation. The remaining factors are set as dictated by economics, status quo, or convenience. At this point you would want to make several (4 to 20) confirmation runs to confirm the previous results and determine a crude estimate of the process capability index, C_{pk}, for the optimal settings. If the confirmation runs did confirm, but did not produce desired results, you might want to see Chapter 7 on how to change your experimental region.

5.5   Variance Reduction Case Study

To avoid the aliasing problem we had in modeling the Table 5.2 data, a resolution IV (or higher) design is recommended. In the example that follows, only 3 factors have been identified for experimentation. Providing the linearity assumption is valid and interactions are possible, a 3-factor (2-level) full factorial is a good design choice. This example is intended to reinforce the previous analysis procedures and to demonstrate one way to handle conflicts in optimizing both the $\hat{y}$ and $\hat{s}$ models.

In a metal casting process for manufacturing jet engine turbine blades, the experimental objective is to determine optimal factor settings to minimize part shrinkage (i.e., difference between mold size and part size) and minimize shrinkage variability. The factors to be tested are mold temperature (A), metal temperature (B), and pour speed (C). The coded test matrix together with the response values and analysis for this experiment are shown in Table 5.12. The $\ln \dfrac{s_{(1)}^2}{s_{(-1)}^2}$ shown in Table 5.12 provides an alternative method for determining the significance of terms for the $\hat{s}$ or $\ln \hat{s}$ prediction equation. Montgomery [10] has shown that $\ln \dfrac{s_{(1)}^2}{s_{(-1)}^2}$ has an approximate standard normal distribution for terms that do not belong in the variation prediction equation. Therefore, the statistical test for each column in the test matrix becomes:

(1) H_0: $\sigma^2_{(+1)} = \sigma^2_{(-1)}$ (i.e., the term representing the design column does not go into the variation prediction equation)

H_1: $\sigma^2_{(+1)} \neq \sigma^2_{(-1)}$ (i.e., the term representing the design column does go into the variation prediction equation)

(2)   Select α (.1, .05, .01, .001).

(3)   Compute $\ln \dfrac{s^2_{(1)}}{s^2_{(-1)}}$.

(4)   Using Appendix D, find $Z(a)$ where $a = (1 - \alpha/2)$.

(5)   If $\left| \ln \dfrac{s^2_{(1)}}{s^2_{(-1)}} \right| \leq Z(a)$ conclude H_0, the column effect is insignificant.

If $\left| \ln \dfrac{s^2_{(1)}}{s^2_{(-1)}} \right| > Z(a)$ conclude H_1, the column effect is significant.

(See Appendix M for simple Rules of Thumb.)

If you can accept $\alpha = .05$, then the rule for determining column effect significance can be simplified to the following:

(i)   If $\left| \ln \dfrac{s^2_{(1)}}{s^2_{(-1)}} \right| \leq 1.96$ conclude H_0.

(ii)   If $\left| \ln \dfrac{s^2_{(1)}}{s^2_{(-1)}} \right| > 1.96$ conclude H_1.

Run	Factors and Interactions							Replicated Response Values				
	A	B	AB	C	AC	BC	ABC	Y_1	Y_2	Y_3	$\bar{y}$	s
1	-1	-1	1	-1	1	1	-1	2.22	2.11	2.14	2.157	0.057
2	-1	-1	1	1	-1	-1	1	1.42	1.54	1.05	1.337	0.255
3	-1	1	-1	-1	1	-1	1	2.25	2.31	2.21	2.257	0.050
4	-1	1	-1	1	-1	1	-1	1.00	1.38	1.19	1.190	0.190
5	1	-1	-1	-1	-1	1	1	1.73	1.86	1.79	1.793	0.065
6	1	-1	-1	1	1	-1	-1	2.71	2.45	2.46	2.540	0.147
7	1	1	1	-1	-1	-1	-1	1.84	1.76	1.70	1.767	0.070
8	1	1	1	1	1	1	1	2.27	2.69	2.71	2.557	0.248
Avg y@ -1	1.735	1.957	1.945	1.993	1.522	1.975	1.913					
Avg y@ 1	2.164	1.942	1.954	1.906	2.377	1.924	1.986					
Δ	0.429	-.015	0.009	-.087	0.855	-.051	0.073					
MSB	1.104	0.001	0.001	0.045	4.386	0.016	0.032					
F	44.36	0.040	0.040	1.810	176.2	0.640	1.280					
Avg s@ -1	0.138	0.131	0.113	0.060	0.145	0.130	0.116					
Avg s@ 1	0.132	0.139	0.157	0.210	0.125	0.140	0.154					
δ	-.006	0.008	0.044	0.150	-.020	0.010	0.038					
$\ln \dfrac{s(1)^2}{s(-1)^2}$	-.090	0.120	0.660	2.500	-.290	-.150	0.570					

$\bar{y} = 1.950 \quad \bar{s} = .135$

$MSE = .02489$

$F(.95;\ 1,\ 16) = 4.49$

$Z(0.975) = 1.96$

Table 5.12 Turbine Blade Casting Example

Based on the information provided in Table 5.12, the two equations for modelling the turbine blade casting process are:

(i) $\hat{s} = .135 + .075C$ *

(ii) $\hat{y} = 1.950 + .2145A + .4275A \cdot C - .0435C$

Optimizing the standard deviation ($\hat{s}$) first indicates that factor C (pour speed) should be set at the low level (coded level -1). Inserting a value of -1 for C in the $\hat{y}$ equation results in

$$\hat{y} = 1.950 + .2145A + .4275(A)(-1) - .0435(-1) \quad \text{or}$$
$$\hat{y} = 1.9935 - .213A$$

Thus, minimizing $\hat{y}$ requires factor A to be set at the high level or coded value of 1. The resulting predicted percent shrinkage for $A_1B_?C_{-1}$ will be

$$\hat{y} = 1.950 + .2145(1) + .4275(1)(-1) - .0435(-1) = 1.7805$$

Since factor B was not important for shifting the average or the variability, it should be set at an appropriate level to reduce costs.

A more in-depth examination of the $\hat{s}$ and $\hat{y}$ equations in this example indicates that, by minimizing $\hat{s}$ first, we have obtained a suboptimal value for $\hat{y}$. If we had instead minimized $\hat{y}$ first, we would have set A at the low level and C at the high level. This type of problem can potentially occur when the $\hat{s}$ and $\hat{y}$ equations contain one or more of the same factors. When both $\hat{s}$ and $\hat{y}$ cannot both be optimized, a tradeoff must take place. Usually, the engineer/researcher will weigh the importance of adjusting the average versus minimizing the variance and use the two equations ($\hat{s}$ and $\hat{y}$) to obtain the best settings.

---

* Ln s could have also been used to model variation. See Appendix M for Rules of Thumb on significant terms in s and ln s models.

An alternative is to use the loss function discussed in Chapter 1. To compute average loss for each run we will use

$$\bar{L} = k\left(s^2 + (\hat{y} - T)^2\right) .$$

For our example, the target percent shrinkage is zero; therefore,

$$\bar{L} = k(\hat{s}^2 + \hat{y}^2) .$$

Using this equation, we can construct a table (see Table 5.13) to aid in selecting appropriate factor-settings. Thus, the best settings to minimize average loss are $A_{-1} B_? C_1$. The B factor will be set based on cost and/or convenience. Of the different strategies discussed, the best one depends on the objective of the experiment.

Factors				
A	C	$\hat{s}$	$\hat{y}$	$\bar{L}$
−1	−1	.060	2.2065	k(4.8722)
−1	+1	.210	1.2645	k(1.6431)*
+1	−1	.060	1.7805	k(3.1738)
+1	+1	.210	2.5485	k(6.5389)
* minimum average loss				

Table 5.13 Tradeoff Analysis Using Average Loss

5.6   Residual Analysis for Replicated or Unreplicated Designs

A third strategy for detecting dispersion effects allows for a reduction in experimental runs and/or an increase in design resolution through utilization of an unreplicated design. It is presented as a modification of the Box-Meyer method [4]. A modified version of their procedure applied to a 2-level design is presented below. This procedure can be used for replicated or unreplicated designs.

(1)   Use an R_{IV} or better design to avoid confounding interaction and dispersion effects.

(2)   Fit the best model to the data.

(3)   Compute the residuals.

(4)   For each level of each factor, compute the standard deviation of the residuals.

(5)   Compute the difference of the largest and smallest standard deviation for each factor.

(6)   Rank order the differences found in (5).

(7)   Use a Pareto chart or $\ln\dfrac{s_{(+)}^2}{s_{(-)}^2}$ to determine the important dispersion effects. (Also see Appendix M for Rules of Thumb.)

(8)   For factors with important dispersion effects, determine the settings for least variance. Set location effect factors to optimize the response average.

(9)   If a factor setting for minimizing variance differs for the setting that optimizes the location effect, use a trade-off study to determine which setting is most crucial to the product quality.

(10)   Set all other factors based on economics, status quo, or convenience.

The example used to illustrate this method contains four factors. The data is generated from the following simulation model:

$$Y_i = 80 + 2A + 6B + 4A \cdot B + (z_i \sigma)$$

(where $\sigma = 3 + 2C$ and $Z_i \sim N(0, 1)$). Therefore, you anticipate location effects for A, B, and $A \cdot B$, a dispersion effect for C, and no effect for a fourth factor D. The generated data appears in Table 5.14. Notice that the run order has been altered from previous 16 run designs. The point to be made here is that starting with − or + is immaterial in generating the orthogonal matrix.

Run	A	B	C	D	Y
1	+	+	+	+	94.50
2	+	+	+	−	102.15
3	+	+	−	+	95.91
4	+	+	−	−	95.66
5	+	−	+	+	76.45
6	+	−	+	−	65.85
7	+	−	−	+	72.79
8	+	−	−	−	72.75
9	−	+	+	+	83.75
10	−	+	+	−	76.15
11	−	+	−	+	79.91
12	−	+	−	−	80.15
13	−	−	+	+	85.00
14	−	−	+	−	74.25
15	−	−	−	+	77.38
16	−	−	−	−	78.08

Table 5.14 Design Matrix and Response Values for Residual Analysis Example

Using least squares regression on the data in Table 5.14 will generate the following prediction equation:

$$\hat{y} = 81.921 + 2.591A + 6.606B + 5.941A \cdot B$$

The computer output from the least squares regression procedure is shown in Table 5.15.

Dependent Variable: y	N: 16	Multiple R: .939	Squared Multiple R: .881				

Adjusted Squared Multiple R: .851   Standard Error of Estimate: 3.927

Variable	Coefficient	Std Error	Std Coef	Tolerance	T	P(2 Tail)
Constant	81.921	0.982	0.000	1.0000000	83.438	0.000
A	2.591	0.982	0.262	1.0000000	2.635	0.022
B	6.606	0.982	0.670	1.0000000	6.724	0.000
A·B	5.941	0.982	0.603	1.0000000	6.056	0.000

Analysis of Variance

Source	Sum-of-Squares	df	Mean-Square	F-Ratio	P
Regression	1370.034	3	456.678	29.609	0.000
Residual	185.032	12	15.423		

Table 5.15 Computer Output for Regression Modeling of Table 5.14 Data

The location effects of A, B, and A·B were correctly identified. Next, the predicted values ($\hat{y}$) and the residuals ($y - \hat{y}$) are calculated and displayed in Table 5.16.

Run	y	$\hat{y}$	$y - \hat{y}$
1	94.50	97.06	−2.56
2	102.15	97.06	5.09
3	95.91	97.06	−1.15
4	95.66	97.06	−1.4
5	76.45	71.96	4.49
6	65.85	71.96	−6.11
7	72.79	71.96	.83
8	72.75	71.96	.79
9	83.75	79.99	3.76
10	76.15	79.99	−3.84
11	79.91	79.99	−.08
12	80.15	79.99	.16
13	85.00	78.68	6.32
14	74.25	78.68	−4.43
15	77.38	78.68	−1.30
16	78.08	78.68	−.60

Table 5.16 Response Values, Predicted Values, and Residuals for Table 5.14 Data

The standard deviation of the residuals for each factor level appears in Table 5.17.

	Factor			
	A	B	C	D
$s_{(+)}$	3.675	3.039	5.035	3.190
$s_{(-)}$	3.596	4.148	.905	3.530
$\ln \dfrac{s_{(+)}^2}{s_{(-)}^2}$	.043	$-.622$	3.432	$-.203$

Table 5.17   Standard Deviation of Residuals for Each Factor in Table 5.14 Data (Also see Appendix M for Rules of Thumb.)

Using the decision rules for $\left| \ln \dfrac{s_{(+)}^2}{s_{(-)}^2} \right|$ from section 5.4, it is evident that the only significant dispersion effect is associated with factor C, i.e., $\left[z \left(1 - \dfrac{\alpha}{2} \right) = z(.975) = 1.96 \right] < 3.432$ implies we are at least 95% confident that factor C is a variance-shifting factor.

5.7   Robust Design Analysis

Robust designs are a type of designed experiment which identify input factor settings that result in products and/or processes robust (insensitive) to noise factors (sources of uncontrolled variation). Taguchi refers to this type of experimentation as parameter design (for more details see Chapter 6 and the Chapter 8 case study, Applying Experimental Design Techniques to Operating Systems and Software). To provide the reader with an introduction to robust design and how the analysis is conducted, consider the following example from an injection molding process.

Statement of the Problem: An injection molding process engineer has been experiencing problems associated with part shrinkage which occurs after curing. This shrinkage problem has contributed to increased part variability and customer dissatisfaction.

Objective: Identify factors which contribute to variability in part shrinkage and determine the best settings to minimize shrinkage.

Response: Part shrinkage is the amount of measured deviation from the desired part size.

Some of the brainstorming in the form of a cause and effect or fishbone diagram is shown in Figure 5.6. The asterisked factors are those determined to be uncontrollable in production; however, with special care, they can be varied in the experiment. These 3 factors are referred to as noise or uncontrollable factors (sources of uncontrolled response variation) and will eventually appear in a separate design array called the outer array. The phrase "uncontrolled factors" indicates that these factors are difficult or impossible to control during production. Their use in the design matrix implies that with additional effort, we can control them during the experiment. A list of controllable and uncontrollable factors is shown in Table 5.18.

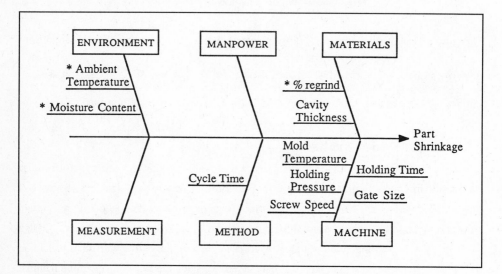

Figure 5.6 Cause and Effect Diagram for Injection Molding Example

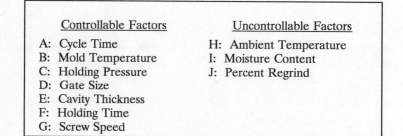

Controllable Factors	Uncontrollable Factors
A: Cycle Time	H: Ambient Temperature
B: Mold Temperature	I: Moisture Content
C: Holding Pressure	J: Percent Regrind
D: Gate Size	
E: Cavity Thickness	
F: Holding Time	
G: Screw Speed	

Table 5.18 List of Controllable and Uncontrollable Factors

Assuming the experimenter is satisfied with a 2-level design and there is no reason to suspect interactions, the resulting robust design matrix will appear as shown in Table 5.19. Notice that the matrix for the uncontrolled factors is also orthogonal. This feature is not a requirement; rather, the combinations for the uncontrolled factors should produce the worst expected variance of the response across the replicates. Our objective is to find a process configuration which, in turn, will minimize the worst expected response variation due to uncontrolled factors.

								J	−	+	+	−		
Control								I	−	+	−	+		
Factors								H	−	−	+	+		
Run	A	B	C	D	E	F	G						$\bar{y}_r$	s_r
1	−	−	−	−	−	−	−		2.6	2.7	2.8	2.8	2.725	.096
2	−	−	−	+	+	+	+		0.8	3.0	0.8	3.2	1.950	1.330
3	−	+	+	−	−	+	+		3.6	1.0	3.3	0.9	2.200	1.449
4	−	+	+	+	+	−	−		2.5	2.4	2.5	2.3	2.425	.096
5	+	−	+	−	+	−	+		3.5	3.6	3.5	3.5	3.525	.050
6	+	−	+	+	−	+	−		2.6	4.7	3.6	1.5	3.100	1.369
7	+	+	−	−	+	+	−		4.5	2.4	2.7	5.1	3.675	1.328
8	+	+	−	+	−	−	+		2.4	2.5	2.3	2.4	2.400	.082

Table 5.19 Robust Design Matrix with Response Values

The marginal mean analysis of the signal ($\bar{y}$) and the noise (ln s) appears in Tables 5.20 and 5.21.

	Analysis of $\bar{y}$						
	Control Factors						
	A	B	C	D	E	F	G
Avg −1	2.325	2.825	2.687	3.031	2.606	2.769	2.981
Avg +1	3.175	2.675	2.812	2.469	2.894	2.731	2.519
Δ	.850	−.150	.125	−.562	.288	−.038	−.462

Table 5.20 Marginal Mean Analysis for Table 5.19 Data

	Analysis of ln s						
	Control Factors						
	A	B	C	D	E	F*	G
ln s Avg −1	−1.01	−1.19	−1.07	−1.17	−1.04	−2.55	−1.02
ln s Avg +1	−1.23	−1.05	−1.16	−1.06	−1.19	.313	−1.21
Δ	−.22	.14	−.09	.11	−.15	2.863	−.19

Table 5.21 Ln s Analysis for Table 5.18 Data

Using the results of Tables 5.20 and 5.21, it is clear that factors A, D, and G shift the average, i.e., these are the location effects and factor F shifts the variability, i.e., it is a dispersion factor. Therefore, to minimize shrinkage and the variability of shrinkage across the uncontrolled factors, the best settings are A_- $B_?$ $C_?$ D_+ $E_?$ F_- G_+. Factors with a question mark are set at levels based on economics, status quo, and/or convenience. If our assumptions in using this design are correct, these settings will minimize shrinkage and make the process robust to uncontrolled variations in percent regrind, ambient temperature, and moisture content.

As a final note on the comparison of dispersion factor identification methods, this data was originally analyzed using signal-to-noise "smaller is better", S/N_S. The results shown in Table 5.22 again failed to identify factor F as an important factor. Had we used signal-to-noise "nominal is best", S/N_N, the results shown in Table 5.23 do point out the importance of factor F; however, more cumbersome analysis is required to determine if F is a location or dispersion factor. In conclusion, the recommended procedure of analyzing the signal $(\bar{y})$ and the noise (ln s) separately is simple and straight forward.

	Analysis Using S/N_S						
	Factor						
	A	B	C	D	E	F	G
Avg −1	−7.90	−9.29	−8.78	−9.90	−8.70	−8.74	−9.60
Avg +1	−10.20	−8.77	−9.28	−8.21	−9.36	−9.33	−8.43
Δ	−2.3	.52	−.50	1.69	−.66	−.59	1.17

Table 5.22 S/N_S Analysis for Table 5.19 Data

* See Appendix M for Rules of Thumb on significance.

	Analysis Using S/N$_N$ Factor						
	A	B	C	D	E	F	G
Avg −	16.03	19.12	17.65	19.63	17.29	30.87	18.28
Avg +	20.57	17.48	18.94	16.96	19.30	5.72	18.32
Δ	4.54	−1.64	1.29	−2.67	2.01	−25.15	.04

Table 5.23 S/N$_N$ Analysis for Table 5.19 Data

5.8 Variance Reduction Using Multi-Variate Charts*

A semiconductor company had been experiencing problems with lot-to-lot, wafer-to-wafer, and within wafer variance of thickness values from a coating process. An engineering team had narrowed the potential contributing factors down to temperature and time. Since only two factors are in question, the team decided to conduct a 2-factor (2-level) full factorial experiment with time (A) settings of 20 and 30 minutes, and temperature (B) settings of 50°C and 70°C. The design matrix, response values, and prediction equations are shown in Table 5.24 To minimize wafer-to-wafer and lot-to-lot variations we would minimize the following equation:

$$\hat{s}_y = 1.648 - .6675B$$

Setting factor B at (+1), i.e., setting temperature at 70°C, will minimize wafer-to-wafer and lot-to-lot variations. Fortunately, setting factor B at (+1) also minimizes the equation for within wafer variation which is measured as within wafer range, R.

$$\hat{R} = 6.602 - 2.73B.$$

To achieve the desired target of 200 microinches, we would perform the following manipulations.

(1) $\hat{y} = 200.9 - 2.46A + 1.04AB - .04B$

(2) $200 = 200.9 - 2.46A + 1.04A(1) - .04(1)$ (set $\hat{y}$ = 200 and B = +1)

(3) $1.42A = 200.9 - 200 - .04$ (solve for A)

(4) $A = .86/1.42 = .606$

(5) Actual time $= \dfrac{x \, d_i}{2} + \bar{t} = \dfrac{.606(30 - 20)}{2} + 25 = 28.03$ minutes

---

* The multivariate chart approach is credited to Dorian Shainin [11].

Response Values:
Wafer Average Thickness and (Wafer Range) in micro inches

Run	A	B	AB	LOT 1				LOT 2				LOT 3				$\bar{Y}$	S_y	$\bar{R}$
				W1	W2	W3	W4	W1	W2	W3	W4	W1	W2	W3	W4			
1	−	−	+	204 (10)	201 (6)	206 (7)	204 (8)	207 (8)	204 (10)	205 (6)	206 (9)	203 (9)	205 (11)	201 (10)	207 (10)	204.4	2.021	(8.667)
2	−	+	−	201 (3)	203 (6)	202 (3)	202 (3)	203 (5)	201 (3)	203 (2)	203 (5)	204 (4)	202 (5)	202 (2)	201 (3)	202.2	0.965	(3.667)
3	+	−	−	194 (8)	200 (10)	197 (7)	199 (9)	201 (9)	194 (10)	197 (7)	196 (10)	197 (9)	199 (11)	194 (9)	201 (8)	197.4	2.610	(8.917)
4	+	+	+	199 (2)	200 (3)	198 (4)	199 (5)	200 (2)	199 (3)	200 (2)	198 (4)	200 (2)	201 (3)	201 (6)	199 (2)	199.4	0.996	(3.000)

	A	B	AB
Avg Y −	203.3	200.9	201.9
Avg Y +	198.4	200.8	199.8
Δ/2	−2.46	−0.04	1.04
F_e	90.52	0.03	16.25

Wafer to Wafer and Lot to Lot Models:

$$\hat{Y} = 200.9 - 2.46A + 1.04AB - 0.04B$$

$$\hat{S}_y = 1.648 - 0.6675B$$

	A	B	AB
Avg R −	6.167	8.792	5.833
Avg R +	5.958	3.333	6.292
Δ/2	−.104	−2.73	−.229
F_e	1.279	191.3	1.35

Within Wafer Variance Model:

$$\hat{R} = 6.062 - 2.73B$$

Table 5.24 Design Matrix and Analysis Summary for Multivariate Chart Example

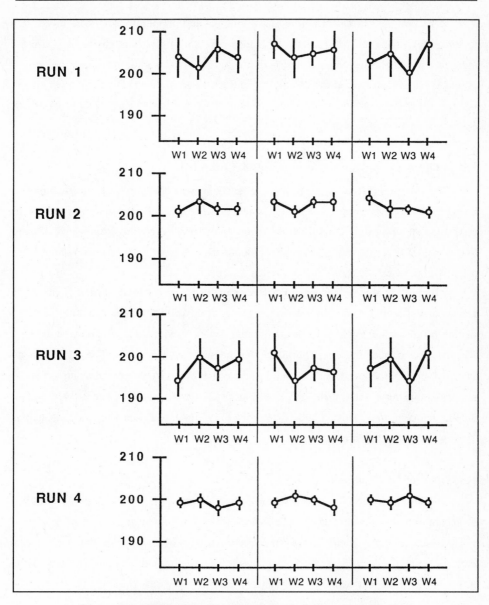

Figure 5.7 Multivariate Charts [11] for Each Run in Table 5.23
(Vertical bars represent within wafer range. Open dots are the wafer
averages. The three sections represent the three lots.)

The optimal settings of time = 28.03 minutes and temperature = 70°C are closely related to experimental run number 4. Notice how the multivariate charts in Figure 5.7 indicate that a high time and high temperature setting will closely approximate the desired optimal. Having completed the experiment and found the best conditions, you would next confirm these findings and monitor production using control charting techniques and/or continued multivariate charting [11].

5.9   Three Level Design Analysis and Comparison

The designs discussed in the previous sections were all of 2 levels thus assuming a linear model is appropriate. If this assumption is incorrect, how does one proceed? This case study will use a simulation process to present 2nd order modeling from 3-level designs. The reader is also referred to Chapter 8, "Identifying a Plasma Etch Process Window", for a real industrial example of a 3-level design.

"Plate 2" is a computer program that simulates a process used to train engineers in Design of Experiments. In this problem, students are asked to optimize an "auto bumper" plating process using thickness as a quantitative response with time, temperature, percent nickel, ph, and percent phosphorous as the independent variables. From background information, it is believed that all of the above independent variables are important in impacting thickness. Additionally, there is concern about strong interactive, as well as non-linear, factors in the process. Crudely following the sequence of steps suggested in Chapter 1, we will address this problem.

Statement of the Problem

We are chartered with characterizing a new process (the plating process). This will be accomplished by developing a mathematical model for predicting plating thickness which will be used for optimizing various products requiring different plating thicknesses. An empirical model will provide the information required for adjusting the settings of the important independent variables in order to obtain the necessary thickness for the various products.

Objectives

(1)   Determine the effect of time, temperature, percent nickel, ph, and phosphorous on plating thickness.

(2)   Develop a "good" mathematical model which characterizes the plating process.

QUALITY CHARACTERISTICS

Response	Type	Anticipated Range	How to measure?
Thickness	Q	0 - 2000	profilometer

FACTORS

Factor	Type	Control or Noise	Range	Interactions?
Time	Q	Control	4 - 12	possible
Temperature	Q	Control	16 - 32	possible
Nickel	Q	Control	10 - 18	possible
ph	Q	Control	2 - 10	possible
Phosphorous	Q	Control	1.40 - 3.88	possible

Appropriate Experiment Design

To resolve the stated problem, different strategies will be demonstrated as described in the following cases. The data has been analyzed using RS software from BBN Software Products, Inc.

CASE I:

A 2-level fractional factorial of resolution V is used to estimate all main effects and 2-factor interactions. The design matrix with the response values is shown in Table 5.25. The computer output for the mathematical model is shown in Table 5.26.

The lack of standard error estimates, t-values, and significance is a result of using all the degrees of freedom for effects and, thus, none remain for estimating the experimental error. As previously demonstrated, a Pareto diagram for the effects can be used to separate the important (few) effects from the trivial (many) effects. The resulting Pareto diagram for

Run #	Time	Temp	Nickel	ph	Phos	Thickness
1	4.00	16.00	10.00	2.00	3.88	113
2	12.00	16.00	10.00	2.00	1.40	756
3	4.00	32.00	10.00	2.00	1.40	78
4	12.00	32.00	10.00	2.00	3.88	686
5	4.00	16.00	18.00	2.00	1.40	87
6	12.00	16.00	18.00	2.00	3.88	788
7	4.00	32.00	18.00	2.00	3.88	115
8	12.00	32.00	18.00	2.00	1.40	696
9	4.00	16.00	10.00	10.00	1.40	99
10	12.00	16.00	10.00	10.00	3.88	739
11	4.00	32.00	10.00	10.00	3.88	10
12	12.00	32.00	10.00	10.00	1.40	712
13	4.00	16.00	18.00	10.00	3.88	159
14	12.00	16.00	18.00	10.00	1.40	776
15	4.00	32.00	18.00	10.00	1.40	162
16	12.00	32.00	18.00	10.00	3.88	759

Table 5.25 Design Matrix (Actual Values) and Response (Thickness)
for Plating Example, Factorial Portion of the CCF

	Term	Coefficient	Std Error	T-Value	Significance
1	1	420.937500	-	-	-
2	~T	318.062500	-	-	-
3	~TE	−18.687500	-	-	-
4	~N	21.812500	-	-	-
5	~P	6.062500	-	-	-
6	~PHO	0.187500	-	-	-
7	~T•TE	−7.062500	-	-	-
8	~T•N	−6.062500	-	-	-
9	~T•P	1.437500	-	-	-
10	~T•PHO	3.812500	-	-	-
11	~TE•N	8.937500	-	-	-
12	~TE•P	2.437500	-	-	-
13	~TE•PHO	−9.937500	-	-	-
14	~N•P	15.187500	-	-	-
15	~N•PHO	12.312500	-	-	-
16	~P•PHO	−10.437500	-	-	-

No. cases = 16   Resid. df = 0   Cond. No. = 1
~indicates factors are transformed

Table 5.26 RS/Discover Computer Output for Coded Table 5.24 Data

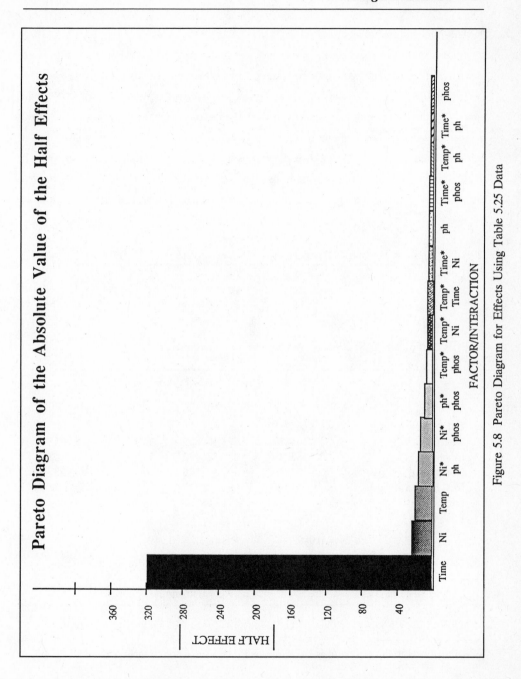

Figure 5.8 Pareto Diagram for Effects Using Table 5.25 Data

the half-effects appears as indicated in Figure 5.8.

Based on the previous Pareto diagram of half-effects, it is obvious that time is an important factor. Nickel and temperature are ranked second and third, followed by everything else. At the outset, we discussed the possibility of non-linear effects. We will now tackle the question of non-linearity through the use of replicated centerpoints. This will be accomplished by complimenting our 16 original data points with 4 center points. For a non-simulated problem, this could be dangerous to do in a sequential fashion especially if there is a possibility that the process has shifted significantly. Therefore, the centerpoints are sometimes added during the 2-level experimental phase. The centerpoint data is as follows:

Run #	Time	Temperature	Nickel	ph	Phos	Thickness
1	8	24	14	6	2.64	351
2	8	24	14	6	2.64	373
3	8	24	14	6	2.64	353
4	8	24	14	6	2.64	321

Analysis of the combined 20 data points is shown in Table 5.27.

Least Squares Summary ANOVA					
Source	df	Sum of Squares	Mean Square	F-Ratio	Significance*
1 Total (Corr.)	19	1662583			
2 Regression	15	1644873	109658	24.77	0.0035
3 Linear	5	1632409	326482	73.74	0.0005
4 Non-linear	10	12464	1246	0.28	0.9527
5 Residual	4	17710	4427		
6 Lack of fit	1	16331	16331	35.53	0.0094
7 Pure error	3	1379	460		

R-sq = 0.9893   R-sq-adj. = 0.9494

Model obeys hierarchy. The sum of squares for linear terms is computed assuming nonlinear terms are removed.
$F(1, 3)$ as large as 35.53 is a rare event => unlikely that model is correct.

Table 5.27 RS/Discover Computer Output for Table 5.24 Data with Centerpoints

* Significance for this software is the same as P(2 tail) discussed earlier from MYSTAT.

Since the "lack of fit" is significant, we need to determine which of our factors are non-linear and develop a mathematical model. As discussed in Chapter 3, an efficient way to accomplish a second order model is through the use of a Box-Wilson (central composite) design. Using $\alpha = (n_F)^{1/4}$ to determine the coded value for the axial points, we obtain $\alpha = 2$ for $n_F = 16$. The lows and highs from the fractional factorial portion of the experiment are, however, already at the extreme conditions. Because of this, $\alpha = 1.0$ is used to construct a Central Composite Faced (CCF) design.

Since ph and phosphorous were screened out during the factorial portion of the experiment, we will not run α-points for these factors. Based on the above considerations, the resultant supplemental runs with associated thickness data are:

Run #	Time	Temp	Nickel	Thickness
1	12	24	14	736.1
2	4	24	14	96.0
3	8	32	14	328.9
4	8	16	14	303.5
5	8	24	18	358.2
6	8	24	10	347.7

By combining the 16 runs from the factorial portion of the design with the four centerpoints and the 6 runs from the axial portion of our design, we obtain the following regression table.

	Term	Coefficient	Std Error	T-Value	Significance
1	1	342.7	11.747		
2	T	318.3	7.092		
3	TE	−15.2	7.092		
4	N	19.9	7.092		
5	T·TE	−7.1	7.522	−0.94	0.36
6	T·N	−6.1	7.522	−0.81	0.43
7	TE·N	8.9	7.522	1.19	0.25
8	T^2	80.1	18.047	4.44	0.00
9	TE^2	−19.7	18.047	−1.09	0.29
10	N^2	17.0	18.047	0.94	0.36

No. cases = 26	R-sq = 0.9923	RMS Error = 30.1
Resid. df = 16	R-sq-adj. = 0.9880	Cond. No. = 5.68

Table 5.28 RS/Discover Computer Output for Table 5.24 Data
with Center Points and Axial Points

Eliminating the unimportant terms leaves us with the final regression table shown below. (The RS software uses backward elimination [12] to remove unimportant terms.)

	Term	Coefficient	Std Error	T-Value	Significance
1	1	342.0	10.498		
2	T	318.3	6.998		
3	TE	−15.2	6.998	−2.17	0.04
4	N	19.9	6.998	2.85	0.00
5	T²	78.4	12.618	6.21	0.00
	No. cases = 26		R-sq = 0.9902		RMS Error = 29.69
	Resid. df = 21		R-sq-adj. = 0.9883		Cond. No. = 3.30

Table 5.29 *RS/Discover* Final Model from Table 5.28 (Using a CCF)

The prediction equation generated from the orthogonally coded CCF design is:

$$\text{Predicted Thickness} = 342.0 + 318.3(\text{time}) - 15.2(\text{temp}) + 19.9(\text{nickel}) + 78.4(\text{time})^2$$

Step XI from Chapter 1 includes making some confirmation runs to verify predicted results. Utilizing RS/Discover software, a 95% confidence interval for the mean response when time = 4, temp = 16, and Nickel = 14 is calculated to be [72.9, 161.7]. This interval provides us something to judge our confirmation runs against. Ten confirmation runs at the above settings resulted in a mean value of 115.5 and a standard deviation of 20.4. Since the mean of our confirmation runs falls within our confidence interval, the model appears to be adequate.

Let's stop briefly and contrast our derived mathematical model with the equation in our simulation package. The actual uncoded equation in our simulation package is:

$$\text{Thickness} = 5(\text{time})^2 - 2.3(\text{temp}) + 4.0(\text{nickel})$$

$$\text{with a standard deviation} = 40$$

When temperature is 16, time is 4, and nickel is 14, the actual model would generate an average value of 99.2 with a standard deviation of 40. The graphic on the next page will contrast the true population distribution with the model distribution for the 10 confirmation runs at temp = 16, time = 4, and nickel = 14.

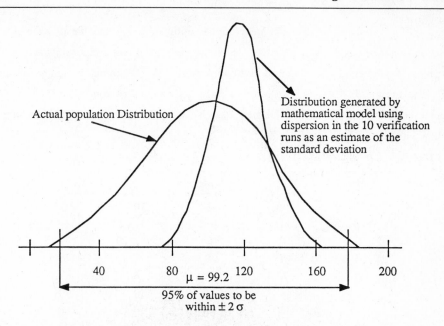

Distribution generated by mathematical model using dispersion in the 10 verification runs as an estimate of the standard deviation

Actual population Distribution

40   80   μ = 99.2   120   160   200

95% of values to be within ± 2 σ

As we can see from the above graphic, our model is off by approximately 20 units on the average and we are nearly a factor of 2 off on the standard deviation. How can we improve on the mathematical model we determine? Large samples are required. A general rule of thumb is: For every term (excluding the constant) in our final prediction equation, we need roughly 10 or more samples. Using this guideline, with the 4 non-constant terms in our model, 40 or more samples in the original experiment would be desired to obtain a "good" model. We used only 26. To estimate standard deviation, the confirmation runs should have 15 or more data points. We used only 10. It is recognized, however, that often times sample size is restricted due to time, money, and resource constraints. In these cases we must simply recognize the limitations of our results.

CASE II:

Instead of using a Central Composite Design to provide us with an empirical model, we could have used other approaches as well. For example, after the screening design was completed, we could have made use of a D-optimal design, a Box-Behnken design, or a Taguchi design.

First, consider a computer generated D-optimal design capable of estimating all linear, quadratic, and 2-way interaction terms for the three most important factors. The minimum number of runs required in this case is 10. The D-optimal design generated by RS Discover is 12 runs, but we ran replicates (2) on the last run. Our 14 run design generated with the applicable thickness data is shown in Table 5.30 and the results are shown in Table 5.31.

Run	Time	Temp	Nickel	Thickness
1	12	32	10	717
2	4	16	10	136
3	12	16	18	787
4	4	32	18	78
5	4	16	18	87
6	4	32	10	80
7	12	16	10	760
8	8	24	10	282
9	8	32	14	318
10	4	24	14	96
11	12	32	18	682
12	12	24	14	747
13	12	24	14	764
14	12	24	14	742

Table 5.30 D-optimal Design (Actual Values) and Response (Thickness)
for Plating Example

From our summary ANOVA table, it appears we have an acceptable fit with the model, but the "T-values" for numerous terms in our coefficient table indicate we can get rid of some terms with little loss in R^2. Our final coefficient table using backward elimination is shown in Table 5.32.

	Term	Coefficient	Std Error	T-Value	Significance
1	1	314.786500	17.517042		
2	~T	322.500000	6.491690		
3	~TE	−26.625000	7.612182		
4	~N	−7.375000	7.612182		
5	~T^2	111.312500	17.517042	6.35	0.0031
6	~T·TE	−10.375000	7.612182	−1.36	0.2446
7	~T·N	5.375000	7.612182	0.71	0.5191
8	~TE^2	29.937500	17.517042	1.71	0.1626
9	~TE·N	−1.875000	7.612182	−0.25	0.8176
10	~N^2	−40.062500	17.517042	−2.29	0.0841

No. cases = 14   R-sq = 0.9986   RMS Error = 21.53

Resid. df = 4   R-sq-adj. = 0.9953   Cond. No. = 7.392

~indicates factors are transformed

Summary of ANOVA

Source	df	Sum of Squares	Mean Square	F-Ratio	Significance
1 Total (Corr.)	13	1283423			
2 Regression	9	1281569	142397	307.20	0.0000
3 Linear	3	1259538	419846	905.70	0.0000
4 Non-linear	6	22030	3672	7.92	0.0325
5 Residual	4	1854	464		
6 Lack of fit	2	1588	794	5.97	0.1435
7 Pure error	2	266	133		

R-sq = 0.9986

R-sq-adj. = 0.9953

Model obeys hierarchy. The sum of squares for linear terms is computed assuming nonlinear terms are first removed. F(2, 2) as large as 5.971 is not a rare event => no evidence of lack of fit.

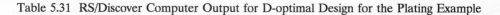

Table 5.31 RS/Discover Computer Output for D-optimal Design for the Plating Example

Term	Coefficient	Std Error	T-Value	Significance
1 1	311.470588	17.084202		
2 ~T	323.657143	6.874237		
3 ~TE	−22.941176	8.053570	−2.85	0.0173
4 ~T^2	107.586555	18.415349	5.84	0.0002

No. cases = 14	R-sq = 0.9957	RMS Error = 23.48
Resid. df = 10	R-sq-adj. = 0.9944	Cond. No. = 5.328

~indicates factors are transformed

Table 5.32 Final Model for D-optimal Design

The prediction equation from the orthogonally coded D-optimal design is:

predicted thickness = 311.47 + 323.66(time) − 22.94(temp) + 107.59(time)^2

Had we chosen a Box-Behnken design for 3 factors it would appear as shown in Table 5.33.

Run	Time	Temp	Nickel	Thickness
1	4	16	14	129
2	12	16	14	792
3	4	32	14	94
4	12	32	14	702
5	4	24	10	37
6	12	24	10	738
7	4	24	18	133
8	12	24	18	714
9	8	16	10	339
10	8	32	10	302
11	8	16	18	319
12	8	32	18	344
13	8	24	14	364
14	8	24	14	342
15	8	24	14	404

Table 5.33 Box-Behnken Design (Actual Values) and Response (Thickness)
for Plating Example

Fitting the coded data with a regression model provided the statistics shown in Table 5.34.

	Term	Coefficient	Std Error	T-Value	Significance
1	1	370.000000	16.123999		
2	~T	319.125000	9.873892		
3	~TE	−17.125000	9.873892		
4	~N	11.750000	9.873892		
5	~T^2	69.375000	14.533976	4.77	0.0050
6.	~T•TE	−13.750000	13.963792	−0.98	0.3700
7	~T•N	−30.000000	13.963792	−2.15	0.0844
8	~TE^2	−10.125000	14.533976	−0.70	0.5171

No. cases = 15   R-sq = 0.9954   RMS Error = 27.93

Resid df = 5   R-sq-adj. = 0.9872   Cond. No. = 4.522

~indicates factors are transformed

Summary of ANOVA

Source	df	Sum of Squares	Mean Square	F-Ratio	Significance
1 Total (Corr.)	14	851473.7			
2 Regression	9	847574.0	94174.9	120.70	0.0000
3 Linear	3	818176.8	272725.6	349.70	0.0000
4 Non-linear	6	29397.2	4899.5	6.28	0.0311
5 Residual	5	3899.8	780.0		
6 Lack of fit	3	1923.7	641.2	0.65	0.6535
7 Pure error	2	1976.0	988.0		

R-sq = 0.9954

R-sq-adj. = 0.9872

Model obeys hierarchy. The sum of squares for linear terms in computed assuming nonlinear terms are first removed. F(3, 2) as large as 0.649 is not a rare event => no evidence of lack of fit.

Table 5.34 RS/Discover Computer Output Using Box-Behnken Design
for Plating Example

From the summary ANOVA, it appears the model is adequate. From the coefficient table, non-important terms can now be consolidated. The result is the coefficients table shown in Table 5.35.

Term	Coefficient	Std Error	T-Value	Significance
1 1	363.769231	14.624921		
2 ~T	319.125000	10.763650		
3 ~N	11.750000	10.763650		
4 ~T^2	70.153846	15.796720	4.44	0.0016
5 ~T·N	−30.000000	15.222100	−1.97	0.0802
6 ~N^2	−33.096154	15.796720	−2.10	0.0656

No. cases = 15   R-sq = 0.9902   RMS Error = 30.44

Resid. df = 9   R-sq-adj. = 0.9848   Cond. No. = 3.513

~indicates factors are transformed

Table 5.35 Final Model for Box-Behnken Design

Thus, the prediction equation generated from the orthogonally coded Box-Behnken design is:

$$\text{predicted thickness} = 363.77 + 319(\text{time}) + 11.75(\text{nickel}) + 70.15(\text{time})^2 -$$
$$30(\text{time} \times \text{nickel}) - 33.09(\text{nickel})^2$$

To model the desired linear, quadratic, and 2-factor interactions, the appropriate 3-level Taguchi design is the L_{27}. In this instance, it is equivalent to the 3^3 full factorial.

The design and the generated thickness data are shown in Table 5.36. After fitting the coded data with the appropriate model and eliminating unimportant terms, our final coefficients table is shown in Table 5.37.

Run #	Time	Temp	Nickel	Thickness
1	4.00	16.00	10.00	113
2	4.00	16.00	14.00	152
3	4.00	16.00	18.00	147
4	4.00	24.00	10.00	65
5	4.00	24.00	14.00	53
6	4.00	24.00	18.00	130
7	4.00	32.00	10.00	83
8	4.00	32.00	14.00	40
9	4.00	32.00	18.00	94
10	8.00	16.00	10.00	339
11	8.00	16.00	14.00	303
12	8.00	16.00	18.00	381
13	8.00	24.00	10.00	348
14	8.00	24.00	14.00	342
15	8.00	24.00	18.00	420
16	8.00	32.00	10.00	327
17	8.00	32.00	14.00	255
18	8.00	32.00	18.00	322
19	12.00	16.00	10.00	745
20	12.00	16.00	14.00	780
21	12.00	16.00	18.00	740
22	12.00	24.00	10.00	772
23	12.00	24.00	14.00	769
24	12.00	24.00	18.00	755
25	12.00	32.00	10.00	735
26	12.00	32.00	14.00	726
27	12.00	32.00	18.00	757

Table 5.36 Full Factorial Design (Actual Values) and Response (Thickness) for Plating Example

Term	Coefficient	Std Error	T-Value	Significance
1 1	321.407407	13.192185		
2 ~T	327.888889	7.225657		
3 ~TE	−20.055556	7.225657	−2.78	0.0016
4 ~N	12.166667	7.225657		
5 ~T^2	87.888889	12.515205	7.02	0.0802
6 ~N^2	24.055556	12.515205	1.92	0.0656

No. cases = 27   R-sq = 0.9902   RMS Error = 30.66

Resid. df = 21   R-sq-adj. = 0.9879   Cond. No. = 4.391

~indicates factors are transformed

Table 5.37 Final Model for Full Factorial (L_{27}) Design

The prediction equation generated from the orthogonally coded L_{27} design is:

$$\text{predicted thickness} = 321.4 + 327.9(\text{time}) - 20.06(\text{temp}) + 12.2(\text{nickel})$$
$$+ 87.89(\text{time})^2 + 24.05(\text{nickel})^2$$

Let's now summarize the models we obtained with the various designs:

Design Type	R^2	Model # of Runs	Total # of Runs (including screening portion)	Model (factors are orthogonally coded)
D-optimal	.9957	14	34	$311.47 + 323.65(\text{time}) + 22.94(\text{temp}) + 107.59(\text{time})^2$
Box-Behnken	.9902	15	35	$363.77 + 319.13(\text{time}) + 11.75(\text{nickel}) + 70.15(\text{time})^2 - 30.0(\text{time} \bullet \text{nickel}) - 33.09(\text{nickel})^2$
CCF	.9902	26	26	$342.0 + 318.3(\text{time}) - 15.2(\text{temp}) + 19.9(\text{nickel}) + 78.4(\text{time})^2$
L_{27}	.9902	27	47	$321.41 + 327.89(\text{time}) - 20.06(\text{temp}) + 12.17(\text{nickel}) + 87.89(\text{time})^2 + 24.06(\text{nickel})^2$

The difference in prediction models is primarily due to large variation in the data coupled with small sample sizes. The actual simulation equation for the uncoded factors is:

$$\text{Thickness} = 5.0(\text{time})^2 - 2.3(\text{temp}) + 4.0(\text{nickel})$$

The program works as follows: You choose a specific uncoded value for time, temp, and nickel and enter it into the program. The simulation will generate a response based on the equation above plus a normal error distribution with a $\sigma = 40$.

One way to test the effectiveness for each of the various designs is to obtain cross-validated R^2 values. Using the simulation equation without the added error, you can obtain the expected value for all 27 combinations (see Table 5.38). Those 27 values will serve as the true response value in the crossvalidation. The cross-validated R^2 is obtained by correlating these true response values with the values obtained from the prediction equation for each design.

Design Type	R^2
D-Optimal	.996
Box-Behnken	.998
CCF	.999
L_{27}	.998

From the above table we see that the cross validated R^2 for all of the models is quite good, with the best R^2 in this situation going with the CCF design. A major reason for the large R^2 terms for each model is that the thickness values correlated against are actual values, without any noise, calculated directly from the base equation. This was done to provide a fair comparison of the prediction ability of each model.

5.10 A Monte Carlo Simulation for Comparing Dispersion Factor Identification Methods [8] (optional)

To identify dispersion factors, the three most commonly used strategies are signal-to-noise ratios, ln s, and analysis of dispersion effects from residuals. Several papers have been written discussing the appropriateness of these strategies from a theoretical standpoint. This section is not intended to add to the collection of theoretical arguments; rather, it is our purpose to put these strategies to the test using simulated data. For purposes of this section, four modeling scenarios are presented and the competing strategies are compared based on $k = 6$ input factors and 32 available experiments. The designs chosen to satisfy the number

	Actual Model			Predicted Values from Models				
Run #	Time	Temp	Nickel	Thickness	D-Opt	B/B	CCF	L_{27}
1	4	16	10	83.2	118.3	39.9	97.3	113.4
2	4	16	14	99.2	118.3	114.8	117.3	101.5
3	4	16	18	115.2	118.3	123.5	137.3	137.7
4	4	24	10	64.8	95.4	39.9	82.1	93.2
5	4	24	14	80.8	95.4	114.8	102.1	81.4
6	4	24	18	96.8	95.4	123.5	122.1	117.6
7	4	32	10	46.4	72.5	39.9	66.9	73.2
8	4	32	14	62.4	72.5	114.8	86.9	61.4
9	4	32	18	78.4	72.5	123.5	106.9	97.6
10	8	16	10	323.2	334.4	318.9	337.3	353.4
11	8	16	14	339.2	334.4	363.8	357.2	341.5
12	8	16	18	355.2	334.4	342.4	377.2	377.7
13	8	24	10	304.8	311.5	318.9	322.1	333.3
14	8	24	14	320.8	311.5	363.8	342.0	321.4
15	8	24	18	336.8	311.5	342.4	362.0	357.6
16	8	32	10	286.4	288.5	318.9	306.9	313.2
17	8	32	14	302.4	288.5	363.8	323.8	301.4
18	8	32	18	318.4	288.5	342.4	346.9	337.6
19	12	16	10	723.2	765.7	738.2	733.9	769.1
20	12	16	14	739.2	765.7	753.0	753.9	757.2
21	12	16	18	755.2	765.7	701.7	773.8	793.5
22	12	24	10	704.8	742.7	738.2	718.7	749.1
23	12	24	14	720.8	742.7	753.0	738.7	737.2
24	12	24	18	736.8	742.7	701.7	758.6	773.4
25	12	32	10	686.4	719.8	738.2	703.5	729.0
26	12	32	14	702.4	719.8	753.0	723.5	717.1
27	12	32	18	718.4	719.8	701.7	743.4	753.4

Table 5.38 Summary of 3-Level Design Analysis Strategies for the Plating Example

of factors and limitations of resources are (1) a Taguchi L_8 (2^{6-3} fractional factorial) with a sample size of four per run, (2) a Taguchi L_{16} (2^{6-2} fractional factorial) with a sample of two per run, and (3) a Taguchi L_{32} (2^{6-1} fractional factorial) with one response per run.

As you can see, the resolution of the designs will increase as you increase the number of combinations. The L_8 design is of resolution III indicating that the 6 main effects will be aliased with 2-way interactions. The resolution of the L_{16} design is IV indicating that the mains are free and clear of 2-way interactions, but each 2-way interaction is aliased with another 2-way interaction. The last design, L_{32}, has resolution V which means that 2-way interactions are no longer aliased with other 2-way interactions. Ideally, the researcher prefers the highest design resolution possible, but he/she is constrained by limited resources and number of factors plus the assumed need to replicate each run to estimate dispersion effects.

Description of Models Tested

It is infeasible to model every different relationship of the response with the input factors while also incorporating several different types of error. However, since the problems encountered in industry are not always associated with additive error relationships having normally distributed error, it is important to include in this section some of the more common models.

ADDITIVE ERROR MODEL:

This model is defined as

$$y = f(\underline{x}_L, \underline{x}_B) + \epsilon(\underline{x}_D, \underline{x}_B)$$

where $\underline{x}_L$ is the set of factors which produce only location effects,

$\underline{x}_D$ is the set of factors which produce only dispersion effects, and

$\underline{x}_B$ is the set of factors which produce both a location and a dispersion effect.

For this model, the strategies will be tested using normally and exponentially distributed error.

The actual model for the simulation is $y = w_0 + w_1 x_1 + w_2 x_2 + w_3 x_1 \cdot x_2 + \epsilon$ where the w_i are generated randomly and each $w_i \geq 3\sigma$. In addition,

$$\epsilon = \begin{cases} z M \sigma & \text{if } x_3 = +1 \\ z \sigma & \text{if } x_3 = -1 \end{cases}$$

where z has a standardized normal distribution in one simulation and an exponential distribution ($\lambda = 0.5$) in the other. Therefore, x_3 is clearly the only dispersion effect and the amount of dispersion is controlled by the multiplier M = 2, 3, 4, and 5.

MULTIPLICATIVE ERROR MODEL:

This model is defined as

$$y = f\left(\underline{x}_L, \underline{x}_B\right) \cdot \epsilon\left(\underline{x}_D, \underline{x}_B\right).$$

In this case the actual simulator model is

$$y = (w_0 + w_1 x_1 + w_2 x_2 + w_3 x_1 \cdot x_2) \cdot \epsilon \text{ where}$$

$$\epsilon = \begin{cases} (4 + z)\sigma M & \text{for } x_3 = +1 \\ (4 + z)\sigma & \text{for } x_3 = -1 \end{cases}$$

and z has a standardized normal distribution. The values of M tested were 3 and 5.

EXPONENTIAL MODEL:

The exponential model appears as

$$y = \exp(w_0 + w_1 x_1 + w_3 x_1 \cdot x_2 + \epsilon)$$

$$\epsilon = \begin{cases} z M \sigma & \text{for } x_3 = +1 \\ z\sigma & \text{for } x_3 = -1 \end{cases}$$

and z has a standard normal distribution. The values of M tested were 3 and 5. A summary of the different tests is displayed in Table 5.39.

Strategies	Designs	Modeling Scenarios

(1) Signal-to-Noise Ratios
  - Smaller is better
  - Larger is better
  - Normal is better
	(1) 4 replications of L_8
	
	(2) 2 replications of L_{16}
	
	(3) 1 replication of L_{32}
		(1) $y = f(\underline{x}) + N(0, 1)$
		
		(2) $y = f(\underline{x}) \cdot N(4, 1)$
		
		(3) $y = \exp[f(\underline{x}) + N(3,.01)]$

(2) $\bar{y}$ plus ln s (3) Modified Box-Meyer		(4) $y = f(\underline{x}) + \exp(\lambda = .5)$

Table 5.39 Summary of Different Simulation Tests

RESULTS:

The simulations were run 50 times for each strategy, model, and design type. The difference in average marginal S/N_L, S/N_S, S/N_N, and ln s for the replicated designs was used to identify the dispersion factor. The difference in residual standard deviations was used to identify dispersion factors for the unreplicated designs. The results for each model follow:

ADDITIVE ERROR MODEL: (Error term is normally distributed.)

In the September 1988 IIE transactions, Pignatiello [7] discusses this model and concludes that maximizing S/N_N is accomplished by maximizing $\bar{y}$ and minimizing s^2. This is verified in that the results for S/N_N are very similar to ln s. As stated in Box [3], S/N_L and S/N_S are based on ideas about location only and this is verified in that S/N_S and S/N_L never identified the dispersion effect. The results for this model are summarized in Figure 5.8. Starting with the L_8 designs, you can see that S/N_N and ln s are about equal, but they dropped substantially in percent, correctly detecting x_3 as a dispersion effect. This result is discussed by Gunter [5] where he indicates that more information is required to identify dispersion effects than location effects. The L_{32} unreplicated method had very impressive results while providing much more information on interactions. Out of curiosity an unreplicated L_{16} was run which again verified Gunter's statement above. As previously stated, S/N_L and S/N_S were ineffective methods for identifying the dispersion effect.

ADDITIVE ERROR MODEL: (Error term exponentially distributed.)

This test produced the following results:

			Percent Correctly Detected				
					Strategy		
			S/N_S	S/N_L	S/N_N	Ln s	Mod B-M
	L_8	M = 3	0%	0%	30%	72%	N/A
Design	L_8	M = 5	0%	0%	75%	90%	N/A
Type	L_{32}	M = 3	N/A	N/A	N/A	N/A	46%
	L_{32}	M = 5	N/A	N/A	N/A	N/A	50%

Table 5.40 Results for Additive Error Model

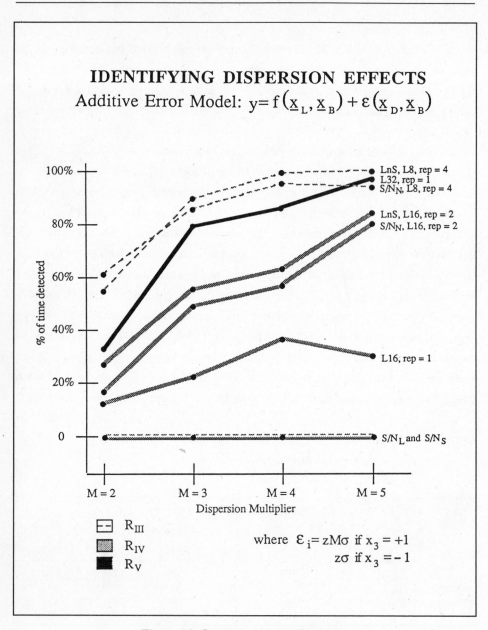

Figure 5.9 Simulation Result Summary

For this method ln s was slightly better than S/N_N and the modified Box-Meyer approach was not as effective as before. The ineffectiveness of S/N_S and S/N_L is still confirmed in this type model.

MULTIPLICATIVE ERROR MODEL:

Pignatiello [7] discussed this model and concluded that S/N_N and ln s would produce similar results. The simulation analysis provided a confirmation to his conclusion. The results are shown below.

			Percent Correctly Detected				
					Strategy		
			S/N_S	S/N_L	S/N_N	Ln s	Mod B-M
	L_8	M = 3	100%	100%	50%	45%	N/A
Design	L_8	M = 5	100%	100%	90%	90%	N/A
Type	L_{32}	M = 3	N/A	N/A	N/A	N/A	0%
	L_{32}	M = 5	N/A	N/A	N/A	N/A	0%

Table 5.41 Results for Multiplicative Error Model

The success of S/N_L and S/N_S was anticipated because, for the multiplicative error model, as the error increases, so does the response. Thus, any dispersion effect is also a location effect. Therefore, the success of S/N_L and S/N_S can be attributed to identifying location effect and not the dispersion effect.

EXPONENTIAL MODEL:

This test produced the results shown below.

			Percent Correctly Detected				
					Strategy		
			S/N_S	S/N_L	S/N_N	Ln s	Mod B-M
	L_8	M = 3	0%	0%	96%	35%	N/A
Design	L_8	M = 5	96%	96%	98%	98%	N/A
Type	L_{32}	M = 3	N/A	N/A	N/A	N/A	0%
	L_{32}	M = 5	N/A	N/A	N/A	N/A	22%

Table 5.42 Results for Exponential Model

Note the difficulty of all strategies except S/N_N in detecting a modest $(M = 3)$ dispersion effect. However, S/N_S, S/N_L, and ln s performed comparably to S/N_N in detecting large dispersion effects. As with the multiplicative error model, it was apparent that the exponential model had transformed non-homogeneous variables into location effects, enabling S/N_S and S/N_L to better detect the dispersion factors.

Of course, if the exponential model, $y = \exp(w_0 + w_1x_1 + w_2x_2 + w_3x_1x_2 + \epsilon)$ is anticipated, then a transformation using the natural logarithm would produce $\ln (y) = w_0 + w_1x_1 + w_2x_2 + w_3x_1x_2 + \epsilon$ which would yield the same results as the additive model. The modified Box-Meyer approach again had only marginal success with the nonlinear model; however, a log transformation as described above would compensate for this ineffectiveness.

CONCLUSIONS

The analysis of dispersion effects due to non-homogeneity of variance has become increasingly important as experimenters strive to optimize a response while minimizing its variance. Taguchi, Box, and others have suggested several techniques for analyzing dispersion effects within an experimental design, either simultaneous with or in addition to location effect analysis. This section compared the dispersion effect detection capabilities of five techniques (three Taguchi signal-to-noise ratios, a response standard deviation analysis, and a modified Box-Meyer technique) by way of a Monte Carlo simulation.

Simulation results clearly indicated that the standard deviation of responses for different factor combinations must typically change by at least a factor of 3 $(M \geq 3)$ to be detected. This is consistent with the non-homogeneity threshold described by Keppel [6] for classical ANOVA analysis. Even with such a large dispersion effect, only S/N_N and ln s proved to be consistent detectors over the range of simulation models. Although comparable performers, ln s is preferred over S/N_N because it is easier to interpret. Because S/N_N simultaneously detects location and dispersion effects, it may be difficult to attribute what kind of effect (dispersion or location) a particular factor has. However, a traditional analysis of the average response for location effects in conjunction with the ln s analysis for dispersion effects forgoes any question as to cause. The modified Box-Meyer method is as good as S/N_N and ln s for comparable number of runs if the underlying process model is linear, and it has the obvious advantages and efficiencies of an unreplicated design.

However, its use should be questioned for process models which are truly unknown and possibly nonlinear unless the appropriate transformation is made. Taguchi's S/N_S and S/N_L are not recommended due to their inability to identify dispersion factors.

Sample size was also found to be a crucial part in finding dispersion factors. Historically, researchers have used 16 data points as a rule of thumb minimum* for estimating a variance (or standard deviation). The 16 run Box-Meyer approach suffered in performance because each set of residuals had 8 values, not enough for a good estimate of variance. The 32 run Box-Meyer approach performed well with 16 residuals per low or high for variance calculations. Thus, sample size for experimentation could be simply tied to the above rule of thumb for variance calculations instead of the rigorous approach in Chapter 4. Using 16 data points as the "ideal" minimum for estimating an average variance (at a factor low or high) in replicated designs would suggest the minimum number of replicates for the following 2-level designs:

2-Level Design	Minimum # of Replicates for Residual Analysis	Modeling Ln s
L_4	8	9
L_8	4	5
L_{16}	2	3
L_{32}	1	2

Realize that when costs, time, resources, etc. prevent you from achieving the above minimum number or replicates, you will still have a chance of detecting dispersion factors. However, the chances for detection are approximately proportional to the ratio of your choice in replicates with the ideal, i.e., replicating an L_8 only 2 times results in an approximate 2/5 (or 40%) chance of detecting a real dispersion factor when using ln s. Another obvious conclusion is that the residual analysis approach will require less in the way of resources.

---

* See Appendix M for Rules of Thumb.

Chapter 5 Problem Set

1. Given the following settings and results:

A	B	C	D = ABC	y_1	y_2	$\bar{y}$	s	ln s
−	−	−		50	70			
−	−	+		47	54			
−	+	−		44	35			
−	+	+		54	71			
+	−	−		57	53			
+	−	+		43	48			
+	+	−		42	39			
+	+	+		60	57			

a) Complete the rest of the table.

b) Find the prediction equation from these results. Only include significant main and 2-way interactions.

c) What settings will minimize the response and the variance of the response?

2. An engineer needs to improve turbine blade quality by reducing thickness variability around a target of 3mm. The brainstorming session identified 4 variables that are likely to affect the thickness. Those variables and their range of values are:

	−1	1
Metal Temperature	20°C	22°C
Mold Temperature	3°C	5°C
Pour Time	1 sec	3 sec
Vendor	A	B

The engineer had been limited to 20 experimental runs, so he decided to run a full factorial of 16 runs, saving 4 runs to confirm his results.

Run #	Metal Temp	Mold Temp	Pour Time	Vendor	Thickness
1	20	3	1	A	2.314
2	20	3	1	B	2.307
3	20	3	3	A	2.321
4	20	3	3	B	2.233
5	20	5	1	A	4.204
6	20	5	1	B	4.260
7	20	5	3	A	4.221
8	20	5	3	B	4.198
9	22	3	1	A	2.687
10	22	3	1	B	2.748
11	22	3	3	A	2.697
12	22	3	3	B	2.728
13	22	5	1	A	4.769
14	22	5	1	B	4.868
15	22	5	3	A	4.787
16	22	5	3	B	4.849

a)   Do the full analysis on this data so the engineer has a prediction equation. Look for main effects, interactions, and dispersion effects.

b)   What settings should the engineer use to meet the goal of 3mm thickness with minimum variability?

3.   A statistical process control analyst is trying to determine which factors have a major effect in a circuit board etching process. She wants to pare down the seven factors brought out in the brainstorming session to maybe two or three. The seven factors are:

		LOW	HIGH
A:	Resist Thickness	.1mm	.5mm
B:	Develop Time	80 sec	90 sec
C:	Develop Concentration	3.1 : 1	2.7 : 1
D:	Exposure	200	240
E:	Develop Temperature	19°C	23°C
F:	Circuit Line Thickness	1mm	3mm
G:	Rinse Time	5 sec	10 sec

She ran the following L_8 design to find the significant effects.

Run	A	B	C	D	E	F	G	Response
1	−	−	−	−	−	−	−	74.48
2	−	−	−	+	+	+	+	70.07
3	−	+	+	−	−	+	+	75.71
4	−	+	+	+	+	−	−	68.79
5	+	−	+	−	+	−	+	84.98
6	+	−	+	+	−	+	−	81.57
7	+	+	−	−	+	+	−	84.03
8	+	+	−	+	−	−	+	80.97

a)   Find out which factors affect the response variable.

b)   If the goal is to maximize the response, what settings should be used? Predict what the response should be.

c)   The analyst ran confirmation runs to see if she had all the effects under control. If the mean of those runs is 86, is there any problem? How about 90? 72?

4.   Reference the Case Study on pages 8-25 through 8-30. In an attempt to get the width of the part to 9.380, the experimenter decided to look for interactions between the top four main effects on width:

D: Mold Temperature
A: Injection Velocity
E: Hold Pressure
B: Cooling Time

Run	D	A	E	B = DAE	y_1	y_2
1	−	−	−		9.34150	9.34160
2	−	−	+		9.36914	9.36916
3	−	+	−		9.34666	9.34664
4	−	+	+		9.36801	9.36809
5	+	−	−		9.36790	9.36800
6	+	−	+		9.34933	9.34937
7	+	+	−		9.36680	9.36690
8	+	+	+		9.35444	9.35446

a)   Analyze his results and create a predictive equation for width. What are the best settings to reach the 9.38 goal for width?

5.   As a reliability engineer, you have been asked to weed out infancy failures in component-populated printed circuit boards. The four factors of interest are:

	LOW	HIGH
A: Stress Temperature	80°C	125°C
B: Thermo Cycle Rate	5°C/min	20°C/min
C: Humidity	15%	95%
D: g level for a 10 minute sinusoid random vibration	3	6

The response is the number of electrical defects per board which has 1000 bonds.

a)   Given the following design matrix and response data, determine the optimal screening method. (The more failures found, the better.)

	A	B	−AB	C	−AC	−BD	D	y_1	y_2	y_3
								Response		
1	−	−	−	−	−	−	−	9	17	12
2	−	−	−	+	+	+	+	21	37	42
3	−	+	+	−	−	+	+	29	35	38
4	−	+	+	+	+	−	−	17	10	15
5	+	−	+	−	+	−	+	32	41	33
6	+	−	+	+	−	+	−	21	17	19
7	+	+	−	−	+	+	−	12	14	18
8	+	+	−	+	−	−	+	33	27	47

6.   A metal casting process for manufacturing turbine blades has four controllable factors:

Metal Temperature
Mold Temperature
Pour Speed
Raw Material

The blades must be 3mm thick; however, the ambient temperature causes the blades to expand and contract. The following experiment was run with ambient temperature as an outer array:

A Metal Temp	B Mold Temp	C Pour Speed	D Raw Material	TempL	TempM	TempH
+	+	+	+	3.06	3.14	3.06
+	+	−	−	3.01	3.00	3.05
+	−	+	−	2.81	2.81	2.80
+	−	−	+	2.80	2.88	3.01
−	+	+	−	2.62	2.61	2.62
−	+	−	+	2.61	2.66	2.72
−	−	+	+	2.42	2.43	2.52
−	−	−	−	2.42	2.43	2.41

a)   Find a combination of settings that will produce the required thickness and is also robust to ambient temperature.

b)   Plot the interactions of:

A with Ambient Temperature
B with Ambient Temperature
C with Ambient Temperature
D with Ambient Temperature

c)   Use the plots from b) to verify the optimal conditions found in a).

Chapter 5 Bibliography

1.   Barker, Thomas B. (1985), *Quality by Experimental Design*, Marcel Dekker.

2.   Box, George E.P., Hunter, William G., and Hunter, J. Stuart (1978), *Statistics for Experimenters*, John Wiley and Sons, Inc., New York.

3.   Box, G.E.P. (1988), "Signal-to-Noise Ratios, Performance Criteria, and Transformation," *Technometrics*, 30, No. 1.

4.   Box, G.E.P, and Meyer, R.D. "Report No. 1, Studies in Quality Improvement: Dispersion Effects From Fractional Designs," Center for Quality and Productivity Improvement, University of Wisconsin, Madison.

5.   Gunter, B. (1988), "Discussion: Signal-to-Noise Ratios, Performance Criteria, and Transformations," *Technometrics*, 30, No. 1.

6.   Keppel, G. (1982), *Design & Analysis, A Researcher's Handbook* (2nd Edition), Prentice-Hall, Englewood Cliffs, NJ.

7.   Pignatiello, J.J. (1988), "An Overview of the Strategy and Tactics of Taguchi," *IIE Transactions*, 20, 247-254.

8.   Schmidt, S.R. and Boudot, J.R. (1989), "A Monte Carlo Simulation Study Comparing Effectiveness of Signal-to-Noise Ratio's and Other Methods for Identifying Dispersion Effects," Rocky Mountain Quality Conference and 1989 ORSA/TIMS Annual Conference.

9.   Taguchi, G. (1987), *System of Experimental Design*, Kraus International Publications, White Plains, New York.

10.   Montgomery, Douglas C. (1990), *Design and Analysis of Experiments* (3rd Edition), John Wiley and Sons, Inc., New York.

11.   Bhote, Keki R. (1988), *World Class Quality*, American Management Association, New York.

12.   Myers, Raymond H. (1989), *Classical and Modern Regression with Applications*, (2nd Edition), Duxbury Press, Boston, MA.

Chapter 6

Taguchi Philosophy, Design, and Analysis

6.1   Introduction

An analysis of Japanese success in the American marketplace and their ability to produce quality products is a matter of current interest. The Japanese did in fact turn around a reputation of manufacturing junk to one of manufacturing quality products that can and still do obtain higher market value than many similar U.S. products. The tools used by the Japanese to accomplish this remarkable turnaround are focused at the product design phase versus quality inspection during manufacturing. By designing quality into the product from its conception, fewer resources are required to produce a quality product. The use of Taguchi Methods in parameter design, known as Quality Engineering, typically results in increased productivity, improved quality, and increased efficiency. The designs used by Taguchi were developed primarily by British and American statisticians as fractional-factorials, Plackett-Burman, Latin Square, and mixed designs. The idea of orthogonality has been around a long time; however, Taguchi's approach is different in that he (1) stresses only a few basic designs most commonly used in industry, (2) provides a table of these designs, and (3) avoids statistical rigor by providing a cookbook approach to the analysis.

The Taguchi approach to quality control emanates from the following definition of quality: "The Quality of a product is the (minimum) loss imparted by the product to society from the time the product is shipped" [11]. From this definition, a loss function is

---

* This chapter contains much of the information found in Warsavage and Schmidt [12].

developed that translates any deviation of a product's quality characteristic from its target value into a financial measure. Any deviation from the target value, either through variability (noise) or by a deviation of the average from target, translates into increased loss. The process of reducing variability and moving the mean to the target is based on designed experiments and emphasizes a phase of product and process design that Taguchi calls parameter design.

Despite much criticism of the Taguchi approach to designed experiments by traditional statisticians, the approach continues to expand. The expansion may be due to the simplicity associated with the presentation of the material by the American Supplier Institute and through the use of success stories such as those from Ford Motor Company, ITT, Digital Equipment Corporation, Xerox, Nippon Telephone and Telegraph Company, Toyota, and Fuji Film. Historically, engineers would vary only one factor at a time or intuitively select the best combination of factor settings based on engineering judgement. This can be extremely inefficient and will rarely lead to an optimal design. Traditional statistics were certainly an alternative; however, the presentation was not always simple and due to its perceived complexity, non-statisticians were less inclined to use it. One positive result of the discussions surrounding the use of Taguchi methods has been a great awakening of interest in industry to numerous experimental design approaches. Many engineers (as compared to a few just a couple of years ago) are now actively learning how to apply various types of orthogonal arrays and analysis approaches. Yet with all the discussions and learning that is taking place, probably fewer than 10% of the experiments done in industry today utilize a "designed experiment" approach. Much remains to be done.

The remainder of this chapter is divided into three parts. Under the topic of Taguchi philosophy and methodology, the loss function and Taguchi's approach to experimental design are discussed. The Taguchi approach to experimental design is presented in the second section. In the last section, the Taguchi approach to experimental design is illustrated.

6.2   Taguchi Philosophy and Methodology

6.2.1 Loss Function

Specification limits are the means by which manufacturers determine the details of a manufacturing product and process. These limits are frequently used in such a way as to imply that if the quality characteristic of a part falls within limits, the product is satisfactory; if it falls outside of these limits, it is defective. In many cases, a target value may be given, but hitting the target has not attained an importance beyond just being within specification. Taguchi suggests that any deviation from target results in incremental loss. He models this philosophy through a loss function which Chernoff and Moses [1] address as a part of statistical decision theory. Taguchi's contribution is the application of the loss function to the measurement of quality.

The quadratic loss function (based on a Taylor Series approximation) will increase as the quality characteristic deviates on either side of the target (see Figure 6.1). It is also symmetric about the target. Loss is minimized by producing all items as close as possible to the target. The old practice of using only the specification limits to determine loss does not hold in practice because a product just within specification is more like an out-of-specification product than a product right on target. As an example, consider a rod built within specification, but at the upper limit, and the sleeve within specification, but at the lower limit. This situation will result in a product with less than optimal performance and reliability. The end result is loss to the purchaser which can be eventually passed on to the producer as loss of market share. Other types of loss functions may be argued more accurate in that a deviation to one side of the target may be more of a problem than a deviation to the other side. The important point is that the development of a loss function results in evaluating costs associated with a deviation from the target and the simplicity of the quadratic loss function results in ease of implementation.

An important aspect of the loss function is that it maps deviations from the target into a financial measure. Everyone understands money, and since it is a common measure, comparisons can be made between products and processes. Operationally, the average loss would be computed using a sample large enough to characterize the process measured on a critical quality characteristic. Using a quadratic loss function, the loss associated with each

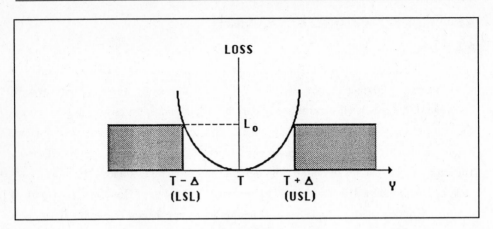

Figure 6.1 Taguchi's Quadratic Loss Function

product would be computed as follows [4]:

$$L_i = k(y_i - T)^2$$

where y_i is the quality characteristic of interest for product i

T is the quality characteristic target

k is a constant that converts deviation to a monetary value

To determine the value of k for a specific target, you need to estimate the loss at some measured value of the characteristic in question. Referring to Figure 6.1, assume you estimate the loss, L_0, associated with a point on the horizontal axis, y_0 (in this case y_0 = LSL). Then using $L_0 = k(y_0 - T)^2$ to represent the quadratic loss function and substituting known values for L_0, y_0, and T, you simply solve for k.

For a specified target T (nominal is best) the average loss, $\bar{L}$, for n products is:

$$\bar{L} = k\sum_i \left(\frac{1}{n}\right)(y_i - T)^2$$

where   n = sample size

y_i = values of the critical parameter

T = target value

The decomposition of the average loss function for all n products is developed as follows:

$$\bar{L} = k\sum \frac{1}{n}(y_i - T)^2$$

$$= \frac{k}{n}\sum \left(y_i - \bar{y} + \bar{y} - T\right)^2$$

$$= \frac{k}{n}\sum [\ (y_i - \bar{y})^2 + (\bar{y} - T)^2 + 2(y_i - \bar{y})(\bar{y} - T)\ ]$$

$$= \frac{k}{n}\sum (y_i - \bar{y})^2 + k(\bar{y} - T)^2 + 2(\bar{y} - T)\frac{k}{n}\sum (y_i - \bar{y})$$

Since the last term results in a value of 0, $\bar{L} = k[\sigma^2_y + (\bar{y} - T)^2]$. Thus, average loss is partitioned into variance of the response and deviation of the average response from the target. To minimize loss, it is clear that not only do you minimize the average response from the target, but you must also minimize response variability.

One illustration of these points is cited by L. Sullivan [10] on the color intensity of television sets. Figure 6.2 illustrates the distribution of the product quality characteristic for Sony Japan and Sony U.S.A. Notice that Sony Japan was producing some products outside the specification limits, but Sony U.S.A. was not. Under conventional methods, Sony U.S.A. would be preferred. Yet, if given a choice, customers chose Sony Japan. A loss function can model this preference, whereas strict adherence to specification limits cannot.

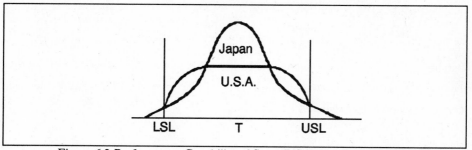

Figure 6.2 Performance Capability of Sony (USA) and Sony (Japan)

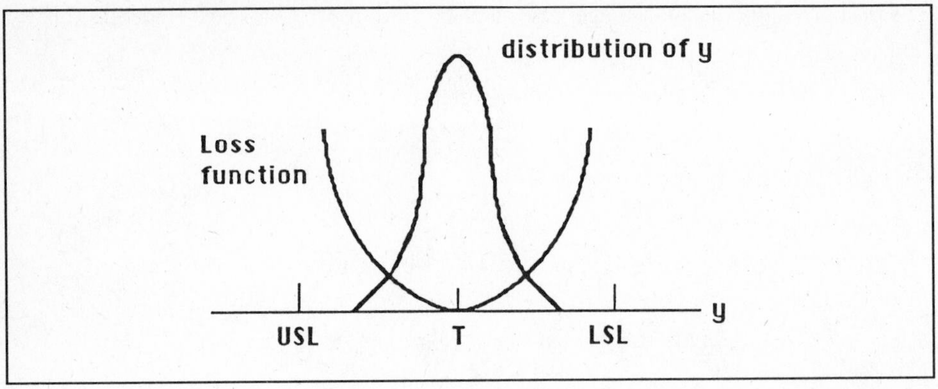

Figure 6.3 Idealized Distribution

Ideally, you would want the variability of the product to appear as shown in Figure 6.3. When the majority of the response measurements occur near the target, the loss is reduced. The use of designed experiments will enable you to determine which factors shift the average (in order to minimize $(\bar{y} - T)^2$) and which factors shift the variability (in order to minimize σ_y^2).

Two quantitative measures which relate σ_y^2 to the specification limits are C_p and C_{pk}.

$C_p = \dfrac{USL - LSL}{6\sigma}$ where $C_p < 1$ is unsatisfactory, $1.0 \leq C_p \leq 1.33$ is marginal, and $C_p > 1.33$ is desired. Since C_p assumes $\bar{y} = T$, which is not always the case, a different measure incorporating the product mean is C_{pk}. To calculate C_{pk}, we use $C_{pk} = \dfrac{\min(USL - \bar{y},\ \bar{y} - LSL)}{3\sigma}$ where $C_{pk} < 1$ is unsatisfactory, $1.0 \leq C_{pk} \leq 1.33$ is marginal and $C_{pk} > 1.33$ is desired. The higher the C_{pk} value, the more the σ^2 portion of the loss function is minimized and the closer $\bar{y}$ is to T. Note that σ is typically estimated with s, the sample data standard deviation.

6.2.2 Quality Engineering

Currently there is a lot of emphasis placed on techniques to control the building of a product once it has entered the manufacturing phase. These techniques are referred to as statistical quality or process control. They were discovered and used by Shewhart in the 1920's and recently made popular by Deming's success with the Japanese. Taguchi refers to these techniques, as well as some innovations of his own, as **On-line Quality Control.**

One of Taguchi's major contributions is to direct attention to the contributions that statistics and quality engineering can make in the design phase of a product. These techniques are thus referred to as Off-line Quality Control [4]. Statistics and a quality oriented approach taken in the design phase can save large amounts of resources in terms of time and effort to manufacture a product. This approach also leads to a reduction of costly scrapping and reworking of the product.

6.2.3 Quality by Design

As previously discussed, the Taguchi approach emphasizes and elaborates on the design phase of a product's life. He subdivides the design phase into three steps or subphases: system design, parameter design, and tolerance design.

System Design. This is the inventive phase where engineers make the prototype. Historically, this is an area in which the U.S. has excelled to the point of leading the world in patents, whereas the Japanese have contributed little. Initially, the Japanese borrowed western product design and then manufactured it better; however, they are currently placing more emphasis on system design. In certain areas, the Japanese are surpassing the U.S.

6.2.4 Parameter and Tolerance Design

Taguchi argues that the Japanese have added or at least emphasized two additional steps between the prototype and the hand-off to manufacturing. The most important step is parameter design. In this step, Taguchi analyzes the effect of each of the controllable factors and determines the best factor settings in relation to the noise factors. Factors that most affect performance are determined, as well as factor settings that are optimal, in terms of performance, cost, ease of manufacture, and reliability. At this stage, one should avoid factor settings that are difficult to maintain and control. If those settings cannot be avoided, then tight tolerances must be set. The setting of those tolerances is addressed in the **tolerance design subphase**. The identification of critical factors and their optimal values are found by experimentation using designed experiment techniques. If tolerances must be determined, these are also determined experimentally.

6.2.5 Controllables and Noise

The factors that affect the performance of a product by increasing variability are referred to as noise or noise factors. Product reliability and ease of manufacture are the result of specifically designing the product to be resistant to noise. In contrast, those factors that are relatively easy to control are called controllables. Some factors can actually fall into both categories. For example, the temperature of a wave solder process can be set closer to 150 degrees than 200 degrees; however, even if 150 degrees is the target, the measured temperature may vary a few degrees within a run or between runs. In the first case, the temperature is considered a controllable factor and in the second, it is a noise factor. The Taguchi strategy investigates the effect of large and small changes. Another noise factor might be ambient humidity which may be systematically varied during experimentation in the parameter design phase, but remains uncontrollable in the operational phase of the process.

Taguchi's consideration of noise results in some changes in the way that an experimental design is set up and analyzed. Noise, a pseudonym for experimental error, is usually considered in statistical thinking as the element of unpredictability of the outcome and is assumed to be independent of the factor settings. In classical experimental design, you would assume that there are no errors in the factor settings. Any small errors, such as they are, are absorbed by randomization. Taguchi recommends that these errors or noise can be directly analyzed using experimental design techniques.

An example which illustrates some of Taguchi's philosophies is based on the characteristic function of a transistor (see Figure 6.4). Consider the following scenario:

(1)   Target output voltage is 100.

(2)   Gain has a large effect on output voltage and is non-linear.

(3)   The circuit engineer has many options in terms of choosing the gain setting.

One approach would be for the circuit engineer to set the transistor gain at 18 so as to obtain the desired output voltage of 100. Please note, however, that small deviations from the target of 18 will result in relatively large variations in our output voltage. A better approach would be for the circuit engineer to select a transistor gain at a much flatter portion of the curve (for example, 30), thus minimizing the resulting variation in output voltage. By using some other component in the circuit (which has a linear effect on output voltage), the engineer can adjust the mean value from 120 to 100. Provided this reconfiguration did not increase the variability in the output voltage, the end result would be much improved.

In general, running transistors near the point where the slope of the output voltage is small may result in a higher quality product. This of course oversimplifies the problem, but it does illustrate a valuable point in that some factor settings can be used to reduce variability and others to obtain the response closest to the target.

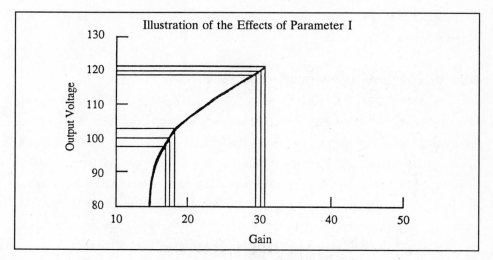

Figure 6.4 Characteristic Function Plot

6.3   Taguchi Strategy of Experimentation

Taguchi's strategy is characterized by three elements: (1) orthogonal arrays based on fractional factorials, Plackett-Burman, or Latin Square experimental designs, (2) design modification that incorporates information about potential noise factors, and (3) analysis of a transformation called signal to noise ratio.

6.3.1 Saturated Fractional Factorial Experimental Designs

A fractional factorial experiment is saturated when the design only allows for the estimation of the main effects. For example, instead of estimating 4 main effects with 16 runs, it is possible to estimate 15 main effects independently of each other with those 16 runs. In this case, all the interactions are aliased with these main effects (see Chapter 3 for details).

Fractional designs, when used as part of a screening process, are effective when there are large numbers of factors. Since large numbers of factors usually include some which do not affect the response measure, the screening process provides a way to remove from consideration (screen out) those unimportant factors as efficiently as possible. There are potentially hundreds of factors in most industrial models and very little may be known about their effects. The screening process is, therefore, a way to start reducing the number of factors in a systematic fashion.

6.3.2 Designs That Assess Information About Noise

Randomization is the key to reaching conclusive results in a time and resource efficient manner. Randomization in factorial designs implies that runs of all different factor combinations will be completed in a random order. There are two reasons for wanting to randomize: (1) the critical assumption of independence is usually ensured through random sampling and randomization, and (2) long-term trends and/or systematic noise factors not included in the design matrix will become part of the error by being spread in an equally likely fashion over the experimental conditions [6]. Since randomization in industrial experiments is frequently time consuming and often impossible, Taguchi does not emphasize the use of randomization, but attempts to compensate for it by incorporating noise in the design. He, therefore, assumes all potential noise factors can be measured and will be included in the design. If this is not the case, there exists a strong possibility that the results will be ambiguous.

Taguchi's robust designs use two sets of factors: (1) an inner array of controllables, and (2) an outer array of noise factors. At each combination of the controllables, an outer array of noise factors is included. Thus, at each combination of the controllables, a separate estimate of the effect of the noise factors can be made. This is similar to a sensitivity analysis by J. S. Hunter, [3]. The primary objective of robust designs is to determine optimal settings for the controllable factors which will result in a desired response which is invariant to noise.

6.3.3 Signal to Noise Ratio

Sensitivity to noise is analyzed through a signal-to-noise statistic which involves both location and dispersion effects. Many different estimates of signal-to-noise have been tested, but typically one of three are used. When a specified target is the criterion, a statistic related to the coefficient of variation is:

$$(S/N) = 10 \ \log_{10} \ \frac{\bar{y}^2}{S^2}$$

$$\text{where} \ \ \bar{y} = \sum \frac{y_i}{n} \ \ \ \text{and} \ \ \ S = \sqrt{\frac{\sum \left(y_i - \bar{y}_i\right)^2}{n-1}}.$$

This transformation converts y to decibel units and is closely related to the conventional signal-to-noise ratio which is $\bar{y}/s$. The optimum factor levels for any process will be found by maximizing (S/N).

In the case of a zero or infinite target, the above formula is replaced by signal-to-noise ratios derived from the formula for average loss, resulting in:

$$(S/N) = -10 \ \log_{10} \ MSD$$

$$\text{where} \ \ \ \ MSD = \frac{1}{n}\sum y_i^2 \ , \ \text{for a zero target (smaller is better)},$$

$$\text{or} \ \ \ \ \ \ MSD = \frac{1}{n}\sum \frac{1}{y_i^2} \ , \ \text{for an infinite target (larger is better)}.$$

A typical Taguchi strategy involves analyzing one of these statistics with or without a graphical analysis of means.

6.4   Example of a Taguchi Experimental Design

In this section, the Taguchi strategy is illustrated with an example of a wave solder process optimization. In a printed circuit board assembly plant, parts are inserted either manually or automatically into a bare board with a circuit printed on it. After most of the parts are inserted, the board is put through a wave solder machine. This process is a means of connecting, both electrically and mechanically, all the parts into the circuit. Boards are placed on a conveyor, and taken through a series of steps. They are bathed in a flux mixture to remove oxide. To minimize warpage, the boards are preheated before the solder is applied. The soldering process takes place as the boards move across the wave of solder. After that, the board usually goes through a device similar to a dishwasher to remove any debris. Because of the many factors used in controlling the process, it is an excellent candidate for experimentation. For wave soldering processes which have not been studied statistically, it is not uncommon for 10-20% of the assembly work force to be involved in solder touchup.

The data used in this example are simulated to illustrate the Taguchi strategy. The objective of the experiment is to minimize the number of solder defects per million joints. Table 6.1 lists the factors and their settings decided upon for this example. The choice of factors to be used must be carefully considered in the brainstorming phase. Experts at other plants with different types of wave solder machines might choose a different set of factors. In our example, the inner array of controllables contained five factors. After brainstorming, it was decided to test each factor at two levels and to include an interaction between solder pot temperature and conveyor speed.

The factors in the noise array are selected as well. Because several different types of assemblies are run through this wave solder process, two different types of assemblies were used. The objective is to find one setting for the wave solder process that is suitable for both types of assemblies. The design will also indicate if assembly type interacts with any of the controllables. In addition to product noise, both the conveyor speed and solder pot temperature will be moved around the initial setting given by the controllables array. This is because it is difficult to set the conveyor speed with any degree of accuracy and it is also difficult to maintain solder pot temperature.

A Designed Experiment for Wave Solder
An Example of the Use of Orthogonal Arrays

Controllable

Factors	Levels	
	Low	High
(1) Solder Pot Temperature (S)	480 °F	510 °F
(2) Conveyor Speed (C)	7.2 ft/m	10 ft/m
(3) Flux Density (F)	.9 °	1.0 °
(4) Preheat Temperature (P)	150 °F	200 °F
(5) Wave Height	0.5"	0.6"

Noise

Factors

(1) Product Noise	Assembly #1, Assembly #2
(2) Conveyor Speed Tolerance	−0.2, +0.2 ft/m
(3) Solder Pot Tolerance	−5 °F, +5 °F

Table 6.1 Factors and Settings for Wave Solder Example

Eight runs will be used to test the effects of the five controllables in a Taguchi L_8 design (see Table 6.2a). Notice that for each factor, there are four runs at the low setting and four runs at the high setting. This balancing is a property of the orthogonality of the design matrix. Table 6.2b lists the array of noise factors to be run at each of the eight settings of the controllables. This is a Taguchi L_4 design. The combination of the inner and outer arrays results in each run of the controllables being repeated over the 4 combinations of the noise factors.

	Controllables Design Inner Array				
Run	Solder Pot Temperature	Conveyor Speed	Flux Density	Preheat Temperature	Wave Height
1	510	10.0	1.0	150	0.5
2	510	10.0	0.9	200	0.6
3	510	7.2	1.0	150	0.6
4	510	7.2	0.9	200	0.5
5	480	10.0	1.0	200	0.5
6	480	10.0	0.9	150	0.6
7	480	7.2	1.0	200	0.6
8	480	7.2	0.9	150	0.5

Table 6.2a Runs Used in Wave Solder Example

Outer Array				
At each combination of the inner array, an outer array of noise factors is run.				
			Run	
Parameter	1	2	3	4
Product Noise	Assembly #1	Assembly #1	Assembly #2	Assembly #2
Conveyor Tolerance	−0.2	+0.2	−0.2	+0.2
Solder Tolerance	−5	+5	+5	−5

Table 6.2b Noise Factors for Wave Solder Example

Table 6.3 presents the results of the experiment. The two far right columns contain the average and signal-to-noise ratio for each of the eight settings of the controllables or inner array. Because the objective is to minimize the response solder defects per million, the formula for signal to noise is $-10 \log \left(\frac{1}{n} \sum y_i^2 \right)$.

		Combined Inner and Outer Arrays									

Combined Inner and Outer Arrays

Results

Run

Noise Factors	1	2	3	4
Product Noise	#1	#1	#2	#2
Conveyor Tolerance	−.2	+.2	−.2	+.2
Solder Tolerance	−5	+5	+5	−5

Controllable Factors   Run

Run	Solder	Conveyor	Flux	Preheat	Wave	1	2	3	4	Mean	S/N
1	510	10.0	1.0	150	0.5	194	197	193	275	215	−46.75
2	510	10.0	0.9	200	0.6	136	136	132	136	135	−42.61
3	510	7.2	1.0	150	0.6	185	261	264	264	244	−47.81
*4	510	7.2	0.9	200	0.5	47	125	127	42	85	−39.51
5	480	10.0	1.0	200	0.5	295	216	204	293	252	−48.15
6	480	10.0	0.9	150	0.6	234	159	231	157	195	−45.97
7	480	7.2	1.0	200	0.6	328	326	247	322	305	−49,76
8	480	7.2	0.9	150	0.5	186	187	105	104	145	−43.59

* Experimental Champion

Table 6.3   Experimental Results for Wave Solder Example

An inspection of the results in Table 6.3 reveals that run number 4 maximized (S/N) and, therefore, our "experimental champion" is S = 510, C = 7.2, F = .9, P = 200, and W = .5. It is quite possible that some of the factors are more influential than others. To determine which factors are important, marginal averages are computed by averaging the four results at a particular level of a factor. For example, the first four runs of the inner array are with the solder pot temperature set at the higher level, the second four are at low solder pot temperature. These sixteen results (4 runs of the inner array each with 4 runs of the outer array) are combined to obtain an average of 170 solder defects per million. The compilation of these results are presented in Table 6.4 and displayed graphically in Figure 6.5. Analysis of the table and graph indicates that low flux density, high solder pot temperature and low wave height reduced the number of solder defects. It was this combination that influenced run number four. The effect of conveyor speed is marginal; however, economic considerations, such as throughput, might influence the decision to run the conveyor at a higher speed. Preheat temperature also appears insignificant with economic considerations favoring the lower temperature. Notice from Figure 6.6 that the anticipated interaction between conveyor speed and solder pot temperature does not appear important. The lines are nearly parallel, indicating that the effect for conveyor speed is the same at each level of solder pot temperature. From the S/N column of Table 6.4, a "paper champion" would be S = 510, C = 10.0, F = .9, P = 150, and W = 0.5, where C and P were set based on economic considerations. These combinations were never tested in the experiment, thus the term "paper champion". To predict S/N at these settings, regression analysis is used to generate a prediction equation for S/N using only the significant effects of S and F.

$$S/N = -45.52 + 1.35(S) - 2.60(F)$$

where coded S and F are (+) for high and (−) for low. Using this equation, the predicted value for our "paper champion" is −41.57.

Parameter	Level	Mean	S/N
Solder Pot Temperature	480	225	−46.87
	510	170	−44.17*
Conveyor Speed	7.2	195	−45.17
	10.0	200	−45.87
Flux Density	0.9	140	−42.91*
	1.0	255	−48.11
Preheat Temperature	150	200	−46.03
	200	194	−45.01
Wave Height	0.5"	174	−44.50*
	0.6"	220	−46.54

* These are the optimum level settings for each significant factor. Factors without an asterisk are not significant and their levels can be based on other considerations.

Table 6.4 Analysis of Results for Wave Solder Example

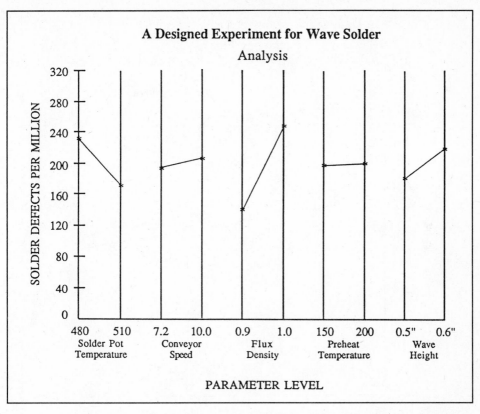

Figure 6.5 Plot of Averages for Wave Solder Example

Although this S/N value for the "paper champion" is a little smaller than that of the "experimental champion," the "paper champion" warrants further investigation because it could possibly cost less to produce. Confirmatory runs should be made for both champions to determine the best combination.

The use of S/N to simultaneously analyze the mean and variability provides a simple approach to the analysis. The reader is referred, however, to Chapter 5 where S/N analysis did not appear sensitive to information on variability. The combining of $\bar{y}$ and s in a S/N fashion may not always be the ideal approach. One should consider separate analysis of $\bar{y}$ and s (or ln s) to supplement S/N analysis (see discussion in Chapter 5).

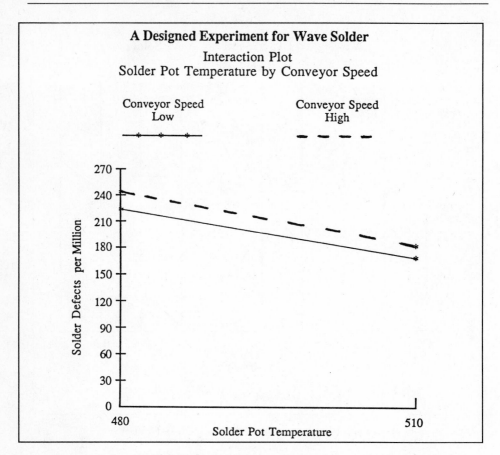

Figure 6.6 Interaction Plot of Temperature vs. Speed for Wave Solder Example

6.5 Comparison of Classical and Taguchi Approaches

<table>
<tr><td align="center">Classical</td><td align="center">Taguchi</td></tr>
<tr><td>1. Brainstorming (as discussed in Chapter 1).</td><td>1. Brainstorming (as discussed in Chapter 1).</td></tr>
<tr><td>2. Effects to be tested typically include main and 2-way interaction effects.</td><td>2. Effects to be tested usually consist of only main effects. Taguchi methods can incorporate 2-way interactions, but</td></tr>
</table>

priority is given to main effects for simplicity.

3. Determine desired levels of α and β in order to estimate the appropriate sample size to measure effects. For recent rules of thumb for numbers of replications see Appendix M.

3. Discussions about α and β normally are not addressed. Sample size, n, is found based upon the number of effects to be tested from (2) and the number of levels of each factor. Little guidance is provided on numbering of replicates.

4. Build an orthogonal design (usually R_{IV} or R_V) to satisfy (2) and (3) above.

4. Orthogonal designs are tabled and because of de-emphasis of interactions are usually of R_{III}.

5. Randomize the order of the runs to spread all unmeasured noise factors evenly across the other effects. This step ensures internal validity of the experiment and allows for causality analysis.

5. Randomization is de-emphasized. Factors are placed in up to 4 groups based on their difficulty level to be reset during experimentation. The hardest factors to reset typically are assigned to columns with only one change of levels (for 2-levels designs) and the easiest to reset are assigned to alternating 1, 2 values. Thus, randomization is not enforced, making the experiment vulnerable to confounding of unsuspecting noise factors. Advocates of Taguchi will argue that if all important controllable and noise factors are in the design, then the confounding problem is resolved. Classical experimenters will still be concerned about factors which were not included due to incomplete brainstorming or inability to be measured.

6. Run the experiment and collect the data.

6. Run the experiment and collect the data.

7. Perform hypothesis tests on all desired effects and classify as significant or not significant using the t or F test. Use replication at some or all points to estimate error and/or pool higher order interaction sums of squares for the error. Recent emphasis is on the use of normal probability plots or Pareto

7. Usually, hypothesis tests are not emphasized; instead, a graphical analysis is conducted or the S/N ratio is used. If ANOVA is conducted, the error estimate is based on the pooling of insignificant sums of squares. Rules for pooling for sums of squares are not rigorous. F values are tested for signifi-

charts for unreplicated designs.

cance per classical procedures; however, it is recommended that final unpooled F ratios greater than 2 not be ignored [4].

8. Build a parsimonious mathematical model to (i) form prediction intervals and (ii) estimate the optimal response through the use of Response Surface Methodology.

8. Important effects are determined graphically to select a "paper champion" or the "experimental champion" based on the best $\bar{y}$ or largest S/N. Prediction equations are generated, but the modeling process is inferior to classical techniques. Prediction intervals are not calculated. The mathematical model is used for prediction, but not to locate the optimal as in Response Surface Methodology.

9. If the optimal response lies outside the sample region, conduct another experiment in the direction of the optimal.

9. No iterative experimentation is used.

10. Find the optimal through an iterative experimental procedure.

10. The true optimal is not really found. Rather, settings for the best response over the experimental region are based on experimental and/or paper champions.

11. Set the process factors at the optimal settings and go on-line. Recent emphasis is also on confirmation runs.

11. The importance of confirmation runs is stressed. Confirmatory runs are made prior to going on-line.

12. Assumptions include normality, independence and equal variability. The F test is robust to minor violations of normality and homogeneity especially for large samples balanced for each experimental condition. If you are concerned about large violations, you can use a Pareto chart approach, i.e., rank order effects based on F values and select the larger ones as being important. More emphasis today is on blending engineering knowledge, common sense, and statistics when drawing conclusions.

12. Assumptions are similar to the classical method. Independence may be a problem due to lack of randomization. Unequal variance is accounted for by the transformation used in S/N calculations.

In addition to the 12 points previously discussed, the Taguchi approach stresses simplicity. The idea of using tabled orthogonal arrays and marginal mean analysis is very appealing to engineers. It is obviously an easy way to get started in designing good experiments. However, engineers should not be content to limit themselves to the basics. Eventually they will want to progress to the more sophisticated designs and analysis techniques discussed in Chapters 3, 4, and 5.

Probably the most important contributions of the Taguchi approach are in the area of the loss function and robust designs. The signal-to-noise statistic does not appear to be the best metric for finding dispersion effects as pointed out in Chapter 5. In addition, the Taguchi approach will infrequently find the optimal; it is widely considered to be 60-80% effective. The Taguchi advocates will maintain that further improvements are so costly that more sophisticated efforts are not cost efficient. One could argue that this is true in many applications; however, there are applications that require more fine tuning in finding the optimal. For 80-100% effectiveness, the classical method of designed experiments and Response Surface Methodology (RSM) techniques should be used.

Chapter 6 Problem Set

1-6. Redo the problems from Chapter 5 using the appropriate S/N formula for those questions. Compare the results using S/N and the results you found originally.

7. When using signal-to-noise, how do you find the optimal levels for a process?

8. Name 3 advantages of Taguchi's methodology. Name 3 shortcomings.

9. Why does Taguchi use a quadratic approximation for the loss function? Into what components does the function partition ?

10. When does a factor belong in an outer array?

Chapter 6 Bibliography

1. Chernoff, H. and Moses, L. E. (1959), *Elementary Decision Theory*, John Wiley and Sons, Inc.

2. Fisher, R. A. (1966), *Design of Experiments*, 8th ed. Hafner (MacMillan).

3. Hunter, J. S. (1985), "Statistical Design Applied to Product Design," *Quality Progress*, (Vol. 17, No. 4).

4. *Introduction to Quality Engineering, 5-Day Seminar Course Manual*, American Supplier Institute, Inc. 1987.

5. Kackar, Raghu N. (1986), "Taguchi's Quality Philosophy: Analysis and Commentary," *Quality Progress*, Dec. 1986.

6. Kirk, Roger E. (1982), *Experimental Design*, Brooks/Cole Publishing Co.

7. Lindquist, E. F. (1953), *Design and Analysis of Experiments in Psychology and Education*, Boston, Haughton Mifflin.

8. Norton, D. W. (1952), "An empirical investigation of some effects of non-normality and heterogeneity of the F-distribution" unpublished doctoral dissertation, State University of Iowa.

9. Rogan, J. C. and Kesselman, H. J. (1977), "Is the ANOVA F-test robust to variance heterogeneity when sample sizes are equal? An investigation via a coefficient of variation." *American Educational Research Journal* 1977, Vol 14, 493-498.

10. Sullivan, L. P. (1984), "Reducing Variability: A New Approach to Quality," *Quality Progress*, Vol. 15, No. 4.

11. Taguchi, G. (1986), *Introduction to Quality Engineering*, Asian Productivity Organization.

12. Warsavage, B. and Schmidt S. R. (1987), "Taguchi Philosophy and Methodology" presented at an IBM Technical Symposium.

Chapter 7

Optimization and Response Surface Methods

7.1   Introduction and Overview

If you have used the technique discussed in Chapter 4 and found factor settings that provide a satisfactory response, then there may be no need for further investigation. Unfortunately, using the simplest optimization techniques may not always lead to desirable results. When your best response does not prove satisfactory, what can you do? There are four things to consider in selecting the next plan of action:

(1)   Were all potential factors and associated 2-way interactions included in the design?

(2)   Is there too much noise in the data?

(3)   Did you select the appropriate number of levels for each factor?

(4)   Did you select the appropriate range for each factor?

If your problem is related to (1) or (2) then you may need to brainstorm again and make sure all relevant effects are measured. If interactions are the only problem, a foldover of the first design may provide the answers you need. If factors were left out of the first design, you will need to build a new design which will consume lots of resources. To counter excessive noise, you can add more important factors to explain the noise or replicate each run to reduce the impact of noise. Once you have encountered these problems due to an inadequate brainstorming process, the consequences involved in correcting the problem should convince you to invest more time in future brainstorming.

If (3) and (4) are contributors to your problem, then you probably need to select a better method to optimize the response. This chapter is designed to compare 3 methods of finding the optimal input factor settings: (1) one-at-a-time designs, (2) orthogonal arrays (Taguchi methods), and (3) response surface methodology (RSM). Although the first two methods are simple and easy to implement, they typically will produce results that are 60–80% of the total improvement to be made. If you need 80–100% of the total improvement, you need another method. To illustrate this point, consider the example that follows.

Given a hypothetical process with 2 input factors, f_1 = temperature ($°$F) where 100 $\leq f_1 \leq$ 400 and f_2 = pressure (psi) where 100 $\leq f_2 \leq$ 200, the objective is to optimize some measure of percent yield, y. Figures 7.1a, 7.1b, 7.2a, and 7.2b present the one-at-a-time approach. In figure 7.1a, the researcher uses brainstorming information to select f_2 = 150 as a place to start his investigation. He sets the pressure knob at 150 psi and varies the temperature over its range to find the optimal temperature setting, which appears to occur at 200$°$F with a corresponding yield of 85.3%.

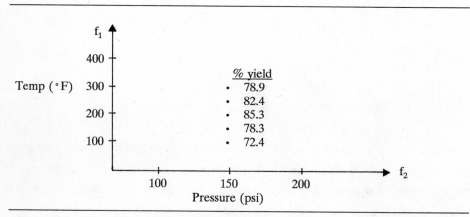

Figure 7.1a Plot of Pressure and Temperature vs. Yield with Pressure Held Constant

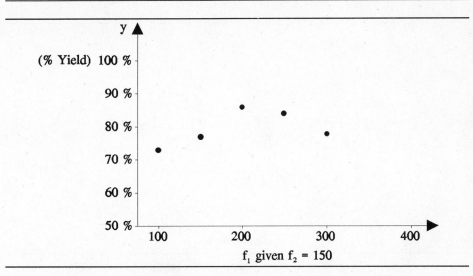

Figure 7.1b Plot of Yield vs. Temperature with Pressure = 150 psi

The temperature knob is now fixed at $200°F$ and pressure is varied over its range as shown in Figure 7.2a and 7.2b. These figures reveal that the best X_2 setting (given $f_1 = 200$) is $f_2 = 150$. At this point, the researcher would conclude his best results occur at ($f_1 = 200$, $f_2 = 150$) and he should make several confirmation runs to ensure the results produce a desired average yield and tolerable variance.

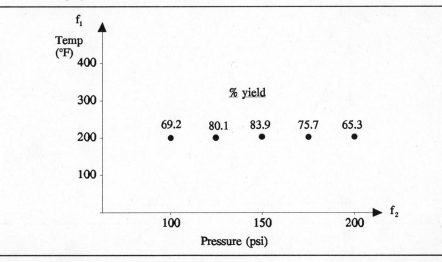

Figure 7.2a Plot of Pressure and Temperature vs. Yield with Temperature held constant

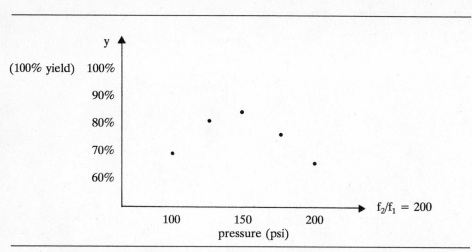

Figure 7.2b   Plot of Yield vs. Pressure with Temperature = 200°F

Unfortunately, if you choose the one-at-a-time approach and your results are less than desirable, it is confusing as to how you might improve the process. In this example, the response contours are displayed in Figure 7.3. The one-at-a-time approach will use up large

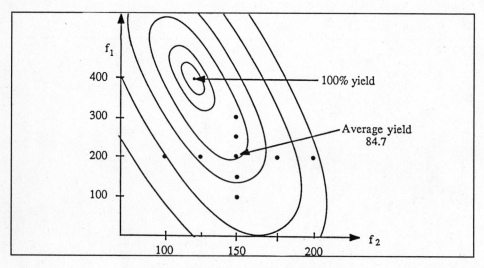

Figure 7.3   Response Surface Contour Plot

amounts of resources and if results are unsatisfactory, there is no efficient way to seek improvements.

Using the Taguchi approach, or any other orthogonal array approach, without RSM would consist of the following:

(1)   The researcher would select an experimental region over which he chooses to investigate the response.
(2)   Since the input factors in this example are both continuous, either a 2 or 3-level design would typically be used to evaluate the response over the experimental region.
(3)   Using analysis techniques as described in Chapter 5, a best setting for each of the factors is obtained.

Referring back to the example, assume the experimental region is decided to be $\{100 \le f_1 \le 200$ and $125 \le f_2 \le 175\}$. The results for a 3-level design appear in Figure 7.4. Based on average marginals, the best settings would be $f_1 = 200$ and $f_2 = 150$ which produces an estimated average response of $\hat{y} = 82.82.*$

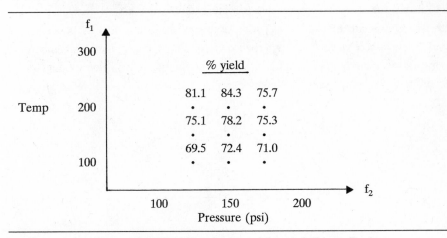

Figure 7.4   Yield Results from 3-Level, 2-Factor Design

$$* \; \hat{y} = \bar{y} + (\bar{f}_{1(200)} - \bar{y}) + (\bar{f}_{2(150)} - \bar{y})$$
$$= 75.844 + (80.366 - 75.844) + (78.3 - 75.844) = 82.82$$

Obviously, the failures of the one-at-a-time and the Taguchi approach associated with this example are due to a poor choice of factor values over which to investigate. As contrived as this data is, the point to be made is that this unfortunate choice of factor ranges can occur with old and new products even from the most experienced researchers, resulting in less than the desired results. When this situation arises, a sequential approach is needed to efficiently lead you in the direction of the optimal. The previous two methods do not have this flexibility and, if modified to produce an optimum search, they will typically be inefficient. What is recommended is an approach based on RSM.

7.2   Response Surface Methodology

The basic idea associated with response surface methodology and experimental design is to build an efficient design to determine if the experimental region contains the optimum. If the optimum lies in the experimental region, build a mathematical model to locate it, else continue experimenting along a gradient of steepest ascent (direction of greatest improvement) to find the center of a new experimental region to be tested. At this new point, build another efficient design in the new experimental region and test whether it contains the optimum. Continue the process until the optimum is located. In order to conduct the process just described, we must be able to represent y as a function of the input factors. If the true statistical relationship is represented by $y = h(x_1, x_2, ..., x_k) + C$, our objective is to estimate $h(x_1, x_2, ..., x_k)$ with as few experimental runs as possible. To accomplish this task, a Taylor Series approximation of $h(x_1, x_2, ..., x_k)$ is recommended. In most cases, a second order Taylor Series is a sufficient approximation. For k input factors, the 2nd order Taylor Series expanded about $\underline{c} = (c_1, c_2, ..., c_k)$ is

$$h(\underline{x}) = h(\underline{c}) + \sum_{1}^{k} (x_i - c_i)\frac{\partial h}{\partial x_i}\bigg|_{\underline{x} - \underline{c}} + \sum_{1}^{k} \frac{(x_i - c_i)^2}{2!}\frac{\partial^2 h}{\partial x_i^2}\bigg|_{\underline{x} - \underline{c}}$$

$$+ \sum_{i>j}\sum (x_i - c_i)(x_j - c_j)\frac{\partial^2 h}{\partial x_i \partial x_j}\bigg|_{\underline{x} - \underline{c}} + R_{\underline{c}}$$

For coded data, i.e., $x_i = 2\dfrac{f_i - \overline{f}_i}{d_i}$, expanding the Taylors about $\underline{c} = \underline{0}$ produces the following model.

$$h(\underline{x}) = h(\underline{0}) + \sum_1^k (x_i)\frac{\partial h}{\partial x_i}\bigg|_{\underline{x} - \underline{0}} + \sum_1^k (x_i^2)\frac{\partial^2 h}{\partial x_i^2}\bigg|_{\underline{x} - \underline{0}} + \sum\sum_{i>j}(x_i)(x_j)\frac{\partial^2 h}{\partial x_i \partial x_j}\bigg|_{\underline{x} - \underline{0}} + R_{\underline{0}}$$

If this 2nd order model is a good fit, then $R_{\underline{0}}$ should be relatively small.

The Taylor Series can be arranged such that it resembles the 2nd order polynomial model below. The weights, b_0, b_i, b_{ii}, and b_{ij} are easily obtained from a multivariable regression software package.

$$y = b_0 + \sum_1^k b_i x_i + \sum_1^k b_{ii} x_i^2 + \sum\sum_{i>j} b_{ij} x_i x_j + R_{\underline{0}}$$

where

$$b_0 = h(\underline{0})$$

$$b_i = \frac{\partial h}{\partial x_i}\bigg|_{\underline{x} - \underline{0}}$$

$$b_{ii} = \frac{\partial^2 h}{\partial x_i^2}\bigg|_{\underline{x} - \underline{0}}$$

$$b_{ij} = \frac{\partial^2 h}{\partial x_i \partial x_j}\bigg|_{\underline{x} - \underline{0}}$$

$R_{\underline{0}}$ is the error term.

The error term, $\underline{e} = R_0$, can be broken into two parts:

(1) pure error   —   estimated through replication.

(2) lack of fit   —   estimated by the difference in the model error and pure error.

   —   This error indicates whether the model type (linear, quadratic, etc.) is appropriate.

Using the central composite design to accomplish RSM, the replications required to estimate pure error will occur at the center point. Consider the previous example where x_1(high) = 200, x_1(low) = 100, x_2(high) = 175, and x_2(low) = 125. The first order design with replicated center points would appear as shown in Table 7.1, where x_i is the transformed value of f_i.

f_1	f_2	x_1	x_2	$x_1 x_2$	y
200	175	+1	+1	+1	75.9
200	125	+1	−1	−1	82.1
100	175	−1	+1	−1	70.1
100	125	−1	−1	+1	69.7
150	150	0	0	0	75.6
150	150	0	0	0	76.2

The next step is to test the model adequacy. Using regression and a 0.10 significance level, the first order model which fits the data is

$$\hat{y} = 75.9 + 4.55x_1 - 1.45x_2 - 1.65x_1x_2$$

The computer output shown in Table 7.1 indicates non-significant lack of fit for a quadratic term.

Dependent Variable: y		N: 6	Multiple R: .999		Squared Multiple R: .998	

Adjusted Squared Multiple R: .991				Standard Error of Estimate: 0.424		

Variable	Coefficient	Std Error	Std Coef	Tolerance	T	P(2 Tail)
Constant	75.900	0.300	0.000	1.0000000	253.000	0.003
x_1	4.550	0.212	0.888	1.0000000	21.449	0.030
x_2	−1.450	0.212	−0.283	1.0000000	−6.835	0.092
$x_1 \bullet x_2$	−1.650	0.212	−0.322	1.0000000	−7.778	0.081
quadratic term	−1.450	0.367	−0.163	1.0000000	−3.946	0.158

Analysis of Variance

Source	Sum-of-Squares	df	Mean-Square	F-Ratio	P
Regression	104.913	4	26.228	145.713	0.062
Residual	0.180	1	0.180		

Table 7.1 Regression Analysis Table for Initial Experimental Region

The insignificant quadratic lack of fit indicates that the optimum lies outside of the initial experimental region. Therefore, the next step is to find a gradient vector which will point us toward the optimum response. This is obtained by taking the derivative of the 1st order model with respect to each of the factors. The gradient vector, g, is

$$ g = \left(\frac{\partial y}{\partial x_1} , \frac{\partial y}{\partial x_2} \right) \Bigg|_{\substack{x_1 = 0 \\ x_2 = 0}} = (4.55, \ -1.45) $$

To experiment along the gradient in reasonable increments, you divide g by the smallest absolute value of the components of g. In this case, we divide by 1.45 indicating that the first new experiment, R_1, should be conducted at (3.14, −1). Subsequent experiments, R_i, will take place at coordinates found by adding 3.14 to the x_1 component and subtracting 1 from the x_2 component. Experimentation along the gradient direction should continue until curvature is detected or until you reach the factor limitations. Then determine the estimated coordinates of the minimum or maximum along the gradient, make these coordinates the center of a new experimental region, conduct a new designed experiment and test for quadratic lack of fit. If the quadratic term is significant, add axial points to estimate the 2nd order model and locate the stationary point.

For our example, the experimental results for R_1, R_2, and R_3 are 91.0, 89.4, and 77.0, respectively. The best response along the gradient appears to occur at R_1, therefore, let the new experimental center point be (3.14, −1. See Figure 7.5). The new design matrix and corresponding response values are shown below.

f_1	f_2	x_1	x_2	y
357	100	4.14	−2.0	87.0
357	150	4.14	0	83.1
257	100	2.14	−2.0	75.7
257	150	2.14	0	86.9
307	125	3.14	−1.0	91.0
307	125	3.14	−1.0	90.1

To convert x_i back to the original values, use the formula

$$x_i = 2\left(\frac{f_i - \bar{f}_i}{f_{max} - f_{min}} \right)$$

For example, $x_1 = 4.14$ results in $4.14 = \dfrac{2(f_i - 150)}{100}$ which simplifies to $f_1 = 357$.

To simplify the model building and analysis for this new experimental region, the design matrix will be altered to reflect R_1 as the center of the experimental region. This is accomplished by the following data table.

w_1	w_2	y
+1	+1	87.0
+1	−1	83.1
−1	+1	75.7
−1	−1	86.9
0	0	91.0
0	0	90.1

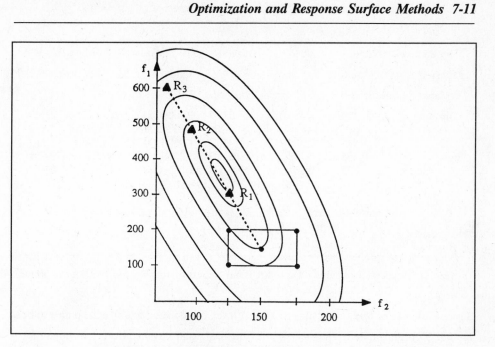

Figure 7.5 Response Surface Contour Plot with Gradient

where

$$w_i = 2\left(\frac{f_i - \bar{f}_i}{f_{i_{max}} - f_{i_{min}}}\right)$$

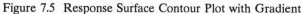

$\bar{f}_i$ = mean for f_i over the new experimental region

$f_{i_{max}}$ = maximum value of f_i over the new experimental region

$f_{i_{min}}$ = minimum value of f_i over the new experimental region

i.e. $\bar{f}_1$ = 307, $f_{1_{max}}$ = 357, $f_{1_{min}}$ = 257

$\bar{f}_2$ = 125, $f_{2_{max}}$ = 150, $f_{2_{min}}$ = 100

Table 7.2 contains the least squares regression analysis of the data.

Variable	Coefficient	Std Error	Std Coef	Tolerance	T	P(2 Tail)
Constant	90.550	0.450	0.000	1.0000000	201.222	0.003
w_1	1.875	0.318	0.299	1.0000000	5.893	0.107
w_2	−1.825	0.318	−0.291	1.0000000	−5.735	0.110
$w_1 \times w_2$	3.775	0.318	0.602	1.0000000	11.864	0.054
quadratic	−7.375	0.551	−0.679	1.0000000	−13.381	0.047

Dependent Variable: y   N: 6   Multiple R: .999   Squared Multiple R: .997

Adjusted Squared Multiple R: .987   Standard Error of Estimate: 0.636

Analysis of Variance

Source	Sum-of-Squares	df	Mean-Square	F-Ratio	P
Regression	156.908	4	39.227	96.857	0.076
Residual	0.405	1	0.405		

Table 7.2 Regression Analysis Table for Final Experimental Region (without α points)

Since the quadratic term is significant at the .1 level, it is decided to add axial points to build the full 2nd order model. The value for α (the axial point distance from the center) is

$$\alpha = (n_F)^{1/4} = (4)^{1/4} = 1.414,$$

where n_F equals the number of runs in the factorial portion of the design. Therefore, the completed data set is shown below.

w_1	w_2	y
1	1	87.0
1	−1	83.1
−1	1	75.7
−1	−1	86.9
0	0	91.0
0	0	90.1
1.414	0	85.4
−1.414	0	77.6
0	1.414	80.5
0	−1.414	77.4

A graphical display is shown in Figure 7.6. The computer output for the fitted model is shown in Table 7.3.

Dependent Variable: y	N: 10		Multiple R: .930		Squared Multiple R: .865	
Adjusted Squared Multiple R: .695				Standard Error of Estimate: 3.019		
Variable	Coefficient	Std Error	Std Coef	Tolerance	T	P(2 Tail)
Constant	90.549	2.135	0.000	1.0000000	42.412	0.000
w_1	2.316	1.068	0.399	1.0000000	2.170	0.096
w_2	−0.365	1.068	−0.063	1.0000000	−0.342	0.750
$w_1 \times w_2$	3.775	1.510	0.460	1.0000000	2.501	0.067
$(w_1)^2$	−3.787	1.412	−0.546	.8164322	−2.681	0.055
$(w_2)^2$	−5.063	1.412	−0.730	.8164322	−3.584	0.023

Analysis of Variance

Source	Sum-of-Squares	df	Mean-Square	F-Ratio	P
Regression	232.776	5	46.555	5.107	0.070
Residual	36.465	4	9.116		

Table 7.3 Region Analysis Table for Final Experimental Region (with α points)

Using a significance level of .10, the parsimonious model is

$$\hat{y} = 90.549 + 2.316w_1 + 3.775w_1 \times w_2 - 3.787(w_1)^2 - 5.063(w_2)^2.$$

The stationary point is found by differentiating $\hat{y}$ with respect to both factors and solving simultaneously for w_1 and w_2.

$$(1) \quad \frac{\partial \hat{y}}{\partial w_1} = 2.316 + 3.775w_2 - 7.574w_1 = 0$$

$$(2) \quad \frac{\partial \hat{y}}{\partial w_2} = -10.126w_2 + 3.775w_1 = 0$$

Now, substituting this value in (2) results in $w_1 = 2.682w_2$.
Substituting this for w_1 in (1) results in

$$3.775w_2 - 7.574(2.682)w_2 = -2.316 \quad \text{or} \quad w_2 = 0.140.$$

Now, substituting this value in (2) results in $w_1 = .375$. Thus, our stationary point is estimated to be at $(w_1, w_2) = (.375, .140)$.

Define B as the second order matrix shown below:

$$B = \begin{bmatrix} b_{11} & \dfrac{b_{12}}{2} \\ \dfrac{b_{21}}{2} & b_{22} \end{bmatrix} \qquad \text{where}$$

b_{11} is the coefficient of $(w_1)^2$

b_{22} is the coefficient of $(w_2)^2$

$b_{12} = b_{21}$ is the coefficient of $(w_1 \times w_2)$

The eigenvalues of B are -6.419 and -2.433. Since both eigenvalues are negative, the stationary point $(w_1 \times w_2) = (.375, .140)$ is a maximum. The untransformed coordinates of the stationary point are (325.75, 128.5). See Figure 7.6 for a summary of the experimentation used to find the optimum.

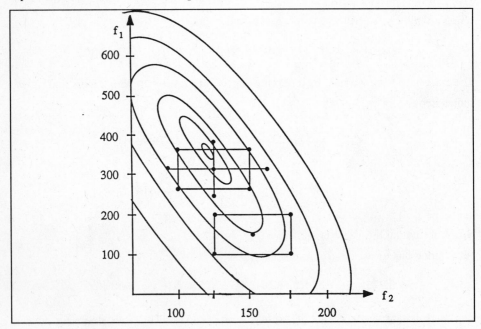

Figure 7.6 Response Surface Contour with Initial and Final Experimental Regions

The following table of possible eigenvalues summarizes the characteristics of stationary points.

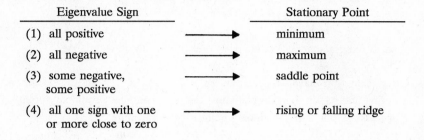

Eigenvalue Sign		Stationary Point
(1) all positive	⟶	minimum
(2) all negative	⟶	maximum
(3) some negative, some positive	⟶	saddle point
(4) all one sign with one or more close to zero	⟶	rising or falling ridge

For a more indepth study of RSM, see Box and Draper [1], Montgomery [2] or Myers [3].

Chapter 7 Problem Set

1. When should you use Response Surface Methodology? What findings in ANOVA indicate that you should use RSM?

2. What is the gradient vector and what does it have to do with RSM?

3. Given that you have searched along the gradient and have selected a new experimental region, how do you know this area contains a stationary point?
    How can you tell if that stationary point is a: maximum?
                                                        minimin?
                                                        saddle point?
                                                        a ridge?

4. State the main reason for using coded variables in regression analysis.

5. If the gradient vector from a 2 term model is:
    $$g = (-2.67, 1.33)$$
    Where should you conduct the first experiment for RSM? Find the second test values as well. If the second test value yields the maximum, what values should you use in a L_4 design?

Chapter 7 Bibliography

1.   Box, G. E. P. and Draper, N. R. (1987), *Empirical Model Building and Response Surfaces*, John Wiley and Sons, Inc., New York.

2.   Montgomery, Douglas C. (1984), *Design and Analysis of Experiments* (2nd ed.), John Wiley and Sons, Inc., New York.

3.   Myers, Raymond H. (1976), *Response Surface Methodology*, Virginia Polytechnic Institute.

Chapter 8

Case Studies

8.1 Introduction

This chapter contains a collection of real industrial case studies, simulated studies, and training type case studies. The complexity of these studies ranges from the fairly simple to the state of the art. The studies have not been altered by the authors of this text and thus it is very possible that many of us would have conducted a similar study using a different approach. The purpose of this chapter is not to critique each case study, but rather to document the types of applications that exist. If the reader has serious concerns about any particular study, please share these concerns with the authors by writing or calling:

AIR ACADEMY PRESS
1155 Kelly Johnson Blvd. Suite 105
Colorado Springs, CO   80920
(719) 531 - 0777
(719) 531 - 0778 (FAX)

The long range plan is to add more case studies to this text and develop a second volume dedicated to applications. If you would like to publish your case study in the next printing of this text, please contact us.

Submitted by: J. Merritt
Honeywell
Colorado Springs, CO

"Results of a First Run Fractional Factorial for the TRE 10X Photo Process on CMOS III Gate Policy"

Background

At the time of this experiment, the author's process was generating an unacceptable rate of rework on the CMOS III 10X stepper process because of poor resolution. In an effort to evaluate this process and determine which process parameters have the most influence on poor resolution, a 2^{8-4} fractional factorial was devised.

Experimental Objective

To detect the relative magnitude of 1st order effects and to screen out less significant factors.

	Controlled Factors	"low" (−1)	"center" (0)	"high" (+1)
(A)	Resist Thickness	12,400 Å	13,700 Å	15,000 Å
(B)	Softbake time	20 min	30 min	40 min
(E)	Develop temperature	19°C	21°C	23°C
(G)	Focus	430	480	530
(C)	Softbake temperature	95°C	100°C	105°
(D)	Develop concentration	3.1:1	3.0:1	2.7:1
(F)	Develop time	80 sec	90 sec	100 sec
(H)	Exposure	200	220	240

Response

The response is a quantitative Critical Dimension (CD). The CD was measured at five points across the wafer for each experimental combination. For this data, the sample mean and sample standard deviation were calculated.

Experimental Array

Design #	A	B	C	D	E	F	G	H
1	−	−	−	−	−	−	−	−
2	+	−	−	−	+	+	+	−
3	−	+	−	−	+	+	−	+
4	+	+	−	−	−	−	+	+
5	−	−	+	−	+	−	+	+
6	+	−	+	−	−	+	−	+
7	−	+	+	−	−	+	+	−
8	+	+	+	−	+	−	−	−
9	−	−	−	+	−	+	+	+
10	+	−	−	+	+	−	−	+
11	−	+	−	+	+	−	+	−
12	+	+	−	+	−	+	−	−
13	−	−	+	+	+	+	−	−
14	+	−	+	+	−	−	+	−
15	−	+	+	+	−	−	−	+
16	+	+	+	+	+	+	+	+
*17	0	0	0	0	0	0	0	0

NOTE: Multiple centerpoints were run so as to obtain an experimental error estimate. This estimate was used to sort out important from unimportant effects for both the mean and standard deviation.

Experimental Conduct

The experiments were all conducted during a 10 hour period. Randomization was conducted where it was possible.

Calculation of Effects

Table I
Mean CD

Factor	Description	Effect (in order of absolute magnitude)
A	Resist Thickness	.504
H	Exposure	−.401
B	Softbake Time	.389
F	Develop Time	−.386
A × C	Resist Thickness × Softbake Temp	−.369
A × B	Resist Thickness × Softbake Time	−.359
C	Softbake Temp	.309
E	Develop Temp	−.301
B × C	Softbake Time × Softbake Temp	−.129
G	Focus	−.121
D × E	Develop Concentration × Develop Temp	.111
C × D	Softbake Temp × Develop Concentration	.069
A × D	Resist Thickness × Develop Concentration	−.059
D	Develop Concentration	.039
B × D	Softbake Temp × Develop Concentration	.024

Calculation of Effects

Table II
Standard Deviation of CD

Factor	Description	Effect (in order of absolute magnitude)
C	Softbake Temp	−.117
A	Resist Thickness	.109
B × C	Softbake Time × Softbake Temp	−.090
H	Exposure	−.084
A × B	Resist Thickness × Softbake Time	−.070
D	Develop Concentration	.066
C × D	Softbake Temp × Develop Concentration	−.063
F	Develop Time	−.052
E	Develop Temp	−.037
A × D	Resist Thickness × Develop Concentration	−.036
G	Focus	.032
B	Softbake Time	.024
D × E	Develop Concentration × Develop Temp	.021
A × C	Resist Thickness × Softbake Temp	−.016
B × D	Softbake Temp × Develop Concentration	.014

Analysis (Mean CD)

From the variation observed in the centerpoints, it was determined that any effect greater than 0.30 would be considered significant. The major surprise for this characteristic was that the Developer Concentration (Factor D) appeared to have little or no effect on the mean CD.

Analysis (standard deviation of CD)

The data in Table II indicates that Softbake Temperature (Factor C) may be a dispersion effect. The sign of the effect indicates smaller variability may result at the higher softbake temperature. Technically, this appears readily explainable since a lower softbake temperature is less efficient in removing solvents from the photoresist.

Conclusions

The experiment was successful in screening out major mean effect factors and interactions, as well as identifying a potential dispersion reduction factor.

Submitted by: Dale Owens
Kurt Manufacturing
Minneapolis, Minnesota

Reduction of Measurement Device Variability Using Experimental Design Techniques

Background

Kurt Manufacturing, headquartered in Minneapolis, Minnesota, is a precision machine shop of considerable size and expertise. Kurt's clients include the nation's elite from aerospace, communications, and computers. Starting in the mid-1980's, Kurt embarked upon an aggressive program to implement gage studies and Process Control techniques across their entire operation. Findings from initial gage studies indicated that in certain situations, the inherent variability of some measurement devices was "large" in relation to the specified engineering tolerances. Kurt's own interval criteria is to use only measurement systems which have an inherent variability of less than 10% of the engineering tolerance.

In what follows, we will learn how Kurt made use of a simple orthogonal array so as to come up with a gaging process with vastly improved variability.

Case Study

When setting up a new process with dedicated gaging, it is not only important to have a capable machine, but a capable gaging device as well. At Kurt, our goal is to have a gage R&R (repeatability and reproducibility) of less than 10% of specification and a process C_{pk} greater than 1.33.

An initial gage R&R study on our device in question yielded a gage error of 80.5% of the characteristic tolerance. Obviously, this was found to be far from acceptable. Accordingly, a team composed of the operator, supervisor, manufacturing engineer, quality engineer, and statistical facilitator met together to determine the variables and levels for a designed experiment.

Extensive brainstorming resulted in the following variables and levels:

(1) Type of indicator

> level 1 = current situation = .0001" increment indicator
>
> level 2 = .00005" indicator

(2) Indicator support

> level 1 = existing special stand
>
> level 2 = height stand

(3) Inspection movement

> level 1 = keep the part stationary and move the indicator
>
> level 2 = keep the indicator stationary and move the part

(4) Type of inspection master

> level 1 = existing part master
>
> level 2 = use gage blocks

(5) Indicating method

> level 1 = bring the indicator straight into the part
>
> level 2 = sweep the indicator into the part from the side

Since this was strictly a screening design, it was decided not to test for two-factor or higher order interactions. After some discussion, the experimental goal was established as "to reduce the amount of gage error." Byproducts of meeting this overall goal would result in reduced cost through a reduction in inspection time and more accurate and precise measurement readings. The experimental objectives of the experiment included determining the effect of factors (1) through (5) on repeatability. Repeatability was quantified through the use of the sample standard deviation.

The team decided to make use of a 2^{5-2} fractional factorial (Taguchi L8). The test plan called for checking one part five times in succession per inspection setup. This approach was used for all eight experimental combinations. The experimental setup was as follows:

Exp #	Increment	Indicator Support	Movement	Master	Method	Result*
1	.0001	tree	part	part	straight	.00004
2	.0001	tree	part	gage	sweep	.00001
3	.0001	stand	indic	part	straight	.00001
4	.0001	stand	indic	gage	sweep	.00005
5	.00005	tree	indic	part	sweep	.00000
6	.00005	tree	indic	gage	straight	.00002
7	.00005	stand	part	part	sweep	.00003
8	.00005	stand	part	gage	straight	.00003

* Result - sample standard deviation of five readings.

By simply "eyeballing" the results, we find that combination #5 is the best. Verification runs by the company showed that this particular combination produced a gage R&R value of 15.2%, a big improvement over previous settings. Efforts were taken by the company to further decrease the gage error.

"Corrosion Study"

The following data was provided to us by an engineer wishing to study an inhibitor of corrosion on a metallic alloy. The questions to be answered in this study were: Is time an important factor? Do corrosion thicknesses change over time? To answer these questions, samples were assessed at zero month, one month, 2 month, and 3 month intervals. Values shown are amounts of material remaining.

Table 1

Zero Months	One Month	Two Months	Three Months
99	104	108	96
114	77	107	68
85	80	69	98
*222	110	sample destroyed	105
127	84	89	104
124	90	124	90

*The value of 222 in column "Zero Months" was discarded for technical reasons. Investigation revealed it was not representative. Because of some additional confusion, only five sample values were available in the "Two Months" column.

Two approaches were taken to answer the question posed. The first approach was to provide a graphical representation of the data. It appears in Figure 1. From this chart we see no clear cut answer to the question. The second assessment approach, a one-way ANOVA resulted in the following statistics:

Source	df	SS	MS	F	"P Value"
Factor	3	1134.9	378.3	1.41	.274
error	18	4844.3	269.1		
total	21	5979.2			

As we can see from this table, at a significance level of .05, there is insufficient evidence to reject the hypothesis that the sample came from populations with equal means.

Based upon the two above analysis approaches, it was concluded there was no major change in corrosion thickness over time.

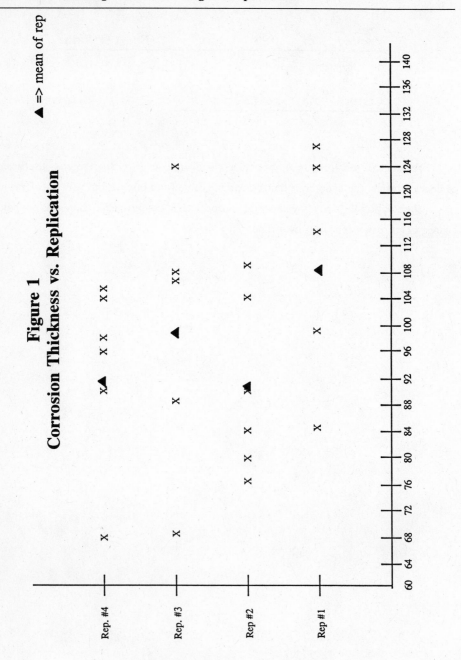

Figure 1
Corrosion Thickness vs. Replication

▲ => mean of rep

Submitted by: Karen Cornwell
                      Digital Equipment Corporation

Linear Motor Design
(Originally presented at the Rocky Mountain Quality Conference, June 1987)

Case Study: The force constant typically drops off approximately 10% on the ends of the motor. To reach optimum seek time, a more linear distribution is required.

Objective: Determine the effect various end cap and motor configurations have on the force constant and leakage flux values.

Analysis: The experiment was conducted using Taguchi's performance parameter (the signal to noise ratio). This made it possible to look at the force constant at both the mean level and variation around the mean. The analysis indicates that the 0.180 groove end cap and standard motor provide the best results.

Results: The improved configuration represents a major improvement over the standard configuration.

I. Introduction:

In an attempt to improve the force constant linearity, the end cap was modified with a groove which created an air gap between the magnet and end cap. In the current configuration, the end cap contacts the magnets. With the addition of the air gap, the magnetic lines of flux will be channeled more toward the center pole rather than through the end cap. The same principle was also attempted with a groove cut at the front of the motor. There is a limited amount of space available for an air gap or additional steel.

The Taguchi analysis technique was employed in developing an optimum motor/end cap configuration which resulted in the maximum force constant with the least variability.

The force constant is measured using a load cell. Current is applied to the coil in the forward and reverse directions. The resulting force is measured on a strain gage. The force constant is simply the measured force divided by the applied current.

II. Experimental Objective

Determine the effect of end cap, front motor cut, and center cut on the dependent variables: force constant and leakage flux.

III. Identification of Dependent Variable(s)

Force Constant

Leakage Flux

IV. Identification of Independent Variables

Noise	Control
Tester Repeatability and Positional Location	A. End Cap
	1. .100 depth groove
	2. .155 depth groove
	3. .180 depth groove
	4. .200 depth groove
	5. .220 depth groove
	6. 44D Stainless Steel
	7. 303 Stainless Steel
	8. Standard
	B. Front Cut Motor
	C. Center Cut Motor

V. Setup of O.A.

We progressed through several experimental stages in conducting the experiment. Briefly, the stages (with applicable results) were as follows:

Stage 1:

We believed that several of the end cap configurations would not produce acceptable results for leakage flux. Because of this, we first tested the leakage flux at six levels for end cap. Results of this experiment were as follows:

End Cap	Leakage Flux (must be < 25)
standard	16
.155	17
.180	18
.200	23
.225	31
standard (303 stainless)	58

From this series of experiments, we found only the first four to have an acceptable leakage flux.

Stage 2:

Using the information from stage 1, we set up an L8 with three control factors. Factor settings for the experiment were:

	Variable	Level 1	Level 2
(A)	end cap	standard	.180
(B)	front	standard	.120
(C)	center	standard	.080

The experimental design with applicable results was:

RUN	A	B	A × B	C	A × C	B × C	A × B × C	S/N*(of force constant)
1	1	1	1	1	1	1	1	11.97
2	1	1	1	2	2	2	2	10.97
3	1	2	2	1	1	2	2	11.08
4	1	2	2	2	2	1	1	10.22
5	2	1	2	1	2	1	2	13.12
6	2	1	2	2	1	2	1	12.10
7	2	2	1	1	2	2	1	12.35
8	2	2	1	2	1	1	2	11.23

*Signal to noise $= 20 \log[\frac{y^2}{s^2}]$, where y^2 and s^2 were calculated from 8 data points.

MARGINAL MEAN RESULTS

FACTOR	LEVEL 1	LEVEL 2	|Δ|
A	11.06	12.20	1.14
B	12.04	11.22	.82
A × B	11.63	11.63	0.00
C	12.13	11.13	1.00
A × C	11.60	11.67	.06
B × C	11.64	11.60	.01
A × B × C	11.66	11.60	.06

From the above, it appears that end cap at 0.180 cut is best, a standard front is best, and a standard center is best. (Note: Leakage flux was acceptable for all combinations tested, i.e.: the average was well below the upper specification of 25.)

Stage 3:

In stage 3, we decided to panic a bit. Perhaps we had been premature in going to

2 levels instead of 4 levels on the end cap. Because of this, we set out to conduct a 4 x 3 full factorial using end cap and motor configuration as factors. The design, with resulting S/N ratios, was as follows:

MOTOR

		Center Cut	Front Cut	Standard
	standard	10.79	11.08	11.83
END	.100	11.24	11.66	12.65
CAP	.155	12.11	12.32	12.69
	.180	12.10	12.35	13.17
	.200	12.09	12.11	12.45

A graphical comparison of the results is provided in Figure 1. As is apparent from the graph, the standard motor with the 0.180 groove is the best performer.

Stage 3A:

After extensive discussion, we decided to re-analyze the raw data using the standard deviation. Our data table appeared as follows:

END CAP	CENTER CUT	FRONT CUT	STANDARD
standard	.041	.039	.030
.100	.037	.034	.025
.155	.031	.028	.025
.180	.031	.026	.022
.200	.031	.026	.026

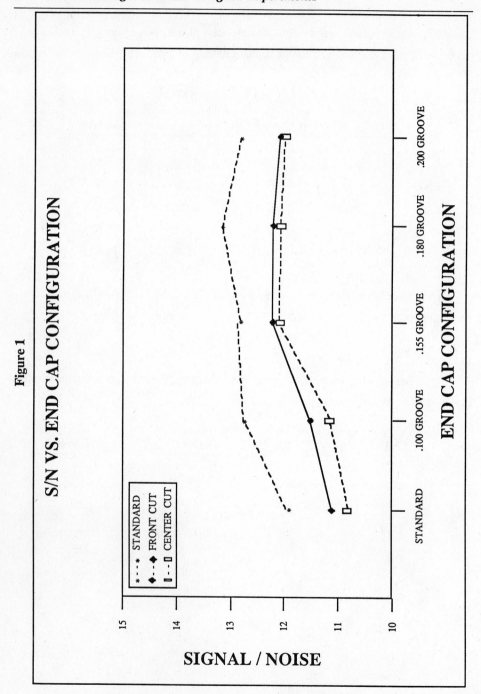

Figure 1

Graphically, these results were as follows:

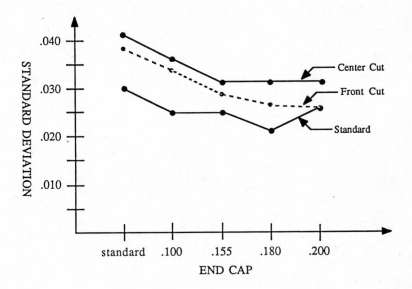

Analysis using the standard deviation indicates that we have the least dispersion with a 0.180 end cap and a standard cut motor.

Stage 4:

We conducted verification testing of the predicted best combinations. Results obtained were in agreement with our predicted best combination.

Conclusions:

Based upon the above findings, it is recommended to go with the standard motor and the 0.180 end cap.

Submitted by: Doug Sheldon
                     Ramtrom Corporation

"Factorial Experiment for Bottom Electrode Stress"

Abstract: Test Lot XXX was run using a 2-level, 3-factor orthogonally designed experiment to characterize Bottom Electrode film stress. The results show that current factor A thickness (200) is too thin for good, controllable adhesion and must be made thicker.

Factors and Levels:

Factor	"Low"	"High"
A	200	1500
B	500	4300
C	no	yes

Responses: Stress and resistance

Experimental Stages:

Stage (1).   Stress and resistance measurements were taken with A and C at all possible combinations (providing us with a 2-level, 2-factor, full factorial design).

Stage (2).   Adding Factor B (3rd process step) at two levels provided us with a 2^3 full factorial design.

Data and Analysis:

Stage (1).   Table I lists the applicable results after deposition of factor A.

TABLE I

	Factor		Stress		Resistance	
Run	A	C	x_i	$\bar{x}$	y_i	$\bar{y}$
1	200	no	−10, −1.26	−5.63	40, 101	70.5
2	1500	no	1.31, 1.23	1.27	6, 9.5	7.75
3	200	yes	7.76, 14.4	11.08	111, 82.5	96.75
4	1500	yes	1.47, 2.43	1.95	7, 10.5	8.750

Parameter	Stress Effect	Resistance Effect
A	−1.115	−75.375
C	8.690	13.625
A × C	−8.015	−12.625

For "stress", note that the A × C interaction effect is the second largest of the three. This means the effect of factor A on stress cannot be considered independently of C. Figure #1 provides a graphical representation of the interaction. At thin values of factor A, the stress varies greatly. At thick values for A, there is less difference in stress regardless of whether C is in or out of the experiment.

For resistance, the most significant factor is A. As factor A thickness increases to 1500, the resistance drops an order of magnitude. The fact that the thin factor A values have a large mean with relatively high dispersion indicates a good factor A layer may not be forming. Resistance values with factor A set at 1500 seems to have much less dispersion. The effect of having factor C present is to increase resistance.

Stage (2).   Once factor A was deposited and measured, factor B was deposited and stress was measured. The applicable data with resultant effects are shown in Table II.

TABLE II

Run	A	B	C	Stress	Parameter	Effect
1	200	500	no	1.67	C	−.22
2	1500	500	no	2.3	B	−.06
3	200	4300	no	.85	C × B	.40
4	1500	4300	no	2.2	A	1.41
5	200	500	yes	.53	A × C	.42
6	1500	500	yes	2.2	A × B	.26
7	200	4300	yes	.72	A × B × C	−.10
8	1500	4300	yes	2.7		

From the above, only factor A appears to stand out. The interaction terms, as well as factors A and C, appear to be of no importance. Figure #2 provides a graphical representation of the effects.

Conclusions:

The current thickness of factor A (200) appears to be too stressful and prone to great dispersion. It is recommended the setting for factor A be increased to 1500. With a setting of 1500 for factor A, the occurrence of factor C is not required. If, however, we go to 1500 for factor A, we will need to reduce factor B. The lower limit of factor B would be where possible factor A interdiffusion into factor B becomes a problem. Results from previous tests indicate it is possible to run with very thin layers of factor B.

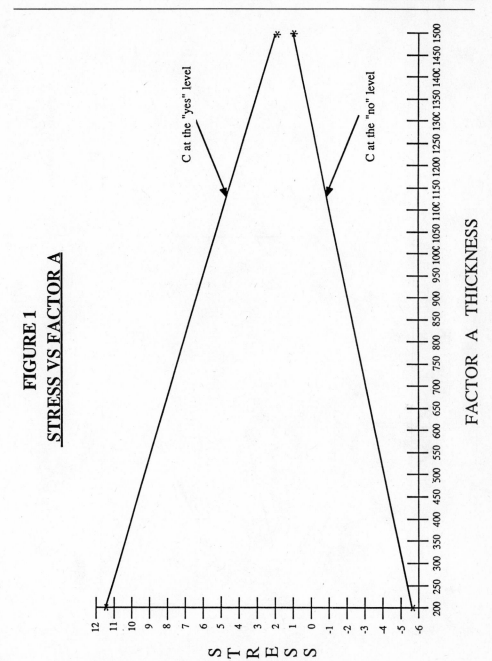

FIGURE 1
STRESS VS FACTOR A

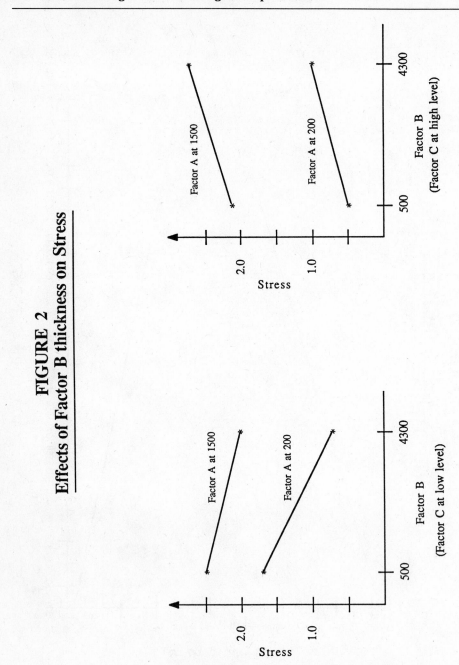

FIGURE 2
Effects of Factor B thickness on Stress

"Experimental Design on an Injection Molded Plastic Part"

Background: Shrinkage is a major concern with some injected molded parts. Typically, a molded die for a part will be built oversize to allow for part shrinkage after it has been produced. In the following situation, a new die had been produced for the production of a part. What are the proper process settings to produce a product with the proper dimensions?

Objective: Determine the effect of injection velocity, cooling time, barrel zone temperature, mold temperature, hold pressure, and back pressure on shrinkage of part XXXX (as defined by the 9.380 and 14.500 dimensions).

Brainstorming: The brainstorming process lasted approximately four hours. Attendees included a process engineer, process supervisor, quality control inspector, quality control supervisor, and a person with some rudimentary training in experimental design. Barrel zone temperatures was a tough one. Barrel zones actually had six different control points. Some discussion was conducted as to whether we should first run a series of experiments about zone temperatures. After a series of discussions, it was concluded to look at only a "low" and "high" group of settings for barrel zone temperatures in the screening design chosen.

Design Type: An eight run fractional factorial at 2 levels (Taguchi L8) was chosen. The controlled factors with applicable levels chosen were:

	Control Factor	Low	High
A	Injection Velocity	1.0	3.1
B	Cooling Time	40 sec	50 sec
C	Barrel Zones	"low temps"	"high temps"
D	Mold Temperature	100	150
E	Hold Pressure	200	1100
F	Back Pressure	50	150

We decided to look only for main effects in this particular design.

Test Plan: We chose to run all experiments during the same shift using raw materials from the same lot and containers. We also chose to use only "fresh" raw materials. No reground product was to be used. Additionally, our test procedure called for cycling in the settings, running until these settings were stable at the specified settings, running 10 shots, and then running the parts for the experiment. Five experimental parts were measured at each design setting with applicable dimensional measurements recorded. Since factor D was a factor which was very time consuming to change, we assigned it to the first column of our array, then randomized the run order within the two levels of factor D.

Design and Resultant Data

Design #	D	C	A	B	E	F	Length
1	1	1	1	1	1	1	14.5000, 14.5005, 14.5000, 14.5000, 14.5005
2	1	1	1	2	2	2	14.5075, 14.5090, 14.5070, 14.5065, 14.5065
3	1	2	2	1	1	2	14.5045, 14.5050, 14.5045, 14.5045, 14.5045
4	1	2	2	2	2	1	14.5100, 14.5105, 14.5105, 14.5110, 14.5105
5	2	1	2	1	2	1	14.5105, 14.5110, 14.5105, 14.5120, 14.5100
6	2	1	2	2	1	2	14.5045, 14.5055, 14.5065, 14.5050, 14.5050
7	2	2	1	1	2	2	14.5150, 14.5140, 14.5155, 14.5150, 14.5145
8	2	2	1	2	1	1	14.5055, 14.5065, 14.5055, 14.5055, 14.5060

Design #	D	C	A	B	E	F	Width
1	1	1	1	1	1	1	9.3575, 9.3560, 9.3570, 9.3585, 9.3590
2	1	1	1	2	2	2	9.3650, 9.3640, 9.3640, 9.3640, 9.3645
3	1	2	2	1	1	2	9.3545, 9.3545, 9.3545, 9.3550, 9.3540
4	1	2	2	2	2	1	9.3630, 9.3630, 9.3625, 9.3635, 9.3635
5	2	1	2	1	2	1	9.3555, 9.3560, 9.3560, 9.3555, 9.3560
6	2	1	2	2	1	2	9.3580, 9.3565, 9.3550, 9.3540, 9.3545
7	2	2	1	1	2	2	9.3600, 9.3580, 9.3585, 9.3590, 9.3585
8	2	2	1	2	1	1	9.3565, 9.3560, 9.3565, 9.3565, 9.3560

Analysis:

Table of Average Marginals
(14.5 - nominal, length)

Level	D	C	A	B	E	F
1	14.5057	14.5059	14.5070	14.5076	14.5040	14.5068
2	14.5092	14.5089	14.5078	14.5072	14.5109	14.5080
|Effect|	.0035	.0030	.0008	.0004	.0069	.0012

Table of Average Marginals
(9.380 - nominal, width)

Level	D	C	A	B	E	F
1	9.3599	9.3583	9.3593	9.3567	9.3560	9.3582
2	9.3566	9.3582	9.3573	9.3598	9.3605	9.3583
|Effect|	.0033	.0001	.0020	.0031	.0045	.0001

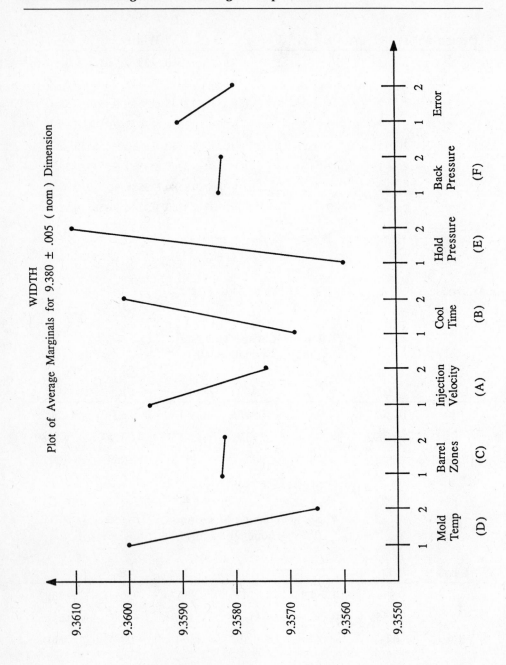

Plot of Average Marginals for 9.380 ± .005 (nom) Dimension

WIDTH

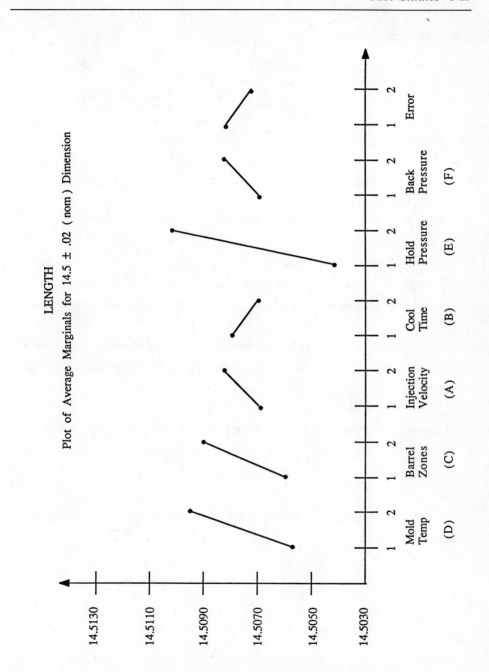

LENGTH

Plot of Average Marginals for 14.5 ± .02 (nom) Dimension

Interpretation of Results

LENGTH	WIDTH
Best settings for 14.5 (nominal)	Best settings for 9.380 (nominal)
Mold temp (1)	Mold temp (1)
Barrel zones (1)	Barrel zones (don't care)
Hold pressure (1)	Hold pressure (2)
Inj. velocity (don't care)	Inj. velocity (1)
Cool time (don't care)	Cool time (2)
Back pressure (don't care)	Back pressure (don't care)

The only major conflict exists on hold pressure. Because of problems with holding the 9.380 (nom) dimension we will select hold pressure as level 2. NOTE: *Under no operating conditions during the experiment were we able to meet the specification on the 9.380 dimension.*

Best conditions appear to happen at mold temperature (1), barrel zones (1), hold pressure (2), injection velocity (1), and cool time (2). Back pressure, with little impact, is best set at level 1.

Findings from Verification Run: Twenty-five verification runs were conducted at the "best conditions" as predicted by our experiment. Under no conditions were we able to meet the criteria established for the 9.380 (nominal).

Conclusions: Based upon the above, it was determined to conduct a redesign of the injection die.

Submitted by: Jack E. Reece, Ph.D.

Consider the Metric
(Are you defining your inputs sensibly?)

1. Introduction

Engineers and scientists eager to try new tools to improve their abilities to solve processing problems readily adapt to the matrix approach to experimentation required by statistical experimental design. Once a person understands the structures of the factorial, fractional-factorial, or various response surface design matrices, setting up an experiment involving common input factors such as time, temperature, power, catalyst type, etc., becomes almost second nature.

However, some particularly troublesome processes for experimental design are those involving combinations of reactive ingredients in a fixed or flowing environment. One example includes the mixture of gases used to deposit a coating or to etch a given layer selectively in the semiconductor industry. Another might be a batch or flow process in a chemical industry involving a mixture of reactants in either a static or flow reactor.

Most texts on experimental design devote most of their attention to the justification of the technique of experimental design and to the statistics and analysis of results. Few devote other than passing reference to assisting the user in defining the metric for these cases. The experimental design literature has generally treated these mixture experiments as a special class and devised special treatments for them (see, for example, "Experiments with Mixtures" by John A. Cornell, New York, Wiley, 1981). The transformations of input metrics described below can accommodate a wide variety of experimental situations.

2. Dealing with Mixtures

2.1 Experiments with Photographic Emulsions:

Conventional photographic emulsion formulations contain light sensitive elements, chemical and photo sensitizers, binders, and other materials. Developing an image in the systems requires treating the exposed film with an external developing chemistry.

At one time the author conducted product development research on a family of photographic emulsions whose entire chemistry (including the developers) resided in the light sensitive coating. Developing an image in an exposed coating for these materials required only heat.

The developing chemistry required a hindered phenolic reducing agent (A) and another agent generally considered a "co-developer" or "development accelerator" (B). The system worked best when the emulsion contained an excess of the phenolic compared to the accelerator. However, too much or too little of the phenolic in the emulsion gave unacceptable image properties; too much of the combination might cause the entire unexposed area to darken upon development; too little gave very slow development with poor image density. Proper balance of these ingredients allowed the "tuning" of the emulsion for specific photographic applications.

In attempting to optimize the performance of this emulsion for a particular application, the author decided to use the amount (or concentration) of the two materials in a fixed batch and their ratio as factors in the design rather than to manipulate the amount of each in the emulsion separately. This approach produced two equations with two unknowns:

$$A + B = \text{Total weight in fixed batch (X1)}$$
$$A/B = \text{Ratio (X2)}; \quad A = B \times X2$$

$$B \times X2 + B = X1$$
$$B = X1/(1 + X2)$$
$$A = X1 - B$$

Where A and B are the amounts of developer and are actually added to the emulsion, and X1 and X2 are the factors included in the design matrix. Using this approach obviously requires knowledge of the likely functional limits of each factor.

This relationship provides the following two-factor matrix in X1 and X2; solving the algebraic expression above provides the actual amounts of each reactant to add:

X1	X2
−1	−1
+1	−1
−1	+1
+1	+1

Thus the design is orthogonal in X1 and X2, but not with respect to the weights of each ingredient actually added to each batch.

The logic of this approach is that it considers the stoichiometry (chemistry) of the system and includes the concentration of these two reactions and their "pseudo-molar" ratios. The practical significance of the approach is that it provided the author a convenient metric for managing these two variables and optimizing the emulsion characteristics.

2.2 Experiments Involving Gas Flows:

2.2.1 Two Reactive Gases:

An extension of the approach described above provides a means for controlling two reactive gases flowing into a reaction chamber to provide a product or to modify (etch, for example) a layer on a substrate. In this example G1 and G2 have flows measured in SCCM (Standard Cubic Centimeters per Minute) or equivalent units. A logical approach to characterizing this system would use a range of the actual flow of each gas in the design matrix.

Based on his experience in optimizing photographic emulsions, the author prefers to use a "chemical" approach to this problem. The usefulness of the method has been demonstrated in actual practice, in that models prepared in this fashion have shown smaller residual sums of squares, less lack of fit sums of squares, and fewer outlier or maverick points in an investigation. Models prepared this way better explain the relationships between inputs and outputs in the system being investigated.

The equations are adaptations of those shown above:

$$G1 + G2 = \text{Total Flow (TF)}$$
$$G1/G2 = \text{Ratio (R)}$$

TF and R become the factors in the experiment, and one calculates the flow settings used to investigate the process as demonstrated above.

2.2.2 More than Two Gases (All Reactive):

Additional gases necessarily complicate the algebra necessary to solve for the actual flow. The example below assumes we must explore three gas flows, all of which play a reactive part in the process.

$$G1 + G2 + G3 = \text{Total Flow (TF)}$$
$$G1/G2 = \text{Ratio1 (R1); } G1 = G2 \times R1$$
$$G3/G2 = \text{Ratio2 (R2); } G3 = G2 \times R2$$

$$G2 \times R1 + G2 + G2 \times R2 = TF$$
$$G2 = TF(1 + R1 + R2)$$

Choosing one of the gases as the denominator in both of the ratio expressions simplifies the algebra somewhat; one convenient approach uses the major component of the gas mixture as that denominator.

2.2.3 Several Reactive Gases and a Carrier (or Diluent):

A further complication occurs when one of the gases in a mixture of three or more is a diluent or carrier. Here we can conveniently use the ratios of reactants as well as the *concentration* of reactants in the stream:

$$G1 + G2 + G3 + D = \text{Total flow (TF)}$$
$$G1/G2 = \text{Ratio1 (R1); } G1 = G2 \times R1$$
$$(G1 + G2 + G3)/TF = \text{Concentration (R3)}$$

$$(G2 \times R1 + G2 + G2 \times R2)/TF = R3$$
$$G2 = R3 \times TF(1 + R1 + R2)$$

Using the expression for the concentration of the reactants both simplifies the algebra and probably better describes the relationship among inputs and outputs.

2.3 Chemical Reactions in Batches:

Consider the case in which a chemist wishes to design experiments to study the reaction of two species in a vessel to produce a particular product. In this case the vessel has a finite capacity, commonly expressed in volumetric or weight measure. Most chemists prefer to consider reaction components in terms of *mole ratios* of the reactants based on some presumed equation for the chemistry of the process. The moles of reactant X1 = k_1 × wt X1; moles of X2 = k_2 × wt X2, where k_1 and k_2 are constants whose values depend on the respective molecular weights of X1 and X2. The mole ratio X1/X2 = k_1 × wt X1/k_2 × wt X2 or k_3 × (wt X1/wt X2). Therefore, one may use the raw weights of each reactant, understanding that multiplying that ratio of weights by k_3 would yield the actual mole ratio.

Quite commonly, chemical reactions in a batch reactor require an inert solvent or diluent -- note that similarity between this problem and the one described in the section above. So, another concern is the concentration of the reactive species in the reactions mixture. Assume for this example that the maximum weight allowed in this reaction vessel is 1,000 units. Then:

$$X1 + X2 + S = 1000$$
$$X1/X2 = \text{Ratio1 (R1)}; \quad X1 = X2 \times R1$$
$$(X1 + X2)/1000 = \text{Concentration (R2)}$$
$$(X2 \times R1 + X2)/1000 = R2$$
$$X2 = R2 \times 1000/(1 + R1)$$
$$X1 = X2 \times R1$$
$$S = 1000 - (X1 + X2)$$

Thus the controlled factors in this experiment are the ratios of the reactants to each other and their concentration in the system.

3. Dealing with Temperature Gradients:

Figure 1 (page 39) represents a tube furnace similar to those used in the semiconductor industry to prepare coatings on silicon wafers, for example. Among the variables one might manipulate in characterizing the performance of this device are the three temperature zones B1, B2, and B3. In this particular furnace configuration reactive gases enter through the ports J in the end cap C, and proceed down the furnace to the exhaust pump outlet B. Because the process depletes the reactive species in the gas flow, "tilting" the furnace temperature from low to high from the entry to the exit is a common practice.

One might decide to investigate this process using entry temperatures that average 20 degrees cooler than the center temperatures, and exit temperatures 20 degrees warmer:

Entry -- 790 to 810

Center -- 810 to 830

Exit -- 830 to 850

The two-level three-factor matrix that results is:

Entry	Center	Exit
790	810	830
810	810	830
790	830	830
810	830	830
790	810	850
810	810	850
790	830	850
810	830	850

While the general trend of the temperature settings is higher toward the exit in each run, the distribution of settings produced contains center temperatures equal to entry and center temperatures equal to exit. The author suggests an alternative metric: consider the center zone the key or master zone and *offset* the other zones with respect to it.

Entry -- −30 to −10
Center -- 810 to 830
Exit   -- 10 to 30

Entry	Center	Exit
-30(780)	810	10(820)
-10(800)	810	10(820)
-30(780)	830	10(840)
-10(800)	830	10(840)
-30(780)	810	30(840)
-10(800)	810	30(840)
-30(780)	830	30(860)
-10(800)	830	30(860)

Now each run contains an upward temperature bias from entry to exit; the severity differs from run to run. The design is orthogonal in the offset settings, but not in the actual temperatures run.

4. Summary:

The metrics described in the preceding sections ultimately have provided useful regression models describing the behavior of one or more outputs as a function of these inputs. These metrics by no means represent all those an investigator might select to use rather than the obvious one.

Experiments and the time necessary to conduct them are expensive. Experiments using the matrix approach of statistical experimental design such as factorials, etc., can make our investigations much more cost effective and provide more reliable information than investigations done by some other method. But the engineer or scientist can improve on the results obtainable from a statistical experimental design if she or he will take the time to step back from a problem and consider alternative, possible novel, metrics for input variables before starting the experiments. Conversion of factors to a new metric *after* all runs are complete may give useful information, but the non-orthogonality of the matrix resulting in that case may cast doubt on the validity of any regression models derived.

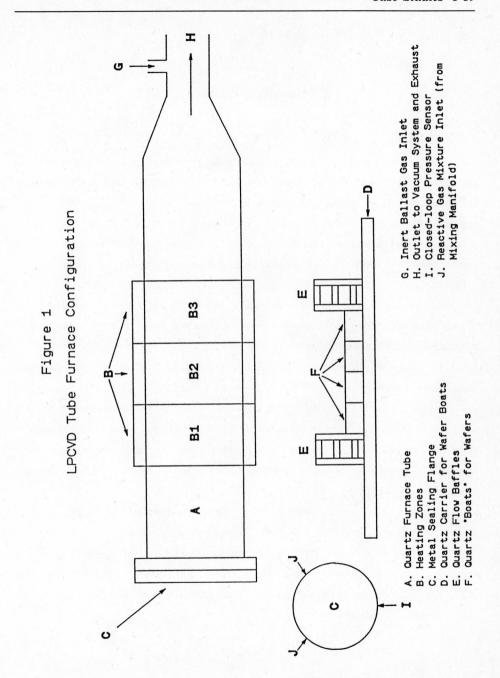

Figure 1

LPCVD Tube Furnace Configuration

A. Quartz Furnace Tube
B. Heating Zones
C. Metal Sealing Flange
D. Quartz Carrier for Wafer Boats
E. Quartz Flow Baffles
F. Quartz "Boats" for Wafers

G. Inert Ballast Gas Inlet
H. Outlet to Vacuum System and Exhaust
I. Closed-loop Pressure Sensor
J. Reactive Gas Mixture Inlet (from Mixing Manifold)

Submitted by: Jack E. Reece, Ph.D.

Minimizing Variability

1. Introduction:

The techniques of statistical experimental design evolved from the work led by Sir Ronald A. Fisher during World War I at Rothamstad Agricultural Experiment Station in London as he and his colleagues strived to increase crop yields. Generally scientists and engineers using these methods concentrated most commonly on the "location" of a process, trying to increase yields, decrease byproducts, or understand a process enough to set and control inputs to maintain a desired result or combination of results (1,2).

Particularly in recent years the widespread teaching of so-called "Taguchi Methodology" has focused attention not only on the location of a process, but also on its dispersion or variability (3). Within that methodology are a number of metrics designed to describe variability and location in one parameter. Classical statisticians have attacked these metrics, creating considerable confusion among "lay practitioners" of the art of statistical experimental design (4).

Engineers tend to focus on the "bottom line". What interests them are methods they can exploit to find answers in terms they and their managers can understand. This paper discusses the analysis of the results of a designed experiment in the semiconductor industry in which the investigators sought to understand the factors controlling coating uniformity during a batch process (5).

The study was successful using one of the metrics described below. Having obtained access to an advanced experimental design package called RS/Discover Version 2.0 (marketed by BBN Software Products Corporation, Cambridge, MA) which supports a variety of "performance statistics" and which contains a sophisticated multiple regression utility, the author elected to re-examine that data to compare predictions of optimal regions from several such statistics.

2. The Process:

Figure 1 (page 39) illustrates the furnace used to prepare the samples for the experimental study.

One inserts a batch of silicon wafers mounted on edge in the boats "F" on the quartz carrier "D" and places them in the furnace. The objects "E" on the carrier are quartz baffles which provide turbulent flow through the furnace and which have thermal mass to preheat the reactive gases before they contact the wafers. From a mixing manifold (not shown) the gas mixture containing two reactive species enters at the positions "J" in the end cap "C" and passes through the furnace tube "A" to the exhaust pump and scrubbing system at "H". Heaters around the tube at "B1", "B2", and "B3" heat the gas stream to about 800 degrees centigrade. Metering an inert ballast gas at the port "G" controls the pressure.

In characterizing the process the engineers manipulated six variables: CLP (Closed Loop Pressure), CTMP (temperature of zone B2), Z1OFF (offset of zone B1 with respect to zone B2), Z3OFF (offset of zone B3 with respect to zone B3), TF (total flow of reactive gases), and RAB (ratio of gas A to gas B). Because they expected complex, nonlinear relationships among the input and outputs of the process, they elected to run a Box-Wilson Central Composite design requiring 53 trials to allow estimation of curvilinear effects. Each trial contained 12 pilot wafers (among the rest used for ballast); and each boat contained 3 pilots. The engineers gathered 10 thickness measurements from each wafer in each run, for a total of 6360 thickness measurements.

The experimental objective was to identify a process window which provided reasonable deposition rates with minimum variation within or among wafers.

3. Performance Statistics:

3.1 Taguchi's "Target is Best":

As stated above, the experimental objective is to get the process on target with minimum variability. Taguchi suggests a parameter called "Target is Best" or SN_T for this purpose. In mathematical terms this parameter is:

$$10\log_{10}(\bar{y}^2/s^2)$$

where $\bar{y}$ is the average value observed in each run and s is the observed standard deviation in each run.

Taguchi uses the parameter to define the optimum operating point for the process in that the combination of factor settings which maximizes this function defines the optimum. Classical statisticians object to the use of this statistic because it blends the mean with a measure of variability. The absence of close correlation between the variability and the mean causes this statistic to lose information compared to separate measurements of mean and variability in that case.

The form of the statistic reflects Taguchi's conversion of these numbers to "decibel" readings because of his engineering background. This statistic is actually a variant of the classical coefficient of variation statistic:

$$CV = s/\bar{y}$$
$$10 \times \log_{10}(CV) = 10 \times \log_{10}(s/\bar{y})$$
$$-20 \times \log_{10}(CV) = 20 \times \log_{10}(\bar{y}/s)$$
$$= 10 \times \log_{10}(\bar{y}^2/s^2)$$

Thus Taguchi's "Target is Best" statistic is the same as $-20 \times \log_{10}CV$.

3.2 Performance Statistic Based on Range:

A common metric used for expressing the "uniformity" of a variety of different kinds of measurements within a group of measurements in the semiconductor industry is:

$$UNIF = range/(2 \times \bar{y}) \times 100$$

This is also a variant of the coefficient of variation CV in that for small samples one may estimate the standard deviation by dividing the range by 4:

$$UNIF = (2/2) \times (range/2) \times (1/\bar{y}) \times 100$$
$$= 2 \times (range/4) \times (1/\bar{y}) \times 100$$
$$= 2 \times s/\bar{y} \times 100 = 2 \times CV \times 100$$

The author recommends against using the statistic based on the range because of the influence outlier-prone data can have -- the standard deviation is far more robust to outliers that the range. Because the natural limit of either statistic is 0, one should transform the response to preclude negative predictions before attempting to prepare a regression model; common transforms are the logarithm, square root, inverse, inverse square root, etc.

3.3 Performance Statistic Based on the Standard Deviation:

Dr. George Box of the University of Wisconsin, Madison, and his colleagues have published a series of papers dealing with modeling variability independent of the mean or location. They recommend:

$$-\log10(S)$$

At once you can see that this is a very simple metric, requiring the calculation of the standard deviation of each set of observations and finding the logarithm of each standard deviation. The negative sign causes this function to follow the "Target is Best" statistic in that one seeks to maximize its value. The results presented below show how well it performs for the data used here in which the standard deviation appears proportional to the mean.

Another form of this statistic is somewhat more difficult to generate:

1. Find the logarithm of each observation in each group.
2. Determine the standard deviation of each group of logarithms of the observations.
3. Find the logarithm of that standard deviation.

For comparison purposes, the author generated this metric and called it "LGSDLGTHK".

3.4 Other Performance Statistics:

All of the metrics described so far assume that the distribution of the observations for each run is normal, although the transforms in each minimize the impact of non-normality. In the particular case described in this paper, only 17 of the sets of observations of the 53 indicated a normal distribution using the Pearson's Chi Square test. Therefore, metrics independent of the nature of the distribution of the observations can be quite useful in some cases.

One such statistic is the <u>I</u>nter<u>Q</u>uartile <u>R</u>ange (IQR). If one orders observations from low to high and defines the end of the first quartile (25% of the observations) and the end of the third quartile (75% of the observations), the difference between the two numbers is the IQR. The range defined by the IQR represents 50% of the observations. Recall that +/− 1 standard deviation from the mean in a normal distribution represents about 2/3 of the area under the normal curve or 2/3 of the data. Thus one may model the IQR of a group of data or its $-\log_{10}$ and obtain results analogous to the metric recommended by George Box and associates.

Another statistic independent of the distribution is the Median. The median is the middle value in the ordered data when the count of the data is odd; or the average of the two middle values in the ordered data when the count is even. The ratio of the IQR to the Median provides a metric quite analogous to the coefficient of variation and is robust to large departures from normality. Another approach to the problem described here went as follows:

Each run contained 4 wafer boats which held three wafers on which we made 10 measurements of thickness.

Because the distribution of the measurements tended to be non-normal, calculating the mean of each boat helped normalize the distribution.

If we then calculated the standard deviation among the boat means and created a form of the coefficient of variation using the standard deviation and the grand mean of each run, we had a metric describing the relative spread of the data in each boat in each run:

CV' = STDEV (MEANS OF BOATS 1 TO 4 PER RUN)/GRAND MEAN PER RUN

To convert this metric to a measure relatable to the +/− percent variation around a central value, we used the following:

$$BTUNIF = 2 \times CV' \times 100$$

This contrived metric is the one the author used initially in helping engineers model the process discussed here. An important part of its usefulness was that it gave "uniformity" values engineers could understand. It and those based more directly on the coefficient of variation translate to a +/− variation about a central point.

4. About the Software:

BBN Software Products Corporation markets RS/Discover™ and its companion multiple regression analysis package MULREG™. Together these packages provide a comprehensive environment for the design, analysis, and interpretation of statistical experimental designs. In its current form, Version 2.0, RS/Discover supports experiments having inner and outer arrays and calculates a variety of performance statistics automatically when requested by the user.

The MULREG package associated with RS/Discover 2.0 is extremely useful in that it incorporates the Box-Cox system of response transformations and can advise the user on which is statistically appropriate. Furthermore it supports both least squares and robust (bisquare) regression and can transform inputs based on variance inflation factors associated with a particular factor's settings. To give each metric the best opportunity possible to describe the variability seen in this study the author used every utility the MULREG routines indicated were statistically appropriate.

The author completed the initial study of this process using RS/Discover 1.2 and its accompanying multiple regression utilities. These products were the state of the art for menu-driven, user-friendly statistical packages in 1987, but did not support the sophistication and convenience available in 1989. Nor did they support experiments with outer arrays.

One could consider the 120 thickness measurements in each experiment part of an outer array. While they are not true "experiments" or "replicates", they provide an estimate of how the thickness of material deposited varied from one end of the furnace to another. The author created a suitable experiment within the software, then inserted the data from the previous experiment into the tables created. The system calculated the performance statistics "Target-is-Best", IQR, and LogS.

The software also allows the user to define custom performance statistics whose metric is more familiar in his environment. Examples of these include the "BTUNIF", IQR/Mean, UNIF, and LGSDLGTHK discussed above. In these cases the system provides space for them in the worksheet generated, but the user must calculate or insert them.

5. Consider the Data:

Space considerations prohibit the display of the entire data base. Rather, Tables 1 and 1a show the conditions of each run, the mean thickness observed in each wafer boat, the grand mean thickness for each run (as well as the median), stdev among boats, minimum and maximum thicknesses, standard deviation of each run, IQR, and standard deviation of the logarithm base 10 of thicknesses within each run.

Figure 2 plots the observed means of the boats versus the run number. The error bars on this graph represent the minimum and maximum thickness observed. From it one can see that normally the wafers decreased in thickness as boat number increased (boat 1 was near the entrance; boat 4 near the exit); but not always. Some of the runs had very tight distribution of thicknesses -- the purpose of regression modeling is to find the simplest expression relating the input factors to the observed distribution of thicknesses.

6. The Regression Models:

We intend to model variability independent of the mean or median of the response. The small table at the bottom of the next page shows the linear correlation coefficients between the mean thickness value observed and the metric created to estimate variability. Note the $-\log_{10}(S)$ statistic correlates at about -0.45 with the mean -- this translates to an R^2 of about .20. With this amount of data, the correlation is probably statistically significant. Note that those metrics involving a ratio of the standard deviation to the mean correlate less well with the mean. We can decide the importance of this correlation when we compare how well each statistic identifies the region of minimum variability in the process. (Incidentally, the previous results using older software identified such a region, and confirming runs proved it.)

Correlation Matrix
Location and Variability Metrics

Term	Mean Thickness
Mean Thickness	1.000
Median Thickness	0.996
-LOG10(S)	-0.449
IQR	0.432
2*SD/Mean*100	0.243
Target is Best	-0.196
LOG10(SD(LOG10(THICK)))	0.190

Table 1
Factor Settings

EXP NUM	CLP	TOT FLOW	RAB	CTMP	Z1 OFF	Z3 OFF	EXP NUM	CLP	TOT FLOW	RAB	CTMP	Z1 OFF	Z3 OFF
1	425	350	5	790	-20	20	28	425	350	5	790	-20	20
2	478	287	4	786	-24	24	29	478	287	6	786	-24	16
3	372	287	6	794	-16	24	30	478	287	6	786	-16	24
4	478	287	6	794	-24	24	31	478	413	4	794	-24	24
5	372	413	6	786	-24	16	32	478	287	4	794	-24	16
6	478	413	6	794	-16	24	33	425	350	3	790	-20	20
7	425	350	5	790	-20	20	34	372	413	4	794	-24	16
8	478	413	6	786	-24	24	35	425	350	5	790	-30	20
9	372	287	4	794	-24	24	36	372	287	4	786	-16	24
10	478	413	4	786	-16	24	37	372	413	4	786	-16	16
11	372	413	6	794	-24	24	38	425	350	5	790	-20	20
12	425	350	5	790	-20	30	39	478	287	6	794	-16	16
13	425	350	5	790	-20	20	40	372	287	4	794	-16	16
14	372	413	4	794	-16	24	41	372	413	4	786	-24	24
15	372	287	6	786	-16	16	42	478	413	4	786	-24	16
16	478	413	6	794	-24	16	43	372	413	6	786	-16	24
17	478	287	4	794	-16	24	44	425	350	5	790	-20	20
18	425	350	7	790	-20	20	45	425	500	5	790	-20	20
19	372	287	4	786	-24	16	46	478	413	6	786	-16	16
20	478	287	4	786	-16	16	47	372	413	6	794	-16	16
21	372	287	6	786	-24	24	48	425	350	5	790	-10	20
22	300	350	5	790	-20	20	49	425	350	5	780	-20	20
23	372	287	6	794	-24	16	50	425	350	5	790	-20	20
24	425	350	5	790	-20	20	51	425	200	5	790	-20	20
25	425	350	5	790	-20	10	52	425	350	5	790	-20	20
26	425	350	5	800	-20	20	53	550	350	5	790	-20	20
27	478	413	4	794	-16	16							

Table 1a
Results

EXP NUM	BOAT1 AVG	BOAT2 AVG	BOAT3 AVG	BOAT4 AVG	BOAT STDEV	GRNDMN	GRNDMED	MIN	MAX	GRAND STDEV	IQR	STDEV LOG10(THK)
1	995	942	901	895	46.71	933.2	923.5	851	1084	49.42	70.0	0.023
2	968	935	919	908	24.67	932.6	922.0	852	1090	44.16	51.5	0.020
3	893	809	749	732	72.28	795.7	781.5	690	936	68.49	115.0	0.037
4	933	872	797	811	68.27	853.2	859.0	705	1017	69.32	102.5	0.036
5	1003	945	884	868	59.37	925.0	911.0	813	1138	76.24	107.0	0.035
6	1183	1070	996	976	94.16	1056.1	1034.0	903	1401	109.81	156.5	0.044
7	972	925	884	880	43.98	915.4	911.0	837	1053	47.68	63.0	0.022
8	963	951	932	944	15.46	947.4	938.0	886	1052	38.18	55.5	0.017
9	950	920	892	903	29.27	916.2	911.0	853	1019	35.39	49.0	0.017
10	1152	1092	1072	1129	41.67	1111.5	1112.5	1009	1301	49.20	96.0	0.025
11	951	931	909	916	21.02	927.1	922.5	867	1004	31.84	47.5	0.015
12	951	919	905	936	23.43	927.8	922.0	872	1037	36.23	53.5	0.017
13	992	928	888	894	52.50	925.4	916.5	838	1177	58.77	70.5	0.027
14	1105	1056	1032	1077	37.42	1067.4	1063.0	980	1198	49.20	75.0	0.020
15	786	716	658	634	63.95	698.5	682.5	611	843	63.05	102.5	0.039
16	1096	1031	946	904	74.78	994.3	986.5	865	1199	85.66	140.0	0.037
17	1055	936	872	871	92.99	933.7	908.5	812	1312	92.84	118.0	0.041
18	876	824	759	730	58.88	797.0	796.0	694	921	63.85	110.5	0.035
19	880	857	821	809	29.76	841.4	843.5	771	957	37.25	58.0	0.019
20	974	890	823	796	75.61	870.7	860.0	757	1091	76.80	110.0	0.038
21	764	739	707	707	28.66	729.0	727.5	673	801	32.26	50.0	0.019
22	890	865	835	843	27.13	858.2	858.5	798	925	30.03	49.5	0.015
23	863	805	729	689	67.00	771.4	762.5	655	916	72.17	129.0	0.041
24	983	932	889	882	46.85	921.5	914.5	842	1061	50.06	64.0	0.023
25	1072	978	883	820	94.68	938.1	919.0	771	1201	105.02	165.0	0.048
26	1018	956	901	883	58.67	939.6	925.5	849	1097	60.70	93.5	0.028
27	1119	1061	1010	1009	54.74	1049.7	1047.0	951	1195	58.65	85.5	0.024
28	902	877	855	870	23.38	876.1	879.5	819	949	28.00	43.0	0.014
29	741	716	675	656	33.16	697.2	696.0	635	775	37.93	62.5	0.024
30	743	697	667	676	38.49	695.6	691.0	639	802	36.60	53.5	0.023
31	1103	1104	1102	1154	0.85	1115.8	1107.0	1049	1270	46.71	66.0	0.018
32	918	890	846	829	36.26	870.7	870.5	794	965	41.35	67.5	0.021
33	1038	1013	993	1029	22.64	1018.4	1018.0	951	1121	38.40	55.0	0.016
34	1111	1102	1078	1089	16.95	1094.7	1092.5	1019	1184	38.94	63.0	0.015
35	889	896	880	882	8.06	886.5	884.5	841	941	22.19	32.0	0.011
36	1095	945	861	846	118	936.8	906.5	781	1322	116.43	151.5	0.051
37	994	955	926	940	33.76	953.9	947.5	890	1072	40.07	59.0	0.018
38	1223	1052	930	884	147	1022.4	981.5	816	1470	153.40	210.0	0.063
39	855	772	689	654	82.87	742.7	722.5	627	919	83.39	147.5	0.048
40	988	909	848	832	70.08	894.0	883.0	793	1092	69.27	100.5	0.033
41	1403	1310	1222	1212	90.58	1286.6	1273.5	1099	1642	120.95	183.0	0.040
42	1434	1309	1179	1111	128	1258.1	1245.0	1009	1714	156.65	232.0	0.053
43	1234	1052	929	884	153	1024.9	991.5	809	1503	160.26	231.0	0.065
44	1205	1028	899	837	153	992.2	944.0	771	1426	157.88	231.0	0.067
45	1040	1029	1016	1047	12.34	1033.0	1035.0	971	1143	34.21	54.0	0.014
46	864	828	789	783	37.37	815.9	817.0	749	903	39.43	60.0	0.021
47	943	894	847	827	47.87	877.8	872.5	792	1009	51.18	82.5	0.025
48	887	831	791	803	48.54	828.1	822.5	751	937	44.65	66.5	0.023
49	781	771	758	781	11.39	772.8	774.0	727	829	22.30	34.0	0.013
50	881	857	829	838	26.37	851.0	851.0	790	914	29.31	47.0	0.015
51	649	613	578	575	35.53	603.4	598.0	551	677	33.24	52.0	0.024
52	1222	1051	930	871	147	1018.2	977.0	817	1423	150.69	217.5	0.062
53	1222	1046	926	873	149	1016.7	969.0	802	1428	151.53	227.0	0.062

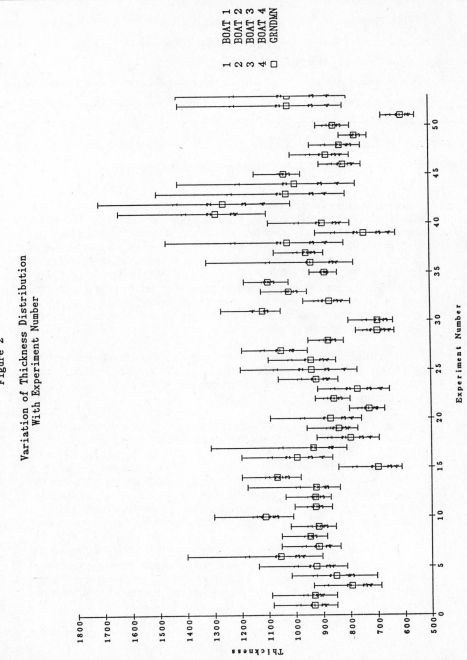

Figure 2

Variation of Thickness Distribution
With Experiment Number

Table 2 summarizes the regression models derived in this study. Note the similarity in structure for locations as well as variation effects. Note also that two interactions appear to control the variability independent of the location.

In performing this analysis we postulated a full 2nd order Taylor Series Model and refined it using stepwise regression which obeyed hierarchy. (Hierarchy rules require that one remove a main effect term only when all higher-order terms involving it have been removed.) This contrasts directly with the methods advocated by G. Taguchi. The quotes below are from a recent paper he published in the ASI Journal (6):

"...In order to prove that there are no interactions, one assigns only main effects to the orthogonal array and finds the optimum conditions. One makes the assumption that there is not interaction and estimates the process average. If there is little interaction, the estimated value of the process average and its confirmatory experiment value will match well. If the effect of interactions is large, these two values will differ greatly and it becomes necessary to take additional measures, such as redoing the experiment on site or bracing oneself for the necessity of improving the design according to marketplace information, or redesigning by rethinking the quality characteristic so that there will no longer be interactions..."

"...The author suggests that, even if there are interactions, only main effects should be assigned to an orthogonal array. The process average is then estimated, and one compares this process average with the value of the confirmatory experiment..."

"...When an interaction is significant, this indicates that the optimum level of a control factor will differ, depending on the levels of other control factors. This also means the optimum conditions we have found will not be useful downstream where conditions differ. This is true because if there are interactions between control factors, there will also be interactions with the manufacturing scale and interactions between laboratory and marketplace conditions. Experience shows that optimum conditions determined after finding interactions do not stand up under large-scale production conditions..."

"... The greatest general difference between Taguchi Methods and traditional design of experiments is the belief that it is useless and counterproductive to assign interactions, whether or not there are interactions. Orthogonal arrays should be used to help us discover experimental failures when interactions exist..."

How in the world anyone would be able to predict a significant interaction is beyond this author. Furthermore, how do we define the "quality characteristic" variability so that interactions are no longer important? The authors experience is that interactions among control factors remain important regardless of the scale of the production. Perhaps new adjustments become necessary if one scales up a process to much larger equipment or output, but control factor interactions do NOT disappear.

This general ignoring of interactions in both optimization and some screening experiments, the general disregard for the importance of randomizing experimental tests, and the tendency to "pick the winner" among existing experimental conditions are the THREE main problems with the "Taguchi Method". Twenty years of doing experimental designs has shown this author that interactions are common, that processes drift with time while we probe them, and that experiments done to satisfy a matrix seldom contain the optimum point.

Modeling a collection of outputs as functions of inputs requires multiple regression analysis. Box and Draper (7) have recently reminded us that all regression models are wrong, but that some might be useful. How useful depends on their ability to predict responses within the limits of our experimental space. Table 3 shows the results of optimization runs done within the software using all the metrics for variability studied here. Regardless of their complexity, they seem to converge on the same general experimental region. Are the differences observed of any consequence?

Within the MULREG program a utility allows us to predict responses from our models, given a set of input parameters. Table 4 summarizes several of these. The ranges of values chosen for use in generating Table 4 represent the ranges seen from the various optimization runs. The confidence intervals show that the minor differences observed among the metrics are inconsequential.

Another utility uses the regression models to create contour plots showing us graphically the region of minimum variability predicted by each. Figures 3 and 4 are the contourplots for all the metrics discussed here. Note how all converge to one region.

Table 2

Regression Models -- Location and Variability

Term	MEAN	MEDIAN	TARGET IS BEST	-LOG10 (STDEV)	LOG10 STDEV	2*SD/MN *100	2*SDBT/GMN *100	THKIQR	IQR/MED *100	-LOG10 IQR
1	0.0329	0.0331	23.939	-1.757	-1.565	0.290	2.302	0.112	0.335	-1.932
CLP	-0.0045	-0.0046	-0.241	-0.060	0.014	-0.018	0.142	-0.006	-0.017	-0.046
TF	0.0027	0.0028	1.042	-0.054	-0.055	0.023	-0.498	-0.005	0.027	-0.052
RAB	-0.0009	-0.0009	-1.781	0.002	0.091	-0.024	0.340	-0.004	-0.040	-0.032
CTMP	0.0007	0.0009	0.148	-0.040	-0.005	-0.011	0.169	-0.011	-0.020	-0.060
ZN1	-0.0010	-0.0010	-3.227	-0.163	0.155	-0.061	0.709	-0.022	-0.072	-0.155
ZN3			1.271	0.025	-0.066	0.020	-0.373	0.011	0.041	0.075
CLP*CTMP			-9.923	-0.327	0.482	-0.095	1.121	-0.037	-0.103	-0.312
TF*CTMP	0.0018	0.0017	6.607	0.337	-0.329	0.090	-0.771	0.041	0.107	0.355
RAB*CTMP	-0.0021	-0.0020	-4.884	-0.312	0.247	-0.083	0.818	-0.038	-0.096	-0.327
ZN1*ZN3			-8.469	-0.324	0.409	-0.091	0.815	-0.045	-0.126	-0.368
TRANSFORM	INVSQRT	INVSQRT	NONE	NONE	NONE	INVSQRT	LOG	INVSQRT	INVSQRT	NONE
VAR. WT.	NONE	NONE	RAB	CTMP	RAB	CTMP	CTMP	CTMP	CTMP	CTMP
ADJ R**2	0.8156	0.8573	0.462	0.296	0.462	0.450	0.502	0.403	0.568	0.403
REL. PRES	0.7760	0.8200	0.314	0.089	0.313	0.282	0.373	0.236	0.449	0.183

Table 3

Optimization Summary -- Variability Metrics

Factor, Resp., Formula	Range	Initial Setting	TARGET IS BEST	-LOGS	LOG10SD LOGTHK	2*SD/MN *100	2*SDBT/GMN *100	THKIQR	IQR/MED *100	-LOG10 IQR
Factors										
CLP	300 to 550	425	300.440	300.000	300.000	300.010	307.260	300.120	300.140	300.000
TOTFLW	200 to 500	350	500.000	499.620	500.000	491.000	490.320	498.780	499.260	499.990
RAB	3 to 7	5	3.006	3.001	3.240	3.003	3.014	3.001	3.002	3.019
CTMP	780 to 800	800	799.920	800.000	799.940	799.990	799.930	799.990	799.950	800.000
ZN1OFF	-30 to -10	-30	-29.230	-28.601	-30.000	-29.184	-29.900	-29.993	-29.874	-29.999
ZN3OFF	10 to 30	30	27.622	28.630	26.538	26.821	26.339	29.824	27.031	29.749
Resp.										
BTUNIF			0.067	0.064	0.078	0.078	0.083	0.046	0.065	0.046
IQRMED			1.270	1.250	1.330	1.326	1.347	1.135	1.265	1.137
LGIQR			-0.522	-0.502	-0.576	-0.568	-0.590	-0.386	-0.517	-0.389
LSDLTH			-3.258	-3.281	-3.207	-3.207	-3.176	-3.404	-3.263	-3.403
SDMNUN			1.803	1.784	1.871	1.869	1.899	1.642	1.791	1.644
THKTGT			58.248	58.701	57.231	57.213	56.596	61.256	58.367	61.222
THKIQR			12.803	12.582	13.396	13.335	13.579	11.371	12.743	11.396
THKLOGS			-0.429	-0.415	-0.468	-0.465	-0.486	-0.315	-0.420	-0.317

Table 4

Predictions from Regression Models (IQR/MED × 100)

Predictions and 95% simultaneous confidence intervals for mean
responses of IQRMED using model TKIQRMD
RAB = 3, CTMP = 800, CLP = 300, TF = 490

ZN3		ZN1 = −30	ZN1 = −29	ZN1 = −28
	Lower	0.657995	0.677811	0.698022
25	Predicted	1.382611	1.427586	1.474793
	Upper	4.563445	4.741089	4.938965
	Lower	0.535777	0.561639	0.588548
30	Predicted	1.145710	1.195903	1.249468
	Upper	3.963159	4.089269	4.238291

Predictions and 95% simultaneous confidence intervals for mean
responses of IQRMED using model TKIQRMD
RAB = 3, CTMP = 800, CLP = 300, TF = 500

ZN3		ZN1 = −30	ZN1 = −29	ZN1 = −28
	Lower	0.640432	0.659353	0.678640
25	Predicted	1.353949	1.397522	1.443234
	Upper	4.541693	4.719994	4.918446
	Lower	0.523517	0.548327	0.574109
30	Predicted	1.124059	1.172821	1.224826
	Upper	3.931724	4.059521	4.210180

7. Summary:

What then is the engineer to do?

This study has shown that where some correlation exists between the variability of a process and the location of a process, one may use that metric to describe that variability that makes the most sense in his environment and is the most convenient.

Without doubt the $\log_{10}(S)$ statistic advocated by Dr. George Box is adequate to the task described here. In this case, however, because engineers in the semiconductor industry are now more familiar with metrics connoting a +/− variation around a central value, we elect to use any one of the several generally based on the coefficient of variation. If one employs a "fudge-factor" to generate values similar to those found using the standard deviation, that metric based on the IQR and the median is probably the one of choice.

If you wish to interpret decibel values from the SN_T metric, then do so. Most importantly, however, instead of the highly-fractionated or saturated designs advocated by Taguchi, use more conservative 1/2-, or 1/4-fraction factorials so that you can estimate interactions. Randomize experimental trials to protect yourself and your results from nuisance variation due to a change in the process as you probe it. Finally model your responses using multiple regression to help identify optimal regions without trying to "pick a winner" among the experimental trials actually done.

Figure 2

BTUNIF, IQRMED, THKTGT, THKLOGS
CLP = 300, CTMP = 800, ZN1OFF = −30, ZN3OFF = 30

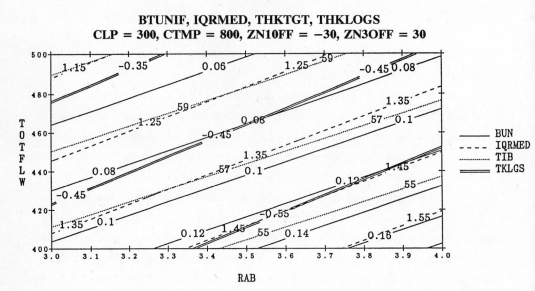

Figure 3

LGIQR, LGSDLGTHK, SDMNUNIF, THKIQR
CLP = 300, CTMP = 800, ZN1OFF = −30, ZN3OFF = 30

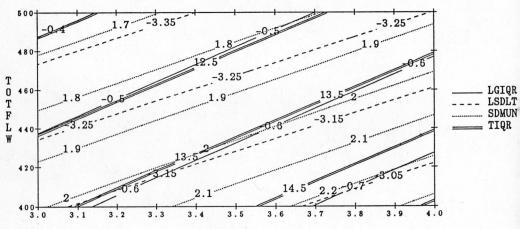

8. References:

1.   Box, G.E.P., Hunter, W.G., and Hunter, J.S., "Statistics for Experimenters," John Wiley & Sons, New York, 1978.

2.   Montgomery, D.C., "Design and Analysis of Experiments," John Wiley & Sons, New York, 1984.

3.   See for example: Taguchi, G., and Phadke, M.S., "Quality Engineering Through Design Optimization," Conference Record Vol. 3, IEEE Globecom 1984 Conference, Atlanta, GA, 1106-1113; Taguchi, G. and Wu, Y., Introduction to Off-Line Quality Control, Central Japan Quality Control Association (1986); and Taguchi, G., Introduction to Quality Engineering: Designing Quality into Products and Processes. Asian Productivity Organization, UNIPUB, Kraus International Publications, White Plains, NY.   See also Phadke, M.S., "Quality Engineering Using Design of Experiments," Proceedings of the American Statistical Association, section on Statistical Education, 11 - 20 (1982).

4.   Box, G.E.P., "Studies in Quality Improvement: Signal to Noise Ratios, Performance Criteria, and Transformation," Report No. 26, University of Wisconsin Center for Quality and Productivity Improvement, Madison, WI, 1987.

5.   Reece, J.E., and Good, M.A., "Process Variability and Response Surface Methodology," *Proceedings of the 1988 Rocky Mountain Quality Conference*, Denver, CO.

6.   Taguchi, G., "The Development of Quality Engineering," ASI Journal 1, (1988).

7.   Box, G.E.P., and Draper, N.R. "Empirical Model-building and Response Surfaces," John Wiley & Sons, New York, 1987.

9. About the Author:

Dr. Jack E. Reece received the Ph.D. in Organic Chemistry from the University of N. Carolina in 1966. He has practiced statistical experimental design techniques since 1970 in all areas of process development and manufacturing in both the manufacture of specialty coatings and semiconductor devices. Currently he is a Senior Principal Process Engineer for a major semiconductor manufacturer and defense contractor. In that position he assists in the integration of applied statistics methods, including experimental design and SPC, in all manufacturing stages and trains co-workers to use the techniques.

Submitted by: Lt Col Mark J. Kiemele
Tenure Associate Professor
United States Air Force Academy

Computer Network Performance Analysis*

4. Introduction

The verification and validation of simulation models is a difficult task [8, 18]. In the absence of real data from the system being simulated, the task seems even close to insurmountable. Action has been taken, however, to gain a high confidence level in the simulator described in Chapter III.

Verification of simulator performance with regard to voice/data transaction arrival patterns was accomplished by Clabaugh [3]. He found that the simulator behavior was within a 95% confidence interval of the true means, where the true means were determined by the use of standard Poisson generators and user run time specifications. Clabaugh also provides additional verification of simulator output by configuring the simulator to correspond to an analytic model and comparing the output results [3]. These verification efforts apply to the modified model as well, since none of the modifications impact Clabaugh's verification work. This chapter documents another step taken toward the verification and validation of the integrated network simulator.

The motivation for this analysis stems not only from the desire to move in the direction of model verification, but also from the need to obtain and analyze performance data from an integrated circuit/packet-switched computer network, for such data is scarce. Fortunately, these two encompassing goals do not diverge.

4.1 Goals and Scope of the Analysis

The specific goals of the analysis are as follows:

(1)   Design an experiment whereby the performance data can be efficiently and economically obtained.

*Reprinted from the author's dissertation, "Adaptive Topological Configuration of an Integrated Circuit/Packet-Switched Computer Network," Texas A&M University, 1984.

(2)　　Determine the effective ranges of network parameters for which realistic and acceptable network performance results.

(3)　　Investigate the sensitivity of performance measures to changes in the network input parameters.　Specifically, determine how network performance is affected by changes in the network traffic load, trunk line or link capacity, and network size.

The scope of the analysis is restricted to those parameters that are closely related to the network topology and the workload imposed on that topology.　With regard to performance sensitivity to workload and link capacity, the following four parameters are investigated:

CS:　Circuit Switch Arrival Rate (voice calls/min)

PS:　Packet Switch Arrival Rate (packets/sec)

SERV:　Voice Call Service Rate (sec)

SLOTS:　Number of Time Slots per Link (a capacity indicator)

The sensitivity to network size is also investigated by varying the number of nodes and links in a network.　In all cases, the performance measures observed are mean packet delay (MPD), fraction of voice calls blocked (BLK), and average link utilization (ALU).

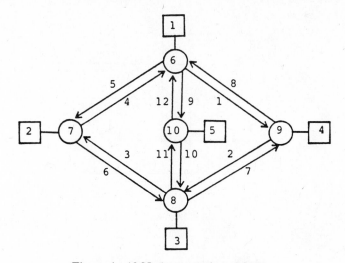

Figure 4.　10-Node network topology

4.2 Design of Experiment

The 10-node network topology shown in Figure 4 was considered sufficiently complex to provide practical performance data without exhausting the computing budget.
The circuit switch (circular) nodes correspond to the major computing centers of Tymshare's TYMNET, a circuit-switched network [14]. The trunk lines are full-duplex (FDX) carriers, each of which is modeled by two independent, half-duplex (HDX) channels.

The preliminary sizing analysis [13] indicates that the effective range of input parameters to be investigated is as follows:

$$
\begin{array}{rl}
\text{CS:} & \text{0-8 (calls/min at each node)} \\
\text{PS:} & \text{0-600 (packets/sec at each node)} \\
\text{SERV:} & \text{60-300 (sec/call)} \\
\text{SLOTS:} & \text{28-52 (a link capacity indicator)}
\end{array}
$$

The fixed parameter settings were such that each slot represents a capacity of about 33 Kbits/sec.

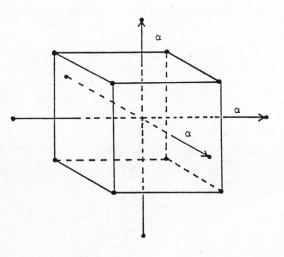

Figure 5. Central composite design (k=3)

The experimental design selected for this analysis is a second order (quadratic), rotatable, central composite design [10, 11]. Such a design for k (number of parameters) = 3 is illustrated in Figure 5. This design was chosen because it reduces considerably the number of experimentation points that would otherwise be required if the classical 3^k factorial design were used. The "central composite" feature of the design replaces a 3^k factorial design with a 2^k factorial system augmented by a set of axial points together with one or more center points. A "rotatable" design is one in which the prediction variance is a function only of the distance from the center of the design and not on the direction. Myers [10] shows that the number of experimental points required for k = 4 is 31. That is, there are 16 factorial points, 8 axial points, and 7 replications at the center point. In order for the design to be rotatable, Myers [10] shows that α must be chosen as $(2^k)^{1/4}$. In this case, α, the distance from the center point to an axial point, is 2. The seven replications at the center point will allow an estimate of the experimental (pure) error to be made; thus, a check for model adequacy is possible [12]. Myers [10] recommends seven replications at the center point simply because it results in a near-uniform precision design. That is, it is a design for which the precision on the predicted value ŷ, given by

$$\frac{N \text{ var } (\hat{y})}{\sigma^2}$$

where N = total number of experimental points, and
σ^2 = the error variance,

is nearly the same at a distance of 1 from the center point as it is at the center point [10].

In this design, each parameter is evaluated at five different levels. The five levels for each of the four parameters are as follows:

CS:	0	2	4	6	8
PS:	0	150	300	450	600
SERV:	60	120	180	240	300
SLOTS:	28	34	40	46	52

The center point is defined as CS / PS / SERV / SLOTS = 4 / 300 / 180 / 40.

Analysis of the results of these 31 runs indicated that the original range of data was too large. Ten of these experimental data points resulted in network performance that would be totally unacceptable, e.g., a MPD of more than 10 seconds. These "outliers" were eliminated and 14 additional points in the moderate to heavy loading of the network were added. Two observations were taken at 8 of these 14 additional points for the purpose of estimating the error variance at points other that the center of the design. Table I shows the 43 observations that form the data set used in the subsequent regression analysis. The first 21 observations in Table I represent those data points retained from the original 31-run design. The remaining 22 observations represent the data points used to augment the original design.

Table I. Experimental Data

OBS	CS	PS	SERV	SLOTS	MPD	ALU	BLK
1	2	150	120	34	0.117	0.262	0.000
2	2	150	120	46	0.116	0.194	0.000
3	2	150	240	34	0.113	0.372	0.000
4	2	150	240	46	0.116	0.275	0.000
5	2	450	120	34	0.430	0.553	0.016
6	2	450	120	46	0.260	0.396	0.000
7	2	450	240	34	0.433	0.676	0.031
8	2	450	240	46	0.449	0.481	0.003
9	6	150	120	34	0.339	0.520	0.012
10	6	150	120	46	0.115	0.386	0.000
11	6	450	120	46	0.462	0.594	0.016
12	0	300	180	40	0.123	0.231	0.000
13	4	0	180	40	0.000	0.325	0.000
14	4	300	180	52	0.153	0.427	0.000
15	4	300	180	40	0.350	0.561	0.016
16	4	300	180	40	0.527	0.571	0.005
17	4	300	180	40	0.502	0.561	0.011
18	4	300	180	40	0.316	0.581	0.020
19	4	300	180	40	0.381	0.562	0.017
20	4	300	180	40	0.798	0.608	0.036
21	4	300	180	40	0.616	0.560	0.007
22	5	400	210	43	1.613	0.739	0.070
23	5	400	210	43	2.754	0.746	0.082
24	5	400	150	37	1.656	0.725	0.067
25	5	400	150	37	1.345	0.725	0.084
26	5	400	150	43	0.192	0.615	0.027
27	5	400	210	40	4.814	0.785	0.106
28	5	400	210	40	5.624	0.782	0.125
29	4	400	210	37	1.604	0.747	0.084
30	4	400	210	37	2.730	0.745	0.101
31	4	400	210	40	2.440	0.695	0.063
32	4	400	210	40	1.199	0.687	0.057
33	4	400	180	40	0.910	0.648	0.035
34	5	400	180	40	1.970	0.722	0.074
35	5	400	180	40	1.130	0.731	0.077
36	5	400	180	37	4.084	0.775	0.110
37	5	400	180	37	4.019	0.778	0.139
38	4	400	180	37	4.817	0.712	0.057
39	4	400	180	37	1.135	0.701	0.065
40	3	400	210	37	0.599	0.652	0.017
41	3	400	210	43	0.293	0.541	0.003
42	3	400	150	37	0.454	0.570	0.003
43	3	400	150	43	0.280	0.476	0.005

4.3 Regression Analysis and Model Selection

The Statistical Analysis System (SAS) [15, 16, 17] has been used to perform a multiple regression analysis for each of the three performance measures. The regression variables are the four input parameters. SAS procedures REG [16] and RSREG [16] were used to analyze the data. The assumption of a quadratic response surface allows for the estimation of 15 model parameters, including the intercept.

Although the model selection procedure can be based on a number of possible criteria [4, 6, 9], the approach taken is as follows. RSREG was used first to check the full (quadratic) model for specification error (lack of fit test) and to determine significance levels for the linear, quadratic, and crossproduct terms. The models for MPD, BLK, and ALU exhibited a lack of fit that was significant at the .14, .05, and .82 levels, respectively. It was found, however, that these significance levels were heavily influenced by the observations at the extremes of the heavy-loading region. For example, by eliminating 13 of the observations in the heavy-loading range of data, the lack of fit for BLK could be raised to the .70 significance level. Hence, although the .05 level for the BLK model borders on statistical significance, the lack of fit was not deemed sufficient to justify a more complex model.

The RSREG results also indicated that some terms were insignificant and possibly could be deleted from the model. Using the RSREG results as a guide, several subsets of the full quadratic models were investigated using SAS procedure REG. In these models, all linear terms were retained because even though a lower order term in a polynomial model may not be considered significant, dropping such a term could produce a misleading model [6]. Several crossproduct and quadratic terms were deleted, however, and the final regression models selected are shown in Table II. Despite their appearance of being insignificant, several of the crossproduct and quadratic terms were retained in the model simply because their retention resulted in a smaller residual mean square than if they had been deleted. The interpretation of these regression models is presented graphically in the following section in terms of a sensitivity analysis.

4.4 Sensitivity Analysis

This section presents data to describe how the various performance measures change with respect to changes in traffic load, link capacity, and network size. All of the graphs presented are obtained from the quadratic response surfaces (regression models) developed in the previous section.

Table II. Regression Models

```
DEP VARIABLE: MPD
```

SOURCE	DF	SUM OF SQUARES	MEAN SQUARE	F VALUE	PROB>F
MODEL	10	61.561986	6.156199	6.874	0.0001
ERROR	32	28.658932	0.895592		
C TOTAL	42	90.220918			
ROOT MSE		0.946357	R-SQUARE	0.6823	
DEP MEAN		1.218093	ADJ R-SQ	0.5831	
C.V.		77.69169			

VARIABLE	DF	PARAMETER ESTIMATE	STANDARD ERROR	T FOR H0: PARAMETER=0	PROB > |T|
INTERCEP	1	12.556905	13.142988	0.955	0.3465
X1 CS	1	-1.654903	1.326989	-1.247	0.2214
X2 PS	1	0.014620	0.012349	1.184	0.2452
X3 SERV	1	-0.026635	0.010541	-2.527	0.0167
X4 SLOTS	1	-0.531166	0.597025	-0.890	0.3803
X11	1	0.232675	0.078122	2.978	0.0055
X12	1	0.001153381	0.0009976058	1.156	0.2562
X13	1	0.013125	0.00331975	3.954	0.0004
X14	1	-0.048214	0.025537	-1.888	0.0681
X24	1	-0.000395179	0.0003018693	-1.309	0.1998
X44	1	0.009018565	0.007302992	1.235	0.2259

```
DEP VARIABLE: ALU
```

SOURCE	DF	SUM OF SQUARES	MEAN SQUARE	F VALUE	PROB>F
MODEL	11	1.123895	0.102172	970.909	0.0001
ERROR	31	0.003262242	0.0001052336		
C TOTAL	42	1.127158			
ROOT MSE		0.010258	R-SQUARE	0.9971	
DEP MEAN		0.581233	ADJ R-SQ	0.9961	
C.V.		1.764929			

VARIABLE	DF	PARAMETER ESTIMATE	STANDARD ERROR	T FOR H0: PARAMETER=0	PROB > |T|
INTERCEP	1	0.022183	0.150510	0.147	0.8838
X1 CS	1	0.068950	0.014713	4.686	0.0001
X2 PS	1	0.001859007	0.0001334741	13.928	0.0001
X3 SERV	1	0.001788392	0.0006903726	2.590	0.0145
X4 SLOTS	1	-0.0094644	0.006917257	-1.368	0.1811
X11	1	-0.000929364	0.0008468003	-1.098	0.2809
X13	1	0.0003897799	.00003664077	10.638	0.0001
X14	1	-0.00127589	0.0002929912	-4.355	0.0001
X24	1	-.0000259887	.00000329748	-7.881	0.0001
X33	1	-.0000024518	.00000146947	-1.669	0.1053
X34	1	-.0000214946	.00000932481	-2.305	0.0280
X44	1	0.0001583866	.00008483328	1.867	0.0714

Table II. (Continued)

```
DEP VARIABLE: BLK
                       SUM OF           MEAN
    SOURCE      DF     SQUARES         SQUARE      F VALUE      PROB>F
    MODEL       12     0.062669     0.005222451     26.743      0.0001
    ERROR       30   0.005858444   0.0001952815
    C TOTAL     42     0.068528
         ROOT MSE       0.013974      R-SQUARE      0.9145
         DEP MEAN       0.038163      ADJ R-SQ      0.8803
         C.V.          36.61764

                     PARAMETER       STANDARD     T FOR HO:
    VARIABLE    DF    ESTIMATE         ERROR     PARAMETER=0    PROB > |T|

    INTERCEP    1     0.197892       0.205845       0.961        0.3441
    X1   CS     1    -0.028562       0.020039      -1.425        0.1644
    X2   PS     1   0.0003439953   0.0002038975     1.687        0.1020
    X3   SERV   1   0.0001307622   0.0005157113     0.254        0.8016
    X4   SLOTS  1    -0.012008       0.00887909     -1.352       0.1863
    X11         1   0.006553606     0.00115578       5.670       0.0001
    X12         1   .00005438269    .00001502268     3.620       0.0011
    X13         1   0.0003339836   0.0000496682      6.724       0.0001
    X14         1   -0.00175778    0.0004020761     -4.372       0.0001
    X22         1   3.79530E-07    1.75917E-07       2.157       0.0391
    X24         1   -.0000158292   .00000450668     -3.512       0.0014
    X34         1   -.0000194136   .00001274852     -1.523       0.1383
    X44         1   0.0002843414   0.0001092323      2.603       0.0142
```

4.4.1 Sensitivity to Traffic Load

The workload imposed upon an integrated network is described by the voice arrival rates (CS) at each circuit switch node and the data packet arrival rates (PS) at each packet switch node, as well as the length of service for each voice call (SERV). It is assumed that, for a given arrival at any particular node, all of the other nodes of the same type are equally likely to be the destination node for that arrival. That is, the workload is said to be uniformly distributed between node pairs. Unless otherwise stated, it is also assumed that SERV = 180 seconds. Besides these three parameters, the link capacity (SLOTS) parameter also affects the network traffic load. If the CS and PS arrival rates are also assumed to be fixed, then the smaller values of SLOTS represent a heavier network load while the larger SLOTS values correspond to a lighter network load.

Figures 6, 7, and 8 depict the MPD, BLK, and ALU performance measures, respectively, as a function of CS for four different SLOTS/PS combinations. Similarly,

Figures 9, 10, and 11 show MPD, BLK, and ALU, respectively, as a function of PS for four load levels defined by combinations of SLOTS and CS. The performance measure sensitivity to the traffic load parameters CS, PS, and SLOTS (for SERV=180) is seen by observing the relative slopes and ordinate values of the four curves in each of the Figures 6 - 11. For example, examination of the curves in Figures 6 and 9 indicates that, within the range of data shown, MPD is more sensitive to CS than to either SLOTS or PS. Additionally, the sensitivity of MPD to both CS and SERV is shown in the three dimensional plot of Figure 12, where the higher levels of one input parameter are seen to magnify the effects of the other parameter.

4.4.2 Sensitivity to Link Capacity

The trunk line carrying capacity is an important design parameter in integrated networks. Hence, in addition to the plots of the performance measures versus number of SLOTS, which are shown in Figures 13, 14, and 15, the confidence intervals for each of the performance measures are also presented. Figures 16, 17, and 18 give the 95% confidence limits for a mean predicted value of MPD, BLK, and ALU, respectively. The voice and data arrival rates for these graphs correspond to a fairly heavy traffic load (CS = 4, PS = 400, SERV = 180).

4.4.3 Sensitivity to Network Size

In addition to checking the performance of the simulator for varying traffic loads and link capacities on a fixed network topology, the analysis also examines a fixed traffic load on varying sized topologies. In particular, a throughput requirement of 2000 data packets per second with a voice call arrival rate of 20 calls per minute was imposed upon three different sized networks. A 10-node, 20-node, and 52-node network were each subjected to the fixed traffic load.

The 10-node network is the TYMNET topology shown previously in Figure 4. Six links interconnect the backbone nodes.

The 20-node network consists of 10 packet switches and 10 circuit switches. The 10 circuit switches forming the backbone of the network are 10 of the major computing centers

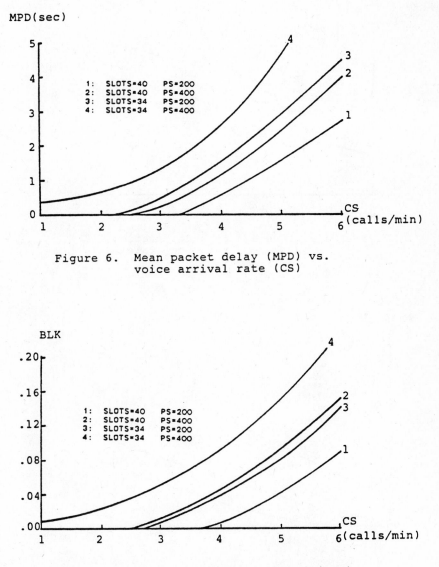

Figure 6.   Mean packet delay (MPD) vs.
             voice arrival rate (CS)

Figure 7.   Fraction of calls blocked (BLK)
             vs. voice arrival rate (CS)

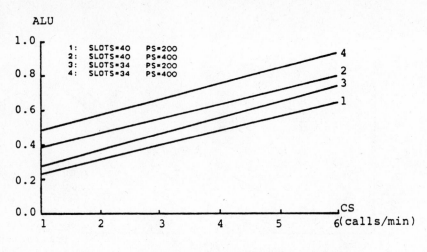

Figure 8.   Average link utilization (ALU)
vs. voice arrival rate (CS)

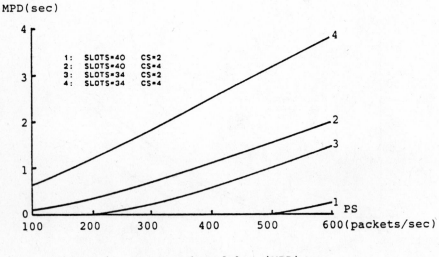

Figure 9.   Mean packet delay (MPD) vs.
data arrival rate (PS)

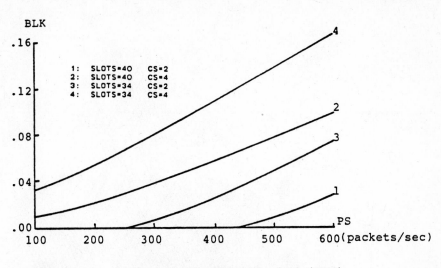

Figure 10. Fraction of calls blocked (BLK)
vs. data arrival rate (PS)

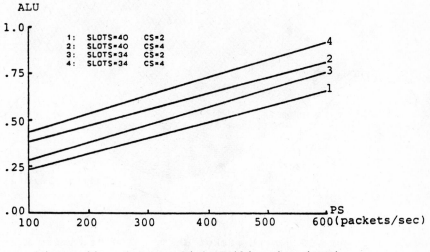

Figure 11. Average link utilization (ALU)
vs. data arrival rate (PS)

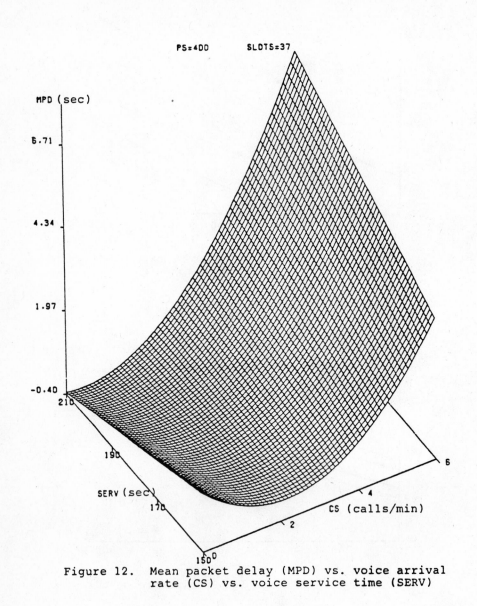

Figure 12.   Mean packet delay (MPD) vs. voice arrival
             rate (CS) vs. voice service time (SERV)

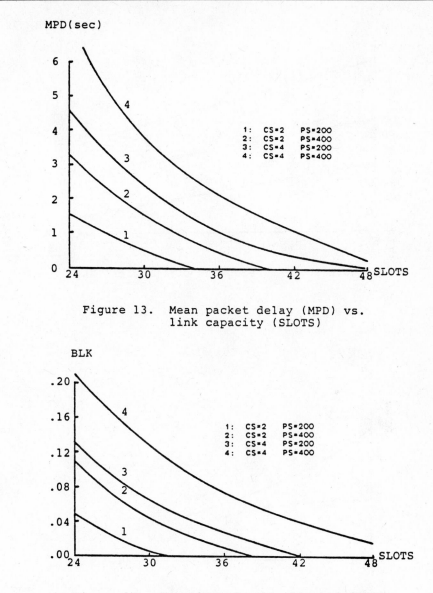

MPD(sec)

1: CS=2   PS=200
2: CS=2   PS=400
3: CS=4   PS=200
4: CS=4   PS=400

Figure 13.   Mean packet delay (MPD) vs.
             link capacity (SLOTS)

BLK

1: CS=2   PS=200
2: CS=2   PS=400
3: CS=4   PS=200
4: CS=4   PS=400

Figure 14.   Fraction of calls blocked (BLK)
             vs. link capacity (SLOTS)

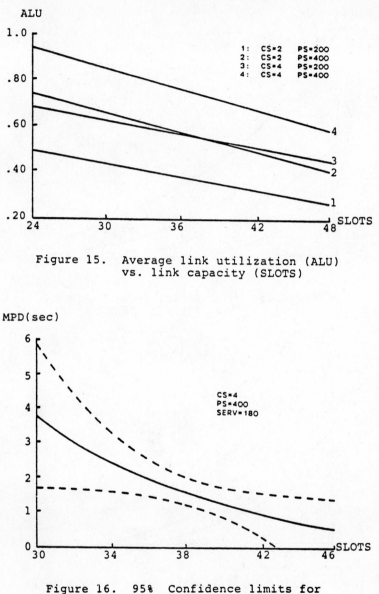

Figure 15.   Average link utilization (ALU)
vs. link capacity (SLOTS)

Figure 16.   95% Confidence limits for
mean packet delay (MPD)

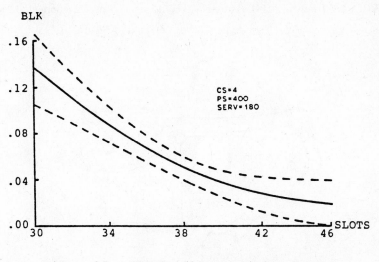

Figure 17. 95% Confidence limits for
            fraction of calls blocked (BLK)

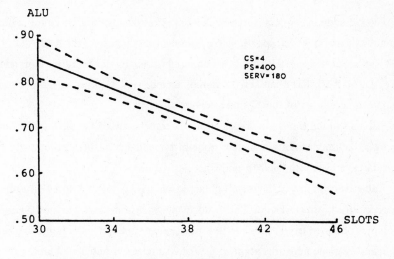

Figure 18. 95% Confidence limits for
            average link utilization (ALU)

in the CYBERNET network [14]. The nodes on the subnet are interconnected by 12 trunk lines. The backbone of the 20-node network is depicted in Figure 19.

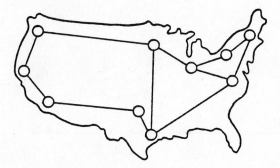

Figure 19. 20-Node network backbone

The 52-node network is comprised of 26 packet switches and 26 circuit switches, with the circuit switch nodes corresponding to a 26-node substructure of the ARPANET network. This 26-node subset of ARPANET is commonly used in the literature for comparative analyses [1, 2, 5, 7]. The subnet is interconnected with 33 links. Figure 20 shows the communications subnet of the 52-node network.

All links in the three topologies have a fixed capacity of 40 slots. Additionally, each of the three communications subnets is node-biconnected (i.e., there are at least two node-disjoint paths between any pair of nodes).

Table III summarizes the results obtained when subjecting the three different network topologies to the given workload. The performance measures given for the 10-node network are averages obtained from a total of 10 simulation runs. Correspondingly, the data for the 20-node and 52-node networks are averages of 7 simulations and 1 simulation, respectively. Despite the fact that there is a reduction in MPD as the number of nodes increases, the relatively smaller decrease in MPD when going from the 20-node network to the 52-node

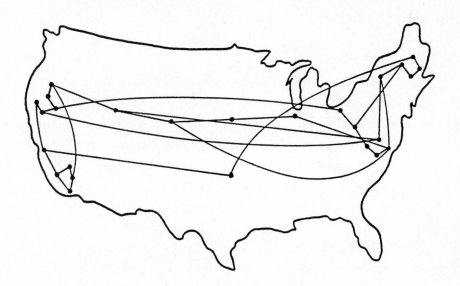

Figure 20. 52-Node network subnet

Table III. Sensitivity to network size

Topology	Subnet Link/Node Ratio	MPD	BLK	ALU
10-Node	1.2	.987	.034	.654
20-Node	1.2	.451	.033	.416
52-Node	1.27	.313	.006	.289

network (as compared to going from the 10-node network to the 20-node network) results from the fact that the backbone of the integrated network is circuit-switched. This implies that there is a fixed switching delay (assumed to be 50 ms in this study) at each node, and each packet incurs this delay at each intermediate node on its route. Hence, if the subnet link/node ratio remains fairly constant, the MPD is expected to decrease to a certain point and then begin to increase as the number of nodes in the network increases. This phenomenon is due to the fact that as more nodes are added to the network, a greater proportion of the delay can be attributed to switching delays, even though the queuing delay steadily decreases. However, as technology improves the switching delays, the effect of this phenomenon on total MPD is reduced.

4.5 Summary

Simulation model validation is a never-ending-process, but a well designed sensitivity analysis of the simulator can increase the user's confidence in the model, as well as his knowledge of the model. In this regard, a sensitivity analysis is an important step in the direction of model validation. This chapter has outlined an approach to analyzing the performance characteristics of an integrated computer-communication network simulation model and has presented the results of the analysis.

The investigation has centered on the integrated network traffic load parameters of packet arrival rate (PS), voice call arrival rate (CS), and voice call service times (SERV), as well as the design parameters of link capacity (SLOTS) and network size. Network performance has been measured in terms of mean packet delay (MPD), a strict data measure; fraction of voice calls blocked (BLK), a pure voice criterion; and average link utilization (ALU), a gauge which tends to combine both the data and voice attributes of an integrated network.

The analysis has determined a range of input parameters for which second-order response surfaces can be used to adequately describe realistic network performance. This range is summarized as follows:

CS: 1-6 calls/min

PS: 100-600 packets/sec

SERV: 150-210 sec

SLOTS: 24-48 slots

In light of the fact that an integrated network with a circuit-switched subnet has not yet been implemented, the term "realistic" performance is admittedly dubious. However, performance criteria for current packet-switched and circuit-switched networks can and have been used as guidelines (e.g., a packet delay of 10 seconds is clearly unacceptable, for the message would automatically time out), although flexibility in the guidelines has been preserved so as to not stifle the range of model applicability.

Furthermore, the graphs presented in this chapter consistently support the theme that the performance measures are more sensitive to the CS (voice) arrival rate than to the other parameters investigated. The "slices" (i.e., graphs) of the response surfaces that correspond to an increased CS level generally have higher ordinate values and steeper slopes than the "slices" that correspond to an increased network loading due to variations in the other parameters. In essence, voice arrival rates tend to dominate the network in the sense that the virtual circuits established by successful call initiations provide the framework of paths by which data packets can "piggyback" the digitized voice.

Additionally, heavily loaded networks tend to intensify the effect of any parameter. Increasing the number of nodes in a network (along with a corresponding increase in the number of links) for a fixed traffic load will decrease link utilization rates and call blocking, but may or may not decrease mean packet delay, depending on what proportion of the delay is switching delay.

The simulation model analyzed in this research is a tool that can be used by integrated network designers and managers alike. As a result of this analysis, the user of the simulation model now has a more precise understanding of the relationship between integrated network performance and the network parameters that influence such performance. In light of this, the model is a more viable tool now than it was prior to the analysis.

REFERENCES

1. Boorstyn, R. R., and Frank, H. Large-scale network topological optimization. *IEEE Trans. on Comm. Com-25*, 1 (Jan 1977), 29-47.

2. Chou, W., and Sapir, D. A generalized cut-saturation algorithm for distributed computer communications network optimization. *IEEE 1982 Int. Conf. on Comm. (ICC-82)*, Philadelphia, PA (June 13-17, 1982), 4C.2.1-4C.2.6.

3. Clabaugh, C. A. Analysis of flow behavior within an integrated computer-communication network. Ph.D. dissertation, Texas A&M University (May 1979).

4. Draper, N. R. and Smith, H. *Applied Regression Analysis (2nd ed.)*. John Wiley & Sons, Inc., New York, NY, 1981.

5. Frank, H., and Chou, W. Topological optimization of computer networks. *Proc. of IEEE 60*, 11 (Nov 1972), 1385-1397.

6. Freund, R. J., and Minton, P. D. *Regression Methods*. Marcel Dekker, Inc., New York, NY, 1979.

7. Gerla, M., and Kleinrock, L. On the topological design of distributed computer networks. *IEEE Trans. on Comm. Com-25*, 1 (Jan 1977), 48-60.

8. Graybeal, W. T., and Pooch, U. W. *Simulation: Principles and Methods*. Winthrop Publishers, Inc., Cambridge, MA, 1980.

9. Montgomery, D. C., and Peck, E. A. *Introduction to Linear Regression Analysis*. John Wiley & Sons, Inc., New York, NY, 1982.

10. Myers, R. H. *Response Surface Methodology*. Allyn and Bacon, Inc., Boston, MA, 1971.

11. Naylor, T. H. (Ed.). *The Design of Computer Simulation Experiments*. Duke University Press, Durham, NC, 1969.

12. Ostle, B., and Mensing, R. W. *Statistics in Research (3rd ed.)*. The Iowa State University Press, Ames, IA, 1975.

13. Pooch, U. W., and Kiemele, M. J. A simulation model for evaluating integrated circuit/packet-switched networks. Presented at Winter Simulation Conf,. Arlington, VA (Dec 12-14, 1983).

14. Pooch, U. W., Greene, W. H., and Moss, G. G. *Telecommunications and Networking*. Little, Brown and Company, Boston, MA, 1983.

15. SAS Institute Inc. *SAS User's Guide: Basics, 1982 Edition*. SAS Institute Inc., Cary, NC, 1982.

16. SAS Institute Inc. *SAS User's Guide: Statistics, 1982 Edition*. SAS Institute Inc., Cary, NC, 1982.

17. SAS Institute Inc. *SAS/Graph User's Guide, 1981 Edition*. SAS Institute Inc., Cary, NC, 1981.

18. Shannon, R. E. *Systems Simulation: The Art and Science*. Prentice-Hall, Inc., Englewood Cliffs, NJ, 1975.

Submitted by: Tom Bingham
Supplier Improvement Manager
Boeing Commercial Airplanes

Optimizing the Anodize Process and Paint Adhesion
for Sheet Metal Parts

Introduction

Sheet metal used for aircraft parts must be finished in such a way that it is resistant to the problems associated with corrosion. The anodizing process and painting of sheet metal are important procedures to insure acceptable corrosion resistance. The problems addressed in this experiment were optimization of the anodizing process and maximization of paint adhesion in order to assure acceptable corrosion resistance on metals exposed to salt spray tests. One of the problems to be overcome was the dilemma caused by conflicting affects of seal times and temperature on anodizing results and paint adhesion. The historical solution was to balance the process so both salt spray tests and paint adhesion failures are in spec. This balancing required slowing down the anodize time and increasing the temperature. Both of these settings produced an expensive solution to the problem. Therefore, it was decided to obtain information through experimental design which could ultimately lead to a more economical solution.

Approach to the Problem

A team of qualified people from diverse backgrounds and levels of expertise was formed to brainstorm the problem. A fishbone diagram generated by the team yielded over 50 possible factors relevant to the anodize process and paint adhesion. The team then reduced the possible factors to 9 probable factors. Two experiments were then designed to model both paint adhesion and salt spray failures. The factors and the levels of interest are shown on the following page.

The response of interest for the anodize process is the number of pits that appeared on the finished metal product after two weeks of exposure to the salt spray test.

Experimental Variables

Factor	Test Values 1	Test Values 2	
A: Deox Time	6 min	20 min	
B: Anodize Time	35 min	55 min	
C: Anodize Temp	91°F	99°F	Controlled
D: Free CrO_3	32%	4.7%	in Anodize
E: Seal Time	23 min	28 min	Process
F: Seal Temp	191°F	199°F	
G: Ph	3.2	3.8	
H: Primer Thickness	.3 mils	1.5 mils	Paint Adhesion
I: Topcoat Thickness	.3 mils	1.5 mils	Specific Factors

Results

The results indicated that anodize time and temperature had average response values as shown in the following graphic.

Salt Spray Pits

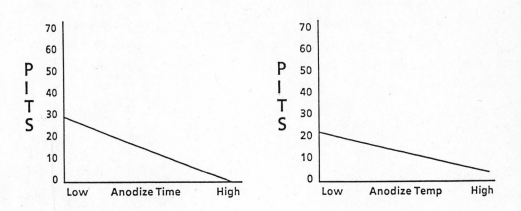

At first glance, it appeared that anodized time and temperature would play a moderate role in optimizing the anodize process. The optimal settings still seemed to favor high time and high temperature which were not the best for reducing costs. However, further investigation through the use of interaction graphs of time with CrO_3 and temperature with CrO_3 provided the following results.

Salt Spray Pits

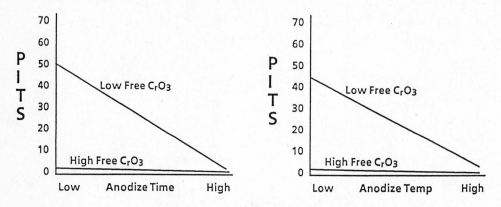

The interaction graphs revealed that when high free CrO_3 is used the time and temperature become unimportant for the anodize process. These two factors can now be set at levels which will speed up the process and reduce costs. The discovery of this very important interaction provided the knowledge required to produce a superior quality product at a lower cost.

Submitted by: Cadet First Class Jim Hecker
Cadet First Class Phil Herre
United States Air Force Academy

Statement of the Problem

The purpose of this experiment is to find the best design and construction of a paper helicopter through the use of experimentation.

Objective of the Experiment

Find a combination of the input factors which will maximize the hangtime.

Start Date: 22 JAN 1989

End Date: 27 JAN 1989

Quality Characteristic

Our response variable will be hangtime − time in the air. Hangtime is a quantitative variable (seconds), with an anticipated range of one to four seconds. The person who launches (drops) each helicopter from a height of ten feet will also time the fall in order to cut down the variability of reaction time.

Factors

Three controllable variables are considered − 1) ratio of blade length to width, 2) type of paper, and 3) the number of paper clips attached.

The ratio of blade length to width will have two levels: 2:1 and 1:2 (length to width). Paper is also a categorical variable with two levels: filler paper and construction paper. The paper clip is a quantitative variable with either one or two as the levels of interest. We expect all the interactions to be worthy of investigation, except for the three-way interaction of ratio-paper-clips.

The noise variables (uncontrollable and unavoidable) that may have an effect are:

1) **wind or drafts** – To control for these, we performed the experiments in a closed room.

2) **fall height** – The helicopter does not start to spin when released, creating a "fall height" where the helicopter does not spin. From observation, we assume the fall height to be random for each helicopter, therefore, not an important factor.

3) **reaction time** – In timing the hangtime, there may be some variance caused by the reaction time of hitting the timer when the helicopter is released and when it hits the ground. To control for this problem, the same person launched the helicopters and timed the fall in order to minimize and randomize reaction time.

Experimental Design

Due to the minimal cost, both in time and price, of a single run, we choose a full-factorial design for this experiment. A full-factorial has the advantage of testing all possible combinations of the factors and accounting for all possible interaction effects.

Since we had three factors at two levels, this meant we needed 8 runs. In order to check our results and to have a variance interval, we duplicated each run, meaning there were 16 total runs.

Here is how the design looked:

Run	Ratio	Paper	R–P	Clip	R–C	P–C	R–P–C	Y_1	Y_2	s^2
1	+	+	+	+	+	+	+	1.28	1.31	.00045
2	+	+	+	–	–	–	–	1.60	1.58	.00020
3	+	–	–	+	+	–	–	1.15	1.10	.00125
4	+	–	–	–	–	+	+	1.19	1.20	.00005
5	–	+	–	+	–	+	–	2.03	2.05	.00020
6	–	+	–	–	+	–	+	2.50	2.42	.00320
7	–	–	+	+	–	–	+	1.32	1.40	.00320
8	–	–	+	–	+	+	–	1.64	1.64	.00000

Ratio (length to width)		Paper		Clips	
+	1:2	+	filler	+	2
–	2:1	–	construction	–	1

Analysis Techniques

The primary method we used to analyze our results is a 2-way ANOVA table (ANalysis Of VAriance). We found the average value at each level of each factor, added the squared differences of each level from the grand mean (1.588) per factor, multiplied by the number that went into each average (4 runs each level × 2 runs) and divided by two. The error was calculated by adding the variances of each run. See Appendix 1 for the results.

Appendix 2 is a graph of the marginal means − the averages at each level per factor. The main factors − ratio, paper, and clips − clearly stand out, as do the three two-way interactions. The three-way interaction is insignificant, as expected.

Appendix 3 is a Pareto Diagram, which shows the half effects − half the difference of the averages. This plot is not as clear as the marginal means; however, it does show the importance of the main factors and the two-way interactions in comparison to the three-way interaction.

Since the variable per run was so small, it did not appear to be worthwhile to investigate dispersion effects.

Conclusions

Based on a simple technique for full-factorials called "pick-the-winner", supported by the marginal means plot, we concluded that the best combination is run six − 2:1 ratio, filler paper, one clip. Five confirmation runs were made of this design, giving an average hangtime of 2.378 seconds and a variance of 0.00868. These ratios are much better than those from any other combination.

Recommendations

It is our professional opinion that in order to make a paper helicopter with the greatest hangtime, a light paper with a small weight attached to the bottom with the blades twice as long as wide will be the best design.

We recommend for anyone else who tries this experiment to use a greater release height in order to create more variance in the results. Also, a three factor, three level design would not be that hard to accomplish: simply create another ratio, like a 1:1 length to width; find a third kind of paper, like onion-skin typing paper; and a third level for the additional weight, i.e., 3 clips or even no clips.

Appendix 1: ANOVA Table

	Ratio	Paper	Clip	R–P	R–C	P–C	R–P–C
Average +	1.301	1.846	1.455	1.477	1.630	1.543	1.578
Average –	1.875	1.330	1.721	1.705	1.546	1.634	1.599

Source	SS	df	MS	F_0
Ratio	1.318	1	1.318	1318
Paper	1.065	1	1.065	1065
Clip	0.283	1	0.283	283
R–P	0.219	1	0.219	219
R–C	0.028	1	0.028	28
P–C	0.033	1	0.033	33
R–P–C	0.002	1	0.002	2*
Error	0.009	8	0.001	–
Total	2.957	15	–	–

$$F_{CRIT} = F_{(0.05, 1, 8)} = 5.32$$
* Not significant because $F_0 < F_{CRIT}$

APPENDIX 2: Marginal Means Plot

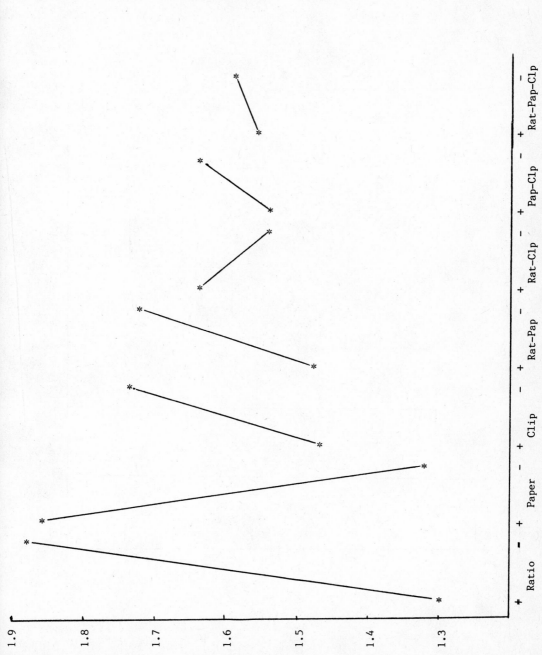

Appendix 3: Pareto Diagram

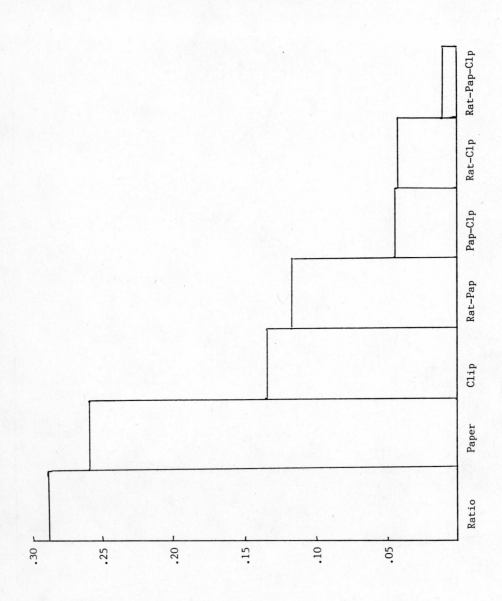

Appendix 4: The Optimal Design

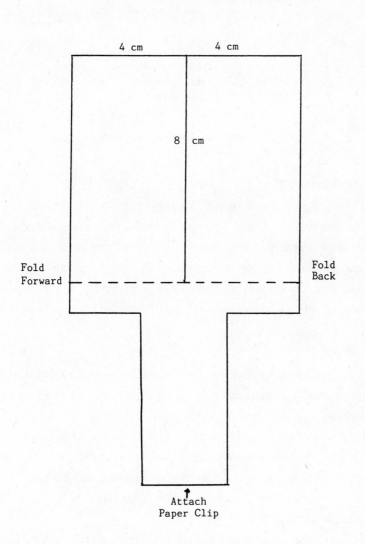

Submitted by: Col. David O. Swint, Professor
United States Air Force Academy
Engineering Design Project
Presented to the American Concrete Institute, Committee 125

Lt. Anderies	Lt. Kleinsmith	Lt. Milam	Lt. Peddycord
Lt. Bennett	Lt. Krause	Lt. Mingus	Lt. Platt
Lt. Bramer	Lt. Leante	Lt. Neulander	Lt. Shackleford
Lt. Finn	Lt. Martin	Lt. Novak	Lt. Smith
Lt. Haggard	Lt. Meyer	Lt. Noyes	

Optimizing Lunar Concrete

Problem Statement

The Spring 1989 Engineering 410 class at the U.S. Air Force Academy took on the project of making lunar concrete. The task at hand was to design, fabricate, test, and analyze concrete mixtures, mixed under ambient and semi-evacuated environments which optimized and maximized engineering properties. The design, construction, and testing of the specimens were performed in accordance with American Society for Testing Materials (ASTM) standards to ensure quality control and to minimize errors (Swint, 1989).

Background

Of the various types of lunar construction materials examined, concrete was found to be one of the most feasible (Lin, 1987). Dr. T. D. Lin has advocated the use of lunar soils and rocks as aggregates because of their quality physical characteristics (Lin, 1987). According to Dr. Lin, lunar materials contain sufficient amounts of silicate, aluminum, and calcium to produce the "cementitious material" required. Finally, when compared to three other major construction materials (aluminum alloy, mild steel, and glass), concrete requires less energy, and thus less cost, for its production. It is also temperature, radiation, and abrasion resistant (Lin, 1987).

A foreseeable problem with lunar construction of concrete lies in the hydration process required in making the concrete (Swint, 1989). Due to the evacuated conditions on the lunar surface, curing of the concrete would have to be accomplished in a pressure chamber to ensure that at least water vapor pressure is maintained. This would prevent the necessary amount of water for optimal concrete strength from being drawn out into a lower pressure environment.

Problem Solving Strategy

With the feasibility of lunar concrete as a construction material established, one is left with solving the problem of determining the best concrete to be used. To accomplish this task, the Engineering 410 group had to determine the best of 80 possible concrete combinations. The group used a D-optimal design to test 18 different combinations of cement, environment, plasticizer, wetting agent, and additive variables. From the results of the testing and the prediction equation, the characteristics of the remaining 62 combinations were estimated. The generation of these 18 test runs will be discussed in the section entitled "Generation of Test Runs."

Concrete Variables

Cement – Two types of cement were used for the experiment, Calcium Aluminate and Portland Type III. The Portland Type III contains approximately 65% calcium oxide, 23% silica, and 4% alumina (Lin, 1986). Calcium Aluminate is a high alumina cement (38% calcium oxide, 10% silica, 52% alumina oxide) and approximates well a cement that could be made from lunar soils. According to Dr. Lin, these proportions are satisfactory for the Portland cement to be considered a cementitious material (Lin, 1986).

Environment – The two mixing environments used during this experiment were ambient and semi-evacuated (mixed at 200 Torr) conditions. This was done to confirm the effect of mixing conditions on the resulting porosity of the concrete.

Plasticizer/Wetting Agent (admixture) – The concrete combinations were made with and without each of these variables. According to Dr. Lin, these variables would be two of the determining factors in the ultimate strength of the concrete produced.

Reinforcements – The reinforcements used were steel wire, aluminum wire, polymer, simulated lunar soil, and no additive. Though aluminum is known to react with the calcium oxide present in the cements, the group was asked to investigate its feasibility as an additive.

If it was a feasible additive, using aluminum already in space would eliminate the cost of shipping it to the moon.

Generation of Test Runs

With the five factors and their possible variations mentioned above, the group had 80 possible combinations of ingredients to test (see Appendix C). However, due to time and monetary constraints, it was not feasible to test all 80 combinations. After consulting with Lt. Col. Stephen R. Schmidt, the class used his work on Experimental Design theory for the project (Schmidt, 1988). Lt. Col. Schmidt consulted Dr. Lin to insure that the most probable interactions, those between environment, plasticizer, and wetting agent, were included in the design generation. These interactions combined with the other variables were used to generate the 18 combinations of ingredients that would provide the information necessary to predict the properties of the other 62 combinations (see "Prediction Process" section). The actual combinations generated can be seen in Appendix A.

Mixing Conditions

For each of the 18 combinations discussed above, one beam and three column specimens were produced. For every concrete batch made, certain mixing conditions were standardized.

Sand — The sand used in the concrete mixtures is Ottawa sand because its particle size distribution is similar to that of a lunar soil sample (Lin, 1987). The grain size distribution was the same proportion as that used by Dr. Lin, 4% of #30 sieved sand, 31% of #40 sieved sand, 45% of #50 sieved sand, and 20% of #100 sieved sand (to the nearest percent). It was sieved according to ASTM procedures, C136 (Lin, 1987).

Water — As with Dr. Lin's experiment, the distilled water was used.

Mix — Though the recommended ASTM mixture for Portland cement is 2.75:1:0.485 of sand-cement-water, the sand-cement ratio was adjusted to achieve the same level of workability and attain the same water-cement ratio as the Calcium Aluminate cement (Lin, 1987). The resulting ratio used by Dr. Lin was 1.75:1:0.485. Due to work from previous classes that produced different results, it was decided that a ratio likely to provide better results would be 1.75:1:0.435. This ratio was used in all concrete combinations poured.

Additives – The additives were added by volume after removing 2% of the sand from the mixtures. This was done because of the different densities of the various additives.

Mixing Time – Each specimen was mixed for 3 minutes before being placed in the molds. According to the Fall 1988 class, a 3-minute mix resulted in the strongest mix and removed a high percentage of air pockets from the specimens.

Curing Time – Dr. Lin's test results confirmed a 3- to 4-day cure of the Calcium Aluminate cement for optimal strength (Lin, 1987). For this reason, the Engineering 410 class allowed the Calcium Aluminate cement mixes to cure for 4 days. The Portland cement cured for 7 days according to standard testing procedures. The first day of curing took place in the molds in a 100% humidity environment, the molds were removed after 24 hours, and the remaining days of curing took place in a 100% humidity environment also.

Dimensions for the cylinder and beam specimens are:

Cylindrical Column: Diameter = 1"   Length = 2"
Square Cross Section Beam: Side dimension = 1.57"   Length = 6.69"

Testing Conditions

From each test run, the specimens were tested for compressive and flexural strength. A three-point load was placed on the beams to be tested (according to ASTM C 78-64) using the Satec Systems 120 WHVL Universal Testing Machine. The load rate used was 0.0025 in/min, consistent with the Spring and Fall 1988's load rate, and was applied until beam failure.

The columns were capped with sulfur (for uniform application of load), and strain gauges were applied with epoxy onto the columns after surface preparation. The wires were connected to a Vishay/Ellis 20 Strain Indicator and a load rate was set at 0.0012 in/min. The lower load rate (as compared with the beam loading) was done so the group could obtain the necessary information from the testing device. As with the beam, the columns were loaded to ultimate failure.

From these tests it was possible to determine maximum flexural strength (modulus of rupture), maximum compressive strength, modulus of elasticity, and Poisson's ratio.

Experimental Data

Two sets of specimens were prepared for each of the 18 experimental conditions shown in Appendix A. Each specimen was tested for compressive strength resulting in two response values (A and B) for each row of the design matrix. The average compressive strength per experimental condition is displayed in Appendix A in the column labeled "Compressive Actual."

Analysis

Once all the specimens were tested, the information from the testing was used to determine the performance of the specimens in terms of four different properties: maximum compressive strength, maximum flexural strength, Poisson's Ratio, and Young's Modulus of Elasticity. Maximum compressive strength is a measure of the compressive load per unit area a specimen can withstand before catastrophic failure occurs. Maximum flexural strength is the product of the bending moment times the distance from the neutral axis to the outer fibers divided by the area moment of inertia. Poisson's Ratio is the ratio of lateral strain to axial strain in the specimen. Young's Modulus of Elasticity is the ratio of stress versus strain which can be determined using the compressive load versus longitudinal deformation curve. It measures the stiffness of the specimen.

To determine the values for these four characteristics, a computer program, CNCRT.PAS was written. The input values for this program were: maximum compressive load, two points defining the load versus deformation curve, maximum flexural load, and the code defining the concrete mixture. The equations used in this program are as follows:

MAXIMUM COMPRESSIVE STRENGTH (see Appendix A)

$$f_c' = P_{max} / \text{Area}$$

POISSON'S RATIO

$$\mu = \frac{\text{LATERAL STRAIN}}{\text{AXIAL STRAIN}}$$

FLEXURAL STRENGTH

$$f_r = \frac{My}{I} = \frac{Pl}{bh^3}$$

YOUNG'S MODULUS OF ELASTICITY

$$E = \frac{STRESS}{STRAIN} = \frac{\sigma}{e}$$

Using the program, these four properties for both runs of the 18 combinations were obtained.

Prediction Process

At this point in the experiment, the group decided to carry on only with the compressive strength for the prediction equation. The reason this was done is that, although standard mixing and testing procedures were followed, the group was not confident about the measurements for the three properties of Poisson's Ratio, maximum flexural strength, and Young's Modulus of Elasticity. The group was, however, confident about the maximum compressive strength because of the simple procedure used to obtain the results. Having the values necessary to determine the maximum compressive strength, the prediction equation below was used to estimate the value of the compressive strength for the other 62 untested combinations.

$$\hat{Y} = \bar{T} + (\bar{C}_i - \bar{T}) + (\bar{E}_i - \bar{T}) + (\bar{P}_i - \bar{T}) + (\bar{A}_i - \bar{T})$$
$$+ [(\overline{P_iW_i} - \bar{T}) - (\bar{P}_i - \bar{T}) - (\bar{W}_i - \bar{T})]$$
$$+ [(\overline{W_iE_i} - \bar{T}) - (\bar{W}_i - \bar{T}) - (\bar{E}_i - \bar{T})]$$
$$+ [(\overline{E_iP_i} - \bar{T}) - (\bar{E}_i - \bar{T}) - (\bar{P}_i - \bar{T})]$$

(Schmidt, 1988)

In this equation, " $\hat{Y}$ " represents the predicted compressive strength for any concrete sample with a given combination of ingredients. " $\bar{C}_i$ " represents the average response when C is at level i. The " i " subscript allows for variation of the type of cement used. For example, $\bar{C}_1$ represents the average response ($f_c{}'$) for specimens made with Portland Type III cement, while $\bar{C}_2$ represents the average response ($f_c{}'$) for samples made with Calcium Aluminate cement. The other factors are interpreted in a similar fashion. The marginal averages can be located in Appendix B.

Just as the single factors provide marginal averages (average response), the double factor combinations provide information on interactions. An interaction is very similar to a marginal average except that instead of averaging the value of the property for experimental specimens containing one specific ingredient, one must average the value of the property for the experimental specimens containing the two ingredients in the interaction. For example, for " $\hat{Y}$ " representing compressive strength, $\overline{E_1P_1}$ is calculated by finding the average of the compressive strength for all of the 18 combinations containing both E_1 and P_1 (see Appendix B). " $\bar{T}$ " is the average value of compressive strength for all 18 of the experimental runs. Applying the prediction equation to any of the 80 possible combinations of ingredients, one would be able to determine the estimated compressive strength for a concrete specimen containing the given combination. For instance, if one wishes to determine the compressive strength for the specimen with the ingredients Calcium Aluminate (C_2), wetting agent (W_1), plasticizer (P_1), and no additives (A_1) mixed under ambient conditions (E_1), the equation would be as follows:

$$\hat{Y} = \bar{T} + (\bar{C}_2 - \bar{T}) + (\bar{E}_1 - \bar{T}) + (\bar{W}_1 - \bar{T}) + (\bar{P}_1 - \bar{T}) + (\bar{A}_1 - \bar{T})$$
$$+ [(\overline{P_1W_1} - \bar{T}) - (\bar{P}_1 - \bar{T}) - (\bar{W}_1 - \bar{T})]$$
$$+ [(\overline{W_1E_1} - \bar{T}) - (\bar{W}_1 - \bar{T}) - (\bar{E}_1 - \bar{T})]$$
$$+ [(\overline{E_1P_1} - \bar{T}) - (\bar{E}_1 - \bar{T}) - (\bar{P}_1 - \bar{T})]$$

In order to speed up the prediction process, two programs were written to implement the equation described above. The first program, 410NIT.FOR, accepts the property values for the 18 runs and generates all the marginal averages, interactions, and $\bar{T}$ value for the given property, i.e., compressive strength (see Appendices A & B).

This data is then inserted into the text of the second program, 410.FOR. This program iterates through all 80 combinations using the prediction algorithm above and gives the predicted $\hat{Y}$ value for each combination (see Appendix C). In this way it is possible to approximate the remaining, untested 62 conditions with the results of only 18 tests.

Optimal Combinations

Using the mathematical model, the group was able to predict the compressive strength for all of the 80 combinations. In determining the optimal combinations, the graphical results of the prediction algorithm were analyzed (see Appendix D). The high points marked the greatest value of compressive strength. From this graph, runs 61, 66, and 71 were of interest and chosen for confirmation testing (see Appendix D).

Confirmation Tests

Once the optimal runs were chosen, they were remixed and tested to confirm the results with those of the prediction algorithm. The combinations of interest were as follows:

1) **#61** – Calcium Aluminate cement, semi-evacuated mixing conditions, wetting agent, plasticizer, no reinforcement.

2) **#66** – Calcium Aluminate cement, semi-evacuated mixing conditions, wetting agent, no reinforcement.

3) **#71** – Calcium Aluminate cement, semi-evacuated mixing conditions, plasticizer, no reinforcement.

One mix, consisting of three cylinders and a beam, was done for each of the combinations of interest. The mixing and testing procedures were carried out exactly as they had been for the original 18 specimens.

The confirmation results (see Appendix D) exceeded our prediction (from 647 psi to 833 psi). Although the difference in predicted and confirmation results is in a direction advantageous to our objective, the group is still concerned as to what has caused this difference. One strong possibility is that an important interaction was not included in this design. To overcome this problem it is recommended that a follow-on experiment be conducted allowing for all two-way interactions to be estimated.

Conclusions

The purpose of this experiment was twofold. It was designed to find the optimum concrete mixture for lunar construction. It also used Lt. Col. Schmidt's prediction equation as a statistical tool in this process. The confirmation data supports the algorithm's predictions and thus shows that the design of experiments approach is a vital tool to reduce the experimenter's work. The group realizes that for actual lunar construction, many more inputs must be analyzed to obtain the optimal combination. However, the group also realizes that the use of the prediction equation is a must to decrease time and money while still obtaining the optimum concrete mixture.

List of References

"Design and Control of Concrete Mixtures." Portland Cement Association. 12th ed., 1979.

ENGR 410 class, Fall 1988. Final Report.

ENGR 410 class, Spring 1988. Display.

Flexural Strength of Concrete (Using Simple Beam with Third-Point Loading), ASTM C 78-64, pp 40-42.

Lin, T. D. "Concrete for Lunar Base Construction." The Portland Cement Association, 1987.

Lin, T. D., H. Love and D. Stark. "Physical Properties of Concrete Made With Apollo 16 Lunar Soil Sample." Construction Technology Laboratories, 1987.

Rader, Stanley P., Maj. (USAF) Consultant. Fall, 1988, Spring, 1989.

Schmidt, Stephen R., Lt. Col. (USAF) and Robert G. Launsby. *Understanding Industrial Designed Experiments*, 1988.

Shah, S. P. "Alternative Reinforcing Materials for Concrete Construction." *International Journal for Development Technology*, vol 1, 3-15 (1983).

Student Manual for Strain Gauge Technology. Raleigh, NC: Measurements Group, Inc., 1983.

Simmerer, Stephen J. (Capt., USA). "Lunar Construction: The Countdown Begins." *The Military Engineer*, Jul 1987, pp 354-357.

Swint, David O., Col. (USAF). Instruction and discussions. Spring, 1989.

APPENDIX A

Mixing Assignments

C_1 - Portland Type III cement P_1 - Plasticizer - lml

C_2 - Calcium Aluminate cement P_2 - No Plasticizer

E_1 - Ambient mixing A_1 - No Reinforcement

E_2 - Semi-Evacuated A_2 - Steel

                             A_3 - Aluminum

W_1 - Wetting Agent - 0.07 ml A_4 - Polymer

W_2 - No Wetting Agent A_5 - Simulated Lunar Soil

COMPRESSIVE STRENGTH (psi)

	Design Matrix					Response		Compressive Strength		Deviation
Run	C	E	W	P	A	A	B	Actual	Predicted	(%)
1	1	1	2	2	1	4132.94	3428.83	3780.89	3893.40	−2.98
2	2	1	1	1	1	5067.50	6042.79	5555.15	4962.02	+10.68
3	1	2	2	1	1	5207.55	4004.34	4605.95	5064.10	−9.95
4	2	2	1	2	1	6944.25	6009.69	6476.97	6119.87	+5.51
5	1	2	1	1	3	3101.61	2892.21	2946.91	2214.17	+24.86
6	1	1	1	2	3	2272.73	2639.43	2456.08	2188.99	+10.87
7	1	1	2	1	4	2317.30	4278.08	3297.69	2702.07	+18.06
8	1	2	2	2	5	3997.97	4507.27	4252.62	4030.06	+5.23
9	2	1	2	2	2	4010.00	3670.00	3840.00	3665.31	+4.55
10	2	2	1	1	4	4260.26	5458.38	4859.32	4989.60	−2.68
11	2	1	2	1	3	1914.95	1960.79	1937.87	2125.84	−9.70
12	2	1	1	2	5	4318.83	5672.28	4995.56	5100.78	−2.11
13	2	2	2	2	4	4870.00	4210.00	4540.00	4993.30	−9.98
14	2	2	2	1	2	5692.65	5692.65	5692.65	4863.01	+15.05
15	1	1	1	1	2	1706.14	2294.38	2000.26	2534.73	−26.72
16	2	2	1	1	5	3496.32	4219.52	3857.92	5125.97	−33.13
17	1	2	1	2	2	3808.30	3342.25	3575.28	3692.58	−3.28
18	2	2	2	2	3	2591.04	3179.28	2885.16	3317.47	−14.98

$$\overline{T} = 3975.35$$

APPENDIX B

MARGINAL AVERAGES

Variable	Compressive Strength
$\bar{C}_1$	3364.46
$\bar{C}_2$	4464.06
$\bar{E}_1$	3482.94
$\bar{E}_2$	4369.28
$\bar{W}_1$	4080.38
$\bar{W}_2$	3870.31
$\bar{P}_1$	3861.52
$\bar{P}_2$	4089.17
$\bar{A}_1$	5104.74
$\bar{A}_2$	3777.05
$\bar{A}_3$	2556.51
$\bar{A}_4$	4232.34
$\bar{A}_5$	4368.70

INTERACTIONS

Interaction	Compressive Strength
$\bar{W}_1, \bar{E}_1$	3751.76
$\bar{W}_1, \bar{E}_2$	4343.28
$\bar{W}_2, \bar{E}_1$	3214.11
$\bar{W}_2, \bar{E}_2$	4395.28
$\bar{E}_1, \bar{P}_1$	3197.74
$\bar{E}_1, \bar{P}_2$	3768.13
$\bar{E}_2, \bar{P}_1$	4392.55
$\bar{E}_2, \bar{P}_2$	4346.01
$\bar{W}_1, \bar{P}_1$	3843.91
$\bar{W}_1, \bar{P}_2$	4375.97
$\bar{W}_2, \bar{P}_1$	3883.54
$\bar{W}_2, \bar{P}_2$	3859.73

APPENDIX C

RUN #	C	E	W	P	A	VALUE
1	1	1	1	1	1	3862.417
2	1	1	1	1	2	2534.727
3	1	1	1	1	3	1314.186
4	1	1	1	1	4	2990.017
5	1	1	1	1	5	3126.379
6	1	1	1	2	1	4737.217
7	1	1	1	2	2	3409.527
8	1	1	1	2	3	2188.986
9	1	1	1	2	4	3864.817
10	1	1	1	2	5	4001.179
11	1	1	2	1	1	3574.465
12	1	1	2	1	2	2246.775
13	1	1	2	1	3	1026.234
14	1	1	2	1	4	2702.065
15	1	1	2	1	5	2838.427
16	1	1	2	2	1	3893.400
17	1	1	2	2	2	2565.710
18	1	1	2	2	3	1345.169
19	1	1	2	2	4	3021.000
20	1	1	2	2	5	3157.362
21	1	2	1	1	1	4762.402
22	1	2	1	1	2	3434.713
23	1	2	1	1	3	2214.712
24	1	2	1	1	4	3890.003
25	1	2	1	1	5	4026.365
26	1	2	1	2	1	5020.270
27	1	2	1	2	2	3692.580
28	1	2	1	2	3	2472.039
29	1	2	1	2	4	4147.870
30	1	2	1	2	5	4284.232
31	1	2	2	1	1	5064.096
32	1	2	2	1	2	3736.406
33	1	2	2	1	3	2515.865
34	1	2	2	1	4	4191.696
35	1	2	2	1	5	4328.059
36	1	2	2	2	1	4766.098
37	1	2	2	2	2	3438.408
38	1	2	2	2	3	2217.867
39	1	2	2	2	4	3893.698
40	1	2	2	2	5	4030.060

APPENDIX C

Run #	C	E	W	P	A	VALUE
41	2	1	1	1	1	4962.018
42	2	1	1	1	2	3634.328
43	2	1	1	1	3	2413.787
44	2	1	1	1	4	4089.618
45	2	1	1	1	5	4225.980
46	2	1	1	2	1	5836.818
47	2	1	1	2	2	4509.128
48	2	1	1	2	3	3288.587
49	2	1	1	2	4	4964.418
50	2	1	1	2	5	5100.780
51	2	1	2	1	1	4674.066
52	2	1	2	1	2	3346.376
53	2	1	2	1	3	2125.835
54	2	1	2	1	4	3801.666
55	2	1	2	1	5	3938.029
56	2	1	2	2	1	4993.001
57	2	1	2	2	2	3665.311
58	2	1	2	2	3	2444.770
59	2	1	2	2	4	4120.601
60	2	1	2	2	5	4256.963
61	2	2	1	1	1	5862.004
62	2	2	1	1	2	4534.313
63	2	2	1	1	3	3313.773
64	2	2	1	1	4	4989.604
65	2	2	1	1	5	5125.966
66	2	2	1	2	1	6119.871
67	2	2	1	2	2	4792.181
68	2	2	1	2	3	3571.640
69	2	2	1	2	4	5247.471
70	2	2	1	2	5	5383.833
71	2	2	2	1	1	6163.697
72	2	2	2	1	2	4836.007
73	2	2	2	1	3	3615.466
74	2	2	2	1	4	5291.297
75	2	2	2	1	5	5427.659
76	2	2	2	2	1	5865.699
77	2	2	2	2	2	4538.009
78	2	2	2	2	3	3317.468
79	2	2	2	2	4	4993.299
80	2	2	2	2	5	5129.661

APPENDIX D

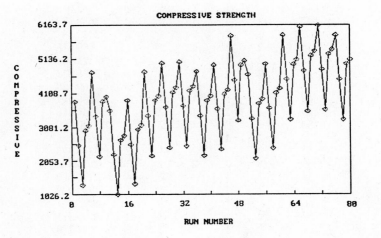

CONFIRMATION RESULTS

	Design Matrix					Response				Compressive Strength	
Test #	C	E	W	P	A	1	2	3	Average (psi)	Predicted	Deviation
61	2	2	1	1	1	4588.6	8225.1	7270.2	6694.6	5862.0	+832.6
66	2	2	1	2	1	6369.8	7109.8	7345.3	6941.6	6119.9	+821.7
71	2	2	2	1	1	5765.3	8641.5	6026.24	6811.0	6163.7	+647.3

Submitted by: Alan Arnholt
Steve Smith
Robert Kaliski

**Design of Experiments
Silane Doping in GaAs**
Statistics Seminar Project - University of Northern Colorado

I. Objective

In the growth of gallium arsenide (GaAs), a material used to fabricate electronic devices, the level of impurities in the crystal has a great influence on the electrical properties. The level of these trace impurities is known as the doping level. Since the doping level determines some of the electrical properties, we desire to have control of this level to tailor the electrical properties to the device or circuit requirements.

II. Response

For the operation of heterostructure bipolar transistors, there are two different doping levels of silicon in GaAs. These levels are 3×10^{16} cm^{-3} and 5×10^{18} cm^{-3}. The goal of this experimental design is to model doping level response, select the proper conditions to obtain the two doping levels, and then confirm the results. The response will be the actual doping level measured on a profiling instrument. This instrument can measure the doping level both accurately and precisely in a range of 10^{14} to 10^{19} cm^{-3}.

III. Factors

In the case of silicon doping in GaAs, three factors may influence the doping level: the silane flow, the arsine flow, and the trimethylgallium (TMGa) flow. There are other factors that could also modify the response, but the machinery is calibrated at these fixed levels. In addition, engineering is only interested in the perturbation of three factors and cannot afford to recalibrate the reactor. The three factors are quantitative and are highly controllable ($+/-$ 2% of set level), and literature and engineering experience suggest that no interactions exist.

IV. Experimental Constraints

The growth parameters selected are to be used in the growth of heterostructure in March. Hence, engineering prefers to have the data in three weeks (end of February) with 15 runs and a maximum of 30 runs including confirmation runs. Although cost is not a concern, each run requires four hours from start to completion.

V. Experimental Plan

Since engineering believes no interaction exists, we will use a fractional factorial design, the Taguchi $L_4(2^3)$, with two replications. The factors with both high and low levels are listed below.

Factors	−	level	+	Total range
silane flow	15		50	0 - 100
TMGa flow	20		35	5 - 50
arsine flow	100		140	90 - 150

The analysis will include a Pareto chart of the three main effects and also a determination of variance. A regression analysis will follow using linear terms. We plan to check the center point for validity of the linear model. If the linear model is inappropriate, we will then examine the response of the axial points and determine the appropriate regression equation. From the empirical regression, we finally will compute the correct levels and make at least two confirmation runs for each target level.

VI. Experiment and Analysis

Results:

Run #	A = Arsine	B = TMGa	C = Silane	Y1	Y2	S	Ln (S)
1	−1	−1	−1	0.70	0.66	0.0282	35.578
2	−1	+1	+1	2.0	1.6	0.282	37.881
3	+1	−1	+1	4.0	3.7	0.212	37.593
4	+1	+1	−1	1.6	1.2	0.282	37.881

Note: Response is in terms of 1×10^{17} cm^{-3}.   T = 1.93

Avg. [Ln(S)] = 37.233

Marginal Means:

	A	B	C
Avg +	2.63	1.60	2.83
Avg −	1.24	2.27	1.04
d	1.39	−0.67	1.79

Factor C is the largest location effect.

	A	B	C
Avg Ln(S.D.) +	37.731	37.881	37.737
Avg Ln(S.D.) −	36.730	36.586	36.730
d	1.007	1.295	1.007

From our limited data, no factor appears to stand out as a dispersion reduction factor.

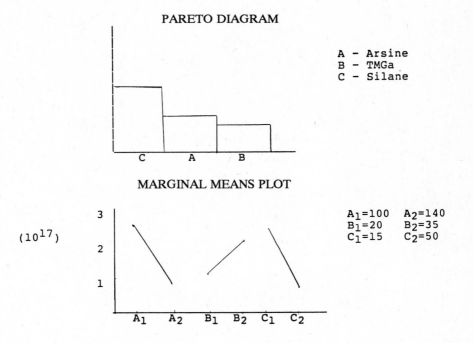

PARETO DIAGRAM

A – Arsine
B – TMGa
C – Silane

MARGINAL MEANS PLOT

(10^{17})

$A_1 = 100 \quad A_2 = 140$
$B_1 = 20 \quad B_2 = 35$
$C_1 = 15 \quad C_2 = 50$

TEST FOR SIGNIFICANCE OF REGRESSION

1) H_0: $A = B = C = 0$
   H_1: $A \neq 0$ or $B \neq 0$ or $C \neq 0$

2) a. F-test
   b. $\alpha = 0.05$
   c. DF = (3, 4)
   d. CV for $F_\alpha = 0.05, 3, 4 = 6.59$

$$
X = \begin{bmatrix} 1 & -1 & -1 & -1 \\ 1 & -1 & 1 & 1 \\ 1 & 1 & -1 & 1 \\ 1 & 1 & 1 & -1 \\ 1 & -1 & -1 & -1 \\ 1 & -1 & 1 & 1 \\ 1 & 1 & -1 & 1 \\ 1 & 1 & 1 & -1 \end{bmatrix}
\quad
X' = \begin{bmatrix} 1 & 1 & 1 & 1 & 1 & 1 & 1 & 1 \\ -1 & -1 & 1 & 1 & -1 & -1 & 1 & 1 \\ -1 & 1 & -1 & 1 & -1 & 1 & -1 & 1 \\ -1 & 1 & 1 & -1 & -1 & 1 & 1 & -1 \end{bmatrix}
\quad
Y = \begin{bmatrix} 0.7 \\ 2.0 \\ 4.0 \\ 1.6 \\ 0.66 \\ 1.6 \\ 3.7 \\ 1.2 \end{bmatrix}
$$

$$
X'X = \begin{bmatrix} 8 & 0 & 0 & 0 \\ 0 & 8 & 0 & 0 \\ 0 & 0 & 8 & 0 \\ 0 & 0 & 0 & 8 \end{bmatrix}
\quad
(X'X)^{-1} = \begin{bmatrix} 0.125 & 0 & 0 & 0 \\ 0 & 0.125 & 0 & 0 \\ 0 & 0 & 0.125 & 0 \\ 0 & 0 & 0 & 0.125 \end{bmatrix}
$$

$$
\underline{B} = (X'X)^{-1}X'Y = \begin{bmatrix} 1.93 \\ 0.63 \\ -0.33 \\ 0.89 \end{bmatrix}
\qquad
Y = 1.93 + 0.69A - 0.33B + 0.89C
$$

$$
\underline{B}X'Y = \begin{bmatrix} 1.93 & 0.69 & -0.33 & 0.89 \end{bmatrix} \begin{bmatrix} 15.46 \\ 5.54 \\ -2.66 \\ 7.14 \end{bmatrix} = 40.892
$$

$$Y'Y = 41.1756 \qquad \frac{(\sum y_i)^2}{n} = 29.876$$

Source	SS	DOF	MS	F_0
Regression	$S_R = \underline{B}X'Y - (\sum y_i)^2 / n = 11.016$	$K = 3$	3.672	51.93
Error	$SS_E = Y'Y - \underline{B}X'Y = 0.2823$	$n - k - 1 = 4$	0.0707	
Total	$SS_T = Y'Y - (\sum y_i)^2 / n = 11.299$	$n - 1 = 7$		

4) Reject H_0, $F_0 = 51.93 > F_\alpha = 0.05, 3, 4 = 6.59$

5) We found evidence that at least $A \neq 0$ or $B \neq 0$ or $C \neq 0$
   $R^2 = SS_R / SS_T = 0.975$

TEST ON INDIVIDUAL REGRESSION COEFFICIENTS

1) a. H_0: $A = 0$ b. H_0: $B = 0$ c. H_0: $C = 0$
        H_1: $A \neq 0$ H_1: $B \neq 0$ H_1: $C \neq 0$

2) a. t-test
    b. $\alpha = 0.05$
    c. $DF = n - k - 1$
    d. CV for $t_{0.05/2,4} = 2.776$

3) Calculations

$$t_0 = \underline{B}_j / \sqrt{\sigma^2 c_{jj}}$$

   a. $t_0 = 0.69 / \sqrt{(0.0707)(0.125)} = 7.339$

   b. $t_0 = -0.33 / \sqrt{(0.0707)(0.125)} = -3.51$

   c. $t_0 = 0.89 / \sqrt{(0.707)(0.125)} = 9.467$

4) Reject H_0 for A, B, and C since $|t_0| > t_{\alpha/2, n - k - 1}$
    a. $|t_0| = 7.339 > t = 2.776$
    b. $|t_0| = 3.51 > t = 2.776$
    c. $|t_0| = 9.467 > t = 2.776$

5) We found evidence that A, B, and C are all non-zero.

From the experimental data, all three factors are significant while no factors have any greater influence on the variability. Since the variability of each factor is approximately equal, we will not concern ourselves with the standard deviation any further.

Using the half-effects of the marginal means, the regression equation is:

$$\underline{Y} = (1.93 + 0.69*A - 0.33*B + 0.89*C) \ 10^{17} \ cm^{-3}.$$

With two runs at the center point A = B = C = 0, the responses are 2.0 and 2.2 (10^{17} cm^{-3}). The average of the center points is 2.1. This value does not appear to be deviate enough from 1.93 to justify the addition of curvature to the regression model.

To reach the lower target of 0.3, we need to set A = −1 and B = 1 by inspection. Solving the regression equation for C, we obtain a coded value for C of −0.67. Uncoding this value by using the inverse of the standard coding transformation yields a silane setting of 21. Two confirmation runs were performed with the three factors set at the specified levels. The responses are 0.27 and 0.30 (10^{17} cm^{-3}) which are acceptable.

VII. Conclusions

To achieve the lower target value, we recommend the arsine be set at 90, the TMGa set at 50, and the silane set at 21. Since the upper target cannot be met with the present machine configuration, we recommend the engineering staff make changes to the machine to achieve this doping level or change their target level.

Submitted by: Randy Boudreaux
Cray Research, Inc.

Bonding Gold Wire to Gallium Arsenide Wafers: Overcoming Pre-Conceived Notions

Statement of the Problem:

The opening of a new Cray plant in Colorado Springs posed an interesting problem. Gold wire bonding to Gallium Arsenide wafers had been optimized at the Wisconsin plant and, using the same parameters, should have optimized the process in Colorado. The solution was not to be that simple. Several months passed while "best guess" and "one-factor-at-a-time experimentation" were used to find an improved process. After hearing and reading the materials in this text, it was decided to try a designed experiment matrix as shown in Chapter 3. Several lessons learned from this first experiment are:

1. Proper experimental designs can be used with little training provided adequate guidance is available.

2. Do not consume all of your resources during the initial experiment – things may go wrong, or the information gained may lead to the need for further experimentation.

3. Preconceived notions are a barrier to process improvements, but they can be overcome through properly designed experiments.

4. Designed experimentation is a much more powerful tool than previously used "best guess" or one-factor-at-a-time approaches.

The Design:

Prior to the actual experimentation, the brainstorming session generated five input factors to be tested along with two responses of interest as shown in Table 1.

INPUT FACTORS	LEVELS
Force	60, 150
Time	40, 100
Ultrasonics	40, 150
Temperature	125, 175
Z-Speed	50, 150
RESPONSE	TYPE
Bond/No Bond	Attribute
Bond Strength	Variable

Table 1

The L_8 design matrix and response values for Bond/No Bond are shown in Table 2.

Results:

As you can see, some errors were made in Z-speed settings for runs 2, 3, and 8. For runs 2 and 8, the problem was associated with machine limitations. Run 3 should have been set at 150, but was inadvertently set incorrectly. These problems will limit our ability to obtain unambiguous results for each factor's effectiveness; however, there still exists a very important finding from comparing the results for each of the 8 runs. For example, prior to collecting the data, it was generally accepted that runs 1, 2, and 6 would be poor performers and were only run because they showed up in the design matrix. Run 6 was obviously a surprise — among other things, preconceived notions had previously prevented this successful combination from being tested.

Conclusions:

Not only did the experimental matrix provide the best combination of input factors, resulting in 100% good bonds, these bonds all exceeded the minimum bond strength requirements. In fact, the bond strength testing device could not knock them off. The knowledge gained in this one experiment was worth many times that of four months of one-at-a-time experiments.

TABLE 2
DESIGN MATRIX AND RESPONSE VALUES FOR BONDING EXPERIMENT

RUN	FORCE	TIME	U/S	TEMP	Z-SPEED	1	2	3	4	5	6	7	8	9	10	11	12	13	14	15	16	17	18	19	20	21	22	23	24
1	60	40	40	125	50	0	0	0	0	0	0	0	0	0	0	0	0	0	0	0	0	0	0	0	0	0	0	0	0
2	60	40	150	175	131	0	0	0	0	0	0	0	0	0	0	0	0	0	0	0	0	0	0	0	0	0	0	0	0
3	60	100	40	125	50	1	1	0	0	0	0	0	0	0	0	0	1	1	1	1	0	0	0	1	1	0	1	1	0
4	60	100	150	175	50	0	0	0	0	0	0	0	1	1	0	0	0	0	0	0	0	0	0	0	0	0	0	0	0
5	60	40	40	175	150	1	1	1	1	1	1	1	0	0	0	0	0	0	0	0	0	1	1	1	1	1	1	0	0
6	150	40	150	125	50	1	1	1	0	1	1	1	1	1	0	1	1	1	1	1	1	1	1	1	1	1	1	1	1
7	150	100	40	175	50	1	1	1	1	0	1	1	1	1	1	1	1	1	1	1	1	1	1	1	1	1	1	0	1
8	150	100	150	125	131	1	1	1	1	1	1	1	1	0	1	1	1	1	1	1	0	1	1	1	1	0	1	1	1

Submitted by: Don F. Wilson
Qinas, Inc.
Littleton, Colorado

Constant Velocity Joints on a Front Wheel Drive
Analysis of Multiple Responses from Designed Experiments

ABSTRACT: This case study presents methods of visualizing and analyzing multiple responses from a designed experiment. This case study is a product development example – elastomeric boots for constant velocity joints on front wheel drive automobiles. Methods presented include use of linear graphs to represent interactive factors, blocking to eliminate noise factors, use of percent contribution to identify strong relationships, analysis of non-parametric (qualitative) data, and compromises to balance performance of several key objectives.

BACKGROUND: The use of constant velocity (CV) joints on front wheel drive automobiles presents a unique challenge for development of elastomeric products. The CV joints are permanently sealed with a boot to contain the grease and to prevent contaminants from entering. If undetected by the customer, the tearing or loosening of the elastomeric boot may require replacement of an entire CV joint, costing the customer in excess of $350.00 per CV joint.

   The boot must be able to flex at angles up to 45 degrees; operate at speeds up to 1900 RPM without more than 7mm of radial growth; maintain resistance to attack by ozone, oxidation, and grease; perform at temperatures below −40 degrees and over 200 degrees; and provide cut resistance to road hazards.

   At the time of this case study, the product dimensions had been determined by the customer and most of the key functional parameters of the CV boots had been met. The primary focus of this study was to improve the resistance of the boot to ozone attack, low temperature properties, and radial growth characteristics at high RPM, while maintaining the performance of previous rubber formulations on other functional tests.

DETERMINING OBJECTIVES: The primary objective of the experiment was to substantially improve the dynamic ozone characteristics of the formulation without adversely affecting the other requirements:

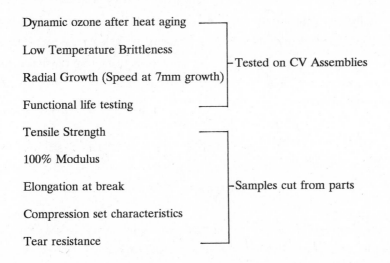

Dynamic ozone after heat aging

Low Temperature Brittleness

Radial Growth (Speed at 7mm growth) — Tested on CV Assemblies

Functional life testing

Tensile Strength

100% Modulus

Elongation at break — Samples cut from parts

Compression set characteristics

Tear resistance

PRIMARY OBJECTIVE: The customer tests the ozone resistance of the formulation on the CV boot, on an accelerated life test preceded by heat aging the samples to simulate internal deterioration (inner noise) of the product. The specification was to show no cracking under 7X magnification after 168 hours of running the CV joint at a 34 degree angle at 23 mph (350 RPM) in an atmosphere of 50 ppm of ozone. At the time of this study, minute cracks were visible at 168 hours (no cracks at 144 hours). Several attempts to "patch" the formulation with one-at-a-time changes had failed. Parts had been marginally passing the radial growth test and low temperature flex requirements. Any change to the rubber formulation might cause problems with these tests.

MEASURING THE OBJECTIVE: The dynamic ozone test at the customer's facility could not be directly duplicated in the development lab because 1) the expense of the test equipment was high, 2) the time necessary to acquire the equipment would delay development, and 3) correlation problems between labs existed. To circumvent these problems, a "predictor" test had been developed in the supplier's development lab to show relative improvements in ozone performance of formulations.

BRAINSTORMING POTENTIAL CAUSES / "KNOBS": The primary cause of deterioration of the rubber boot is the ozone level in the atmosphere. However, the intent of the experiment is not to control the root cause (ozone), but to make the boots "robust" or insensitive to the "outer" noise of ozone. The major factors to improve ozone resistance relate to either reducing the stress level of the rubber in operation or using protective chemicals in the formulation.

PRIORITIZING THE INVESTIGATIONS: Since the product design had been finalized, and modulus requirements were necessary, the focus of the experiment was on modifying the rubber formulation to improve chemical resistance to ozone.

The rubber formulation consists of thirteen ingredients, but the chemist believed only four ingredients were potential major contributors to improving the ozone performance.

SELECTING FACTORS FOR THE EXPERIMENT: The four factors were to be characterized by both type and amount, creating seven Control Factors for the experiment:

Neoprene Type	(N)
Antiozonant Type	(Z)
Antiozonant Level	(ZL)
Antioxidant Type	(X)
Antioxidant Level	(XL)
Wax Type	(W)
Wax Level	(WL)

One Noise Factor was considered to be dominant in the ozone testing — the variability of the ozone concentration from test to test.

Interactions of the three protective ingredients were considered to be strong possibilities, creating seven suspected interactions:

Z×ZL	(Antiozonant Type × Antiozonant Level)
X×XL	(Antioxidant Type × Antioxidant Level)
W×WL	(Wax Type × Wax Level)
Z×X	(Antiozonant Type × Antioxidant Type)
Z×W	(Antiozonant Type × Wax Type)
X×W	(Antioxidant Type × Wax Type)
ZL×XL	(Antiozonant Level × Antioxidant Level)

SELECTING LEVELS FOR EACH FACTOR: Two types of neoprene (GNA, GW) were selected to investigate differences in performance on the three major responses: Dynamic Ozone, Low Temperature, and Radial Growth. Technical literature indicated that stress levels might be significantly different in the neoprene types to improve performance characteristics in all three areas.

The selection of the protective ingredients were based on technical literature and previous lab screenings. The levels of the antioxidant and antiozonant were selected to produce "adequate" to "improved" protection, but below the maximum solubility of the materials in the formulation. The levels of the wax types were varied from 0 to 4 parts per hundred (pph) of base polymer to examine a published theory that way may improve static ozone resistance, but inhibit dynamic ozone resistance.

CHOOSING THE APPROPRIATE DESIGN: With the seven factors listed above and selecting two levels per factor, all combinations of seven factors at two levels each would require 128 formulations. Running the 128 trials would produce information on all interactions of the factors. However, resources for the experiment were limited. The dynamic ozone chamber in the development lab would test only 8 samples at a time, and each test required approximately 2 weeks (32 weeks for 128 combinations). The cost of testing each formulation was $250.00, or $32,000 for 128 combinations.

The minimum use of resources to examine the effects of each factor would be 8 trials, using a one-at-a-time strategy, or an L8 design (a fully saturated fractional factorial). A one-at-a-time strategy would not permit analysis of interactions and would be very sensitive to individual test results. The L8 Design would confound all two-way interactions with the factor effects (3 2-way interactions with each factor), but factor effects would be based on averages of 4 data points each. Larger fractional factorial designs (L16, L32, L64) would reduce confounding of interactions at the expense of spending more resources. Selection of non-geometric orthogonal designs (L12, L20, L24, L28) would prevent total confounding of interactions with factors, but would not permit analysis of the interactions.

The chemist decided to select a design that would prevent confounding of the seven factors and seven interactions that were suspected, but allowing the other 14 potential interactions to be confounded with factors and/or interactions. Using the linear graph method developed by Genichi Taguchi, the experiment can be accomplished using the L16 design:

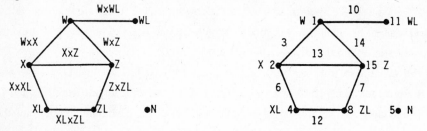

REVIEWING THE DESIGN: The linear graph does not show the confounding of 14 other potential two-way interactions, or higher order interactions. Using an interaction table, the following confounding situations were identified:

FACTOR			POTENTIAL 2-WAY INTERACTIONS	
1	W	WAX TYPE	N×XL	
2	X	ANTIOX TYPE		
3	W×X	WAX × ANTIOX	ZL×WL	
4	XL	ANTIOX LVL	W×N	WL×Z
5	N	NEOPRENE	W×XL	
6	X×XL			
7	Z×ZL		X×N	
8	ZL	ANTIOZ LVL		
9	(not assigned)		W×XL	X×WL
10	W×WL		X×ZL	N×Z
11	WL	WAX LEVEL	XL×Z	
12	XL×ZL			
13	X×Z		N×ZL	
14	W×Z		N×WL	
15	Z	ANTIOZ	XL×WL	

The two-way interactions that were confounded were not considered to be any of the "vital few" effects that would significantly alter ozone performance.

A review of the formulation combinations showed all combinations to be feasible for the experiment:

Design No.	WAX TYPE	ANTIOX TYPE	ANTIOX LEVEL	NEOPRENE TYPE	ANTIOZ LEVEL	WAX LEVEL	ANTIOZ TYPE
1	PARAFIN	X1	1 PHR	GNA	1 PHR	0 PHR	Z1
2	PARAFIN	X1	1 PHR	GNA	3 PHR	4 PHR	Z2
3	PARAFIN	X1	3 PHR	GW	1 PHR	0 PHR	Z2
4	PARAFIN	X1	3 PHR	GW	3 PHR	4 PHR	Z1
5	PARAFIN	X2	1 PHR	GNA	1 PHR	4 PHR	Z2
6	PARAFIN	X2	1 PHR	GNA	3 PHR	0 PHR	Z1
7	PARAFIN	X2	3 PHR	GW	1 PHR	4 PHR	Z1
8	PARAFIN	X2	3 PHR	GW	3 PHR	0 PHR	Z2
9	POLY.	X1	1 PHR	GW	1 PHR	4 PHR	Z2
10	POLY.	X1	1 PHR	GW	3 PHR	0 PHR	Z1
11	POLY.	X1	3 PHR	GNA	1 PHR	4 PHR	Z1
12	POLY.	X1	3 PHR	GNA	3 PHR	0 PHR	Z2
13	POLY.	X2	1 PHR	GW	1 PHR	0 PHR	Z1
14	POLY.	X2	1 PHR	GW	3 PHR	4 PHR	Z2
15	POLY.	X2	3 PHR	GNA	1 PHR	0 PHR	Z2
16	POLY.	X2	3 PHR	GNA	3 PHR	4 PHR	Z1

DETERMINING THE SEQUENCE OF EXPERIMENTING: The test samples could be prepared in random sequence (to reduce the chances of a nuisance variable affecting the results). However, the variability of the dynamic ozone test (8 tests/setup) was likely to affect the test results substantially. Factor 9 was used to block the order of testing to isolate the effect of the ozone test variation.

L16 DESIGN - SETUP TABLE
SORTED BY TEST ORDER

Trial Order	Design No.	Wax Type	Antiox Type	Antiox Level	Neoprene Type	Antioz Level	Test Block	Wax Level	Antioz Type
1	3	Parafin	X1	3 PHR	GW	1 PHR	1st	0 PHR	Z2
2	10	Poly.	X1	1 PHR	GW	3 PHR	1st	0 PHR	Z1
3	7	Parafin	X2	3 PHR	GW	1 PHR	1st	4 PHR	Z1
4	14	Poly.	X2	1 PHR	GW	3 PHR	1st	4 PHR	Z2
5	12	Poly.	X1	3 PHR	GNA	3 PHR	1st	0 PHR	Z2
6	5	Parafin	X2	1 PHR	GNA	1 PHR	1st	4 PHR	Z2
7	1	Parafin	X1	1 PHR	GNA	1 PHR	1st	0 PHR	Z1
8	16	Poly.	X2	3 PHR	GNA	3 PHR	1st	4 PHR	Z1
9	15	Poly.	X2	3 PHR	GNA	1 PHR	2nd	0 PHR	Z2
10	13	Poly.	X2	1 PHR	GW	1 PHR	2nd	0 PHR	Z1
11	6	Parafin	X2	1 PHR	GNA	3 PHR	2nd	0 PHR	Z1
12	11	Poly.	X1	3 PHR	GNA	1 PHR	2nd	4 PHR	Z1
13	9	Poly.	X1	1 PHR	GW	1 PHR	2nd	4 PHR	Z2
14	8	Parafin	X2	3 PHR	GW	3 PHR	2nd	0 PHR	Z2
15	4	Parafin	X1	3 PHR	GW	3 PHR	2nd	4 PHR	Z1
16	2	Parafin	X1	1 PHR	GNA	3 PHR	2nd	4 PHR	Z2

PERFORMING EXPERIMENTS/OBTAINING RESULTS: As anticipated, the dynamic ozone results (hours to first cracks observed) changed dramatically from the first test group to the second, averaging 342 hours and 174 hours, respectively. The chemist, on noting the difference, requested a supplier (DuPont) to perform a "second opinion" test at their laboratories, using a different apparatus. Results were compiled in a data table:

Trial Order	Design No.	Dynamic Ozone	DuPont Ozone	Ozone Rating	Modulus 100%	Tensile PSI	Elong %	Hi Speed Spin	Low Temp
7	1	384	4	2B	380	1870	325	2250	−49
16	2	168	5	2B+	390	1930	350	2300	−47
1	3	360	5	2B+	300	2150	375	2100	−45
15	4	192	2	1B+	315	2070	350	2100	−46
6	5	288	7	3B	360	1970	300	2250	−48
11	6	168	5	2B+	420	1930	325	2350	−48
3	7	240	7	3B	300	2020	375	2050	−46
14	8	168	5	2B+	340	2060	400	2200	−47
13	9	168	5	2B+	295	2140	325	2150	−44
2	10	408	3	2B−	340	2180	375	2150	−46
12	11	216	2	1B+	370	2030	325	2150	−48
5	12	408	4	2B	415	2060	300	2300	−49
10	13	168	7	3B	310	2200	450	2100	−47
4	14	336	6	3B−	325	2170	425	2100	−45
9	15	144	7	3B	390	2080	300	2300	−48
8	16	312	6	3B−	395	2050	350	2250	−47

ANALYZING DATA: The data for each response was analyzed using DESIGN CUBE software. Analysis of Variance (ANOVA) tables and Level Effects reports were generated for each response. The analysis for the Dynamic Ozone responses are shown on the following page. To determine significant effects, factors and interactions with less importance were "pooled" into error terms.

The results on Dynamic Ozone show the TEST BLOCK (the difference between the first and second test groups) to be the largest effect on ozone performance (79.9% contribution), followed by ANTIOXIDANT TYPE (10.1%), and WAX LEVEL (3.6%). The pooling process may have created "significant" results out of insufficient information by pooling the smaller sources of variation preferentially. Thus, the Probability column (1−alpha) may not be accurate. Reviewing the tables of Level Means and Level Effects shows the preferred Antioxidant is X1. The preferred wax level is 0 PHR. The interaction plots show a preference for 3 PHR of Antioxidant if X1 is selected.

The Dynamic Ozone results, however, were questionable, considering the large variation between tests. Therefore, a "second opinion" test was conducted by DuPont on a different method of dynamic ozone testing. All samples were run for 336 hours in the ozone chamber and removed. Unfortunately, the DuPont measurement criteria was not a continuous quantitative scale as with the "time to cracking" criteria. The data was qualitative, using a visual rating scale, from 1B− (very slight cracks under 7X magnification) to 3B+ (almost ready to break). The data from the DuPont Test was reviewed in two ways: 1) By sorting the trials from low to high ratings and 2) by assigning a numerical rating scale to the visual ratings (1B = 1 . . . 3B+ = 8).

CV BOOT COMPOUNDING STUDY   L16 DESIGN

Grp.	Factor	Name	Level 1	Level 2	2 way interactions
I	1 W	WAX TYPE	PARAFIN	POLY.	2x3 4x5 6x7 8x9 10x11 12x13 14x15
II	2 X	ANTIOX.	X1	X2	1x3 4x6 5x7 8x10 9x11 12x14 13x15
II	3 WxX				1x2 4x7 5x6 8x11 9x10 12x15 13x14
III	4 XL	ANTIOX LVL	1 PHR	3 PHR	1x5 2x6 3x7 8x12 9x13 10x14 11x15
III	5 N	NEOPRENE	GNA	GW	1x4 2x7 3x6 8x13 9x12 10x15 11x14
III	6 XxXL				1x7 2x4 3x5 8x14 9x15 10x12 11x13
III	7 ZxZL				1x6 2x5 3x4 8x15 9x14 10x13 11x12
IV	8 ZL	ANTIOZ LVL	1 PHR	3 PHR	1x9 2x10 3x11 4x12 5x13 6x14 7x15
IV	9 B	TEST BLOCK	1st	2nd	1x8 2x11 3x10 4x13 5x12 6x15 7x14
IV	10 WxWL				1x11 2x8 3x9 4x14 5x15 6x12 7x13
IV	11 WL	WAX LEVEL	0 PHR	4 PHR	1x10 2x9 3x8 4x15 5x14 6x13 7x12
IV	12 XLxZL				1x13 2x14 3x15 4x8 5x9 6x10 7x11
IV	13 XxZ				1x12 2x15 3x14 4x9 5x8 6x11 7x10
IV	14 WxZ				1x15 2x12 3x13 4x10 5x11 6x8 7x9
IV	15 Z	ANTIOZ.	Z1	Z2	1x14 2x13 3x12 4x11 5x10 6x9 7x8

Trial Order	Design Order	1	2	3	4	5	6	7	8	9	10	11	12	13	14	15	DYNAMIC OZONE
7	1	1	1	1	1	1	1	1	1	1	1	1	1	1	1	1	384
16	2	1	1	1	1	1	1	1	2	2	2	2	2	2	2	2	168
1	3	1	1	1	2	2	2	2	1	1	1	1	2	2	2	2	360
15	4	1	1	1	2	2	2	2	2	2	2	2	1	1	1	1	192
6	5	1	2	2	1	1	2	2	1	1	2	2	1	1	2	2	288
11	6	1	2	2	1	1	2	2	2	2	1	1	2	2	1	1	168
3	7	1	2	2	2	2	1	1	1	1	2	2	2	2	1	1	240
14	8	1	2	2	2	2	1	1	2	2	1	1	1	1	2	2	168
13	9	2	1	2	1	2	1	2	1	2	1	2	1	2	1	2	168
2	10	2	1	2	1	2	1	2	2	1	2	1	2	1	2	1	408
12	11	2	1	2	2	1	2	1	1	2	2	1	2	1	2	1	216
5	12	2	1	2	2	1	2	1	2	1	1	2	1	2	1	2	408
10	13	2	2	1	1	2	2	1	1	2	2	1	1	2	2	1	168
4	14	2	2	1	1	2	2	1	2	1	1	2	2	1	1	2	336
9	15	2	2	1	2	1	1	2	1	2	2	1	1	2	1	2	144
8	16	2	2	1	2	1	1	2	2	1	1	2	2	1	1	1	312

Grand Average   258.00   Averages   258.00   0.00

ANOVA TABLE

ERROR (e)	SOURCE	NAME	df	SUM OF SQUARES	MEAN SQ.	% CONTR.	F	Prob.
	1 W	WAX TYPE	1	2304.00	2304.00	1.54	18.67	.9960**
	2 X	ANTIOX.	1	14400.00	14400.00	10.12	116.67	.9998**
e	3 WxX		1	0.00	0.00	0.00		
e	4 XL	ANTIOX LVL	1	144.00	144.00	0.00		
e	5 N	NEOPRENE	1	144.00	144.00	0.00		
	6 XxXL		1	1296.00	1296.00	.83	10.50	.9858 *
e	7 ZxZL		1	144.00	144.00	0.00		
	8 ZL	ANTIOZ LVL	1	2304.00	2304.00	1.54	18.67	.9960**
	9 B	TEST BLOCK	1	112896.00	112896.00	79.91	914.67	.9999**
	10 WxWL		1	576.00	576.00	.32	4.67	.9341
	11 WL	WAX LEVEL	1	5184.00	5184.00	3.59	42.00	.9993**
e	12 XLxZL		1	144.00	144.00	0.00		
	13 XxZ		1	1296.00	1296.00	.83	10.50	.9858 *
e	14 WxZ		1	144.00	144.00	0.00		
e	15 Z	ANTIOZ.	1	144.00	144.00	0.00		
		Replications	0	0.00	0.00	0.00		
		Error	7	864.00	123.43	1.31		
		Total	15	141120.00		100.00		

Standard Error   11.11

CV BOOT COMPOUNDING STUDY
L16 DESIGN

FACTOR	NAME	LEVEL	LEVEL MEANS	LEVEL EFFECTS	TOTAL EFFECTS
1 W	WAX TYPE	1 PARAFIN	246.00	-12.00	24.00**
		2 POLY.	270.00	12.00	
2 X	ANTIOX.	1 X1	288.00	30.00	-60.00**
		2 X2	228.00	-30.00	
3 WxX		1	258.00	0.00	0.00
		2	258.00	0.00	
4 XL	ANTIOX LVL	1 1 PHR	261.00	3.00	-6.00
		2 3 PHR	255.00	-3.00	
5 N	NEOPRENE	1 GNA	261.00	3.00	-6.00
		2 GW	255.00	-3.00	
6 XxXL		1	249.00	-9.00	18.00 *
		2	267.00	9.00	
7 ZxZL		1	261.00	3.00	-6.00
		2	255.00	-3.00	
8 ZL	ANTIOZ LVL	1 1 PHR	246.00	-12.00	24.00**
		2 3 PHR	270.00	12.00	
9 B	TEST BLOCK	1 1st	342.00	84.00	-168.00**
		2 2nd	174.00	-84.00	
10 WxWL		1	264.00	6.00	-12.00
		2	252.00	-6.00	
11 WL	WAX LEVEL	1 0 PHR	276.00	18.00	-36.00**
		2 4 PHR	240.00	-18.00	
12 XLxZL		1	261.00	3.00	-6.00
		2	255.00	-3.00	
13 XxZ		1	267.00	9.00	-18.00 *
		2	249.00	-9.00	
14 WxZ		1	255.00	-3.00	6.00
		2	261.00	3.00	
15 Z	ANTIOZ.	1 Z1	261.00	3.00	-6.00
		2 Z2	255.00	-3.00	

Grand Average   258.00

DYNAMIC OZONE
INTERACTION PLOTS

ANTIOXIDANT TYPE x ANTIOXIDANT LEVEL

ANTIOX LVL

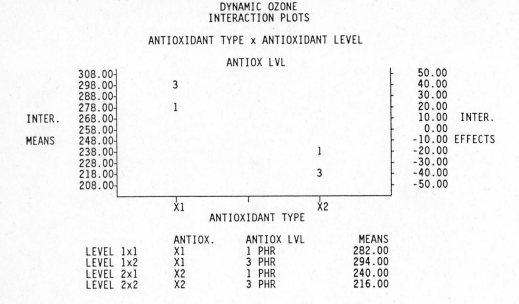

```
        308.00-     3                                    50.00
        298.00-                                          40.00
        288.00-                                          30.00
        278.00-     1                                    20.00
INTER.  268.00-                                          10.00   INTER.
        258.00-                                           0.00
MEANS   248.00-                                         -10.00  EFFECTS
        238.00-                          1              -20.00
        228.00-                                         -30.00
        218.00-                          3              -40.00
        208.00-                                         -50.00
               X1                       X2
                    ANTIOXIDANT TYPE
```

	ANTIOX.	ANTIOX LVL	MEANS
LEVEL 1x1	X1	1 PHR	282.00
LEVEL 1x2	X1	3 PHR	294.00
LEVEL 2x1	X2	1 PHR	240.00
LEVEL 2x2	X2	3 PHR	216.00

WAX TYPE x WAX LEVEL INTERACTION

WAX LEVEL

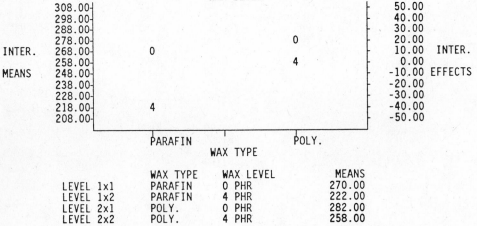

```
        308.00-                                          50.00
        298.00-                                          40.00
        288.00-                                          30.00
        278.00-                          0              20.00
INTER.  268.00-     0                                    10.00   INTER.
        258.00-                          4               0.00
MEANS   248.00-                                         -10.00  EFFECTS
        238.00-                                         -20.00
        228.00-                                         -30.00
        218.00-     4                                   -40.00
        208.00-                                         -50.00
            PARAFIN                      POLY.
                    WAX TYPE
```

	WAX TYPE	WAX LEVEL	MEANS
LEVEL 1x1	PARAFIN	0 PHR	270.00
LEVEL 1x2	PARAFIN	4 PHR	222.00
LEVEL 2x1	POLY.	0 PHR	282.00
LEVEL 2x2	POLY.	4 PHR	258.00

The results from the DuPont Test tended to confirm Antioxidant X1 as the largest contributor in the formulation for ozone protection, but did not confirm the effects of the Wax Level. ANOVA Analysis using the Numerical rating scale appeared to show an effect from the ANTIOXIDANT TYPE (59.5% contribution), the ANTIOZONANT TYPE (9.5%), the ANTIOZONANT LEVEL (9.5%), and an ANTIOXIDANT TYPE / ANTIOZONANT TYPE interaction (9.5%).

DUPONT OZONE RESULTS
SORTED BY VISUAL RATING

Des No	WAX TYPE	ANTIOX TYPE	ANTIOX LEVEL	NEOPRENE TYPE	ANTIOZ LEVEL	TEST BLOCK	WAX LEVEL	ANTIOZ TYPE	DUPONT VISUAL RATING	DUPONT NUMER. RATING
11	POLY.	X1	3 PHR	GNA	1 PHR	2nd	4 PHR	Z1	1B+	2
4	PARAFIN	X1	3 PHR	GW	3 PHR	2nd	4 PHR	Z1	1B+	2
10	POLY.	X1	1 PHR	GW	3 PHR	1st	0 PHR	Z1	2B-	3
12	POLY.	X1	3 PHR	GNA	3 PHR	1st	0 PHR	Z2	2B	4
1	PARAFIN	X1	1 PHR	GNA	1 PHR	1st	0 PHR	Z1	2B	4
6	PARAFIN	X2	1 PHR	GNA	3 PHR	2nd	0 PHR	Z1	2B+	5
3	PARAFIN	X1	3 PHR	GW	1 PHR	1st	0 PHR	Z2	2B+	5
9	POLY.	X1	1 PHR	GW	1 PHR	2nd	4 PHR	Z2	2B+	5
8	PARAFIN	X2	3 PHR	GW	3 PHR	2nd	0 PHR	Z2	2B+	5
2	PARAFIN	X1	1 PHR	GNA	3 PHR	2nd	4 PHR	Z2	2B+	5
14	POLY.	X2	1 PHR	GW	3 PHR	1st	4 PHR	Z2	3B-	6
16	POLY.	X2	3 PHR	GNA	3 PHR	1st	4 PHR	Z1	3B-	6
7	PARAFIN	X2	3 PHR	GW	1 PHR	1st	4 PHR	Z1	3B	7
13	POLY.	X2	1 PHR	GW	1 PHR	2nd	0 PHR	Z1	3B	7
5	PARAFIN	X2	1 PHR	GNA	1 PHR	1st	4 PHR	Z2	3B	7
15	POLY.	X2	3 PHR	GNA	1 PHR	2nd	0 PHR	Z2	3B	7

ANALYSIS OF MULTIPLE RESPONSES: Each response can be analyzed separately by ANOVA to determine the effect of each factor on each response. To gain a better understanding of the relationships of each factor to each response, Multiple Response Summary Tables can be constructed to show a matrix of the factors and product characteristics on a single page. The Summary Tables may display the Level Means, Level

Effects, or Percent Contribution. The Percent Contribution Summary Table shows the relative effect of each factor on each product characteristic and can be used to quickly identify the "vital few" factors, versus the "trivial many." The Percent Contribution Summary Table for this experiment shows the importance of antioxidant selection for Ozone Protection, and the importance of Neoprene selection for Modulus, Tensile, Elongation, High Speed Spin, and Low Temperature properties. The Wax selection appears important to maintain tensile strength. Factors of lesser significance are also identified.

The Percent Contribution Table does not indicate the preferred level for each factor — only the relative importance. Summarizing information by Level Effects shows which level is preferred for any particular response. The Level Effect is obtained by subtracting the Grand Average from the Mean Response of each Factor Level. Thus, a prediction can be made by adding the Level Effects for any combination of Factor Levels to the Grand Average. For example, the Tensile prediction for the combination of Poly Wax with GNA Neoprene would be: 2056.88 (Grand Average) + 56.88 (Poly Wax Level Effect) − 66.88 (GNA Neoprene Level Effect) = 2046.88, assuming other Factor effects to be insignificant.

The Level Effects Summary Table shows the preferred Levels of Neoprene Type to be GW Neoprene for Tensile and Elongation Properties, but GNA Neoprene for High Speed Spin and Low Temperature. The selection of Neoprene Type is dependent on the relative importance of the response criteria. For factors that have no substantial relationship to any responses, the factor level with the lower cost would be the choice.

CV BOOT COMPOUNDING STUDY
SUMMARY TABLE - % CONTRIBUTION

FACTOR	NAME	LEVEL	DYNAMIC OZONE	DUPONT OZONE	MODULUS 100%	TENSILE PSI	ELONG. %	HI SPEED SPIN	LOW TEMP BRITTLE.
1 W	WAX TYPE	1 PARAFIN 2 POLY.	1.63	0.00	.28	36.56	.52	.48	.79
2 X	ANTIOX.	1 X1 2 X2	10.20	59.52	.28	.11	8.38	.48	.79
3 WxX		1 2	0.00	2.38	.28	.75	8.38	.48	.79
4 XL	ANTIOX LVL	1 1 PHR 2 3 PHR	.10	2.38	.00	.75	2.09	1.93	.79
5 N	NEOPRENE	1 GNA 2 GW	.10	0.00	81.05	50.55	52.36	69.56	63.78
6 XxXL		1 2	.92	2.38	.05	2.76	.52	1.93	.79
7 ZxZL		1 2	.10	0.00	.05	2.34	13.09	1.93	7.09
8 ZL	ANTIOZ LVL	1 1 PHR 2 3 PHR	1.63	9.52	12.64	.00	2.09	7.73	0.00
9 B	TEST BLOCK	1 1st 2 2nd	80.00	2.38	.05	.04	0.00	1.93	0.00
10 WxWL		1 2	.41	2.38	.00	.53	.52	0.00	3.15
11 WL	WAX LEVEL	1 0 PHR 2 4 PHR	3.67	0.00	4.81	.99	.52	7.73	7.73
12 XLxZL		1 2	.10	0.00	.14	.22	.52	.48	3.15
13 XxZ		1 2	.92	9.52	.00	.11	.52	.48	3.15
14 WxZ		1 2	.10	0.00	.28	2.34	8.38	.48	0.00
15 Z	ANTIOZ.	1 Z1 2 Z2	.10	9.52	.05	1.95	2.09	4.35	3.15

☐☐ = STRONG CONTRIBUTOR TO RESPONSE

☐ = LESSER CONTRIBUTOR TO RESPONSE

CV BOOT COMPOUNDING STUDY
SUMMARY TABLE - LEVEL EFFECTS

FACTOR	NAME	LEVEL	DYNAMIC OZONE	DUPONT OZONE	MODULUS 100%	TENSILE PSI	ELONG. %	HI SPEED SPIN	LOW TEMP BRITTLE.
1 W	WAX TYPE	1 PARAFIN	-12.00	0.00	-2.19	-56.88	-3.12	6.25	-.12
		2 POLY.	12.00☺	0.00	2.19	56.88●	3.12	-6.25	.12
2 X	ANTIOX.	1 X1	30.00●	-1.25●	-2.19	-3.12	-12.50	-6.25	.12
		2 X2	-30.00	1.25	2.19	3.12	12.50☺	6.25	-.12
3 WxX		1	0.00	.25	-2.19	8.12	12.50☺	-6.25	.12
		2	0.00	-.25	2.19	-8.12	-12.50	6.25	-.12
4 XL	ANTIOX LVL	1 1 PHR	3.00	.25	-.31	-8.12	6.25	12.50	.12
		2 3 PHR	-3.00	-.25	.31	8.12	-6.25	-12.50	-.12
5 N	NEOPRENE	1 GNA	3.00	0.00	37.19	-66.88	-31.25	75.00●	-1.12●
		2 GW	-3.00	0.00	-37.19	66.88●	31.25●	-75.00	1.12
6 XxXL		1	-9.00	.25	.94	-15.62	-3.12	12.50	.12
		2	9.00	-.25	-.94	15.62	3.12	-12.50	-.12
7 ZxZL		1	3.00	0.00	.94	-14.38	15.62☺	-12.50	-.38☺
		2	-3.00	0.00	-.94	14.38	-15.62	12.50	.38
8 ZL	ANTIOZ LVL	1 1 PHR	-12.00	.50	-14.69	.62	-6.25	-25.00	0.00
		2 3 PHR	12.00☺	-.50☺	14.69	-.62	6.25	25.00☺	0.00
9 B	TEST BLOCK	1 1st	84.00?	.25	-.94	1.88	0.00	-12.50	0.00
		2 2nd	-84.00?	-.25	.94	-1.88	0.00	12.50	0.00
10 WxWL		1	6.00	-.25	.31	-6.88	3.12	0.00	.25
		2	-6.00	.25	-.31	6.88	-3.12	0.00	-.25
11 WL	WAX LEVEL	1 0 PHR	18.00☺	0.00	9.06	9.38	3.12	25.00☺	-.50☺
		2 4 PHR	-18.00	0.00	-9.06	-9.38	-3.12	-25.00	.50
12 XLxZL		1	3.00	0.00	-1.56	-4.38	-3.12	6.25	-.25
		2	-3.00	0.00	1.56	4.38	3.12	-6.25	.25
13 XxZ		1	9.00	-.50☺	-.31	-3.12	-3.12	-6.25	-.25
		2	-9.00	.50	.31	3.12	3.12	6.25	.25
14 WxZ		1	-3.00	0.00	2.19	-14.38	-12.50	6.25	0.00
		2	3.00	0.00	-2.19	14.38	12.50☺	-6.25	0.00
15 Z	ANTIOZ.	1 Z1	3.00	-.50☺	.94	-13.12	6.25	-18.75	-.25
		2 Z2	-3.00	.50	-.94	13.12	-6.25	18.75	.25
Grand Average			258.00	5.00	352.81	2056.88	353.12	2193.75	-46.88

● = STRONG PREFERENCE FOR LEVEL SETTING

☺ = MILD PREFERENCE FOR LEVEL SETTING

CV BOOT COMPOUNDING STUDY
SUMMARY TABLE - LEVEL MEANS

FACTOR	FACTOR	LEVEL	DYNAMIC OZONE	DUPONT OZONE	MODULUS 100%	TENSILE PSI	ELONG. %	HI SPEED SPIN	LOW TEMP BRITTLE.
1 W	WAX TYPE	1 PARAFIN	246.00	5.00	350.62	2000.00	350.00	2200	-47.00
		2 POLY.	270.00	5.00	355.00	2113.75	356.25	2188	-46.75
2 X	ANTIOX.	1 X1	288.00	3.75	350.62	2053.75	340.62	2188	-46.75
		2 X2	228.00	6.25	355.00	2060.00	365.62	2200	-47.00
3 WxX		1	258.00	5.25	350.62	2065.00	365.62	2188	-46.75
		2	258.00	4.75	355.00	2048.75	340.62	2200	-47.00
4 XL	ANTIOX LVL	1 1 PHR	261.00	5.25	352.50	2048.75	359.38	2206	-46.75
		2 3 PHR	255.00	4.75	353.12	2065.00	346.88	2181	-47.00
5 N	NEOPRENE	1 GNA	261.00	5.00	390.00	1990.00	321.88	2269	-48.00
		2 GW	255.00	5.00	315.62	2123.75	384.38	2119	-45.75
6 XxXL		1	249.00	5.25	353.75	2041.25	350.00	2206	-46.75
		2	267.00	4.75	351.88	2072.50	356.25	2181	-47.00
7 ZxZL		1	261.00	5.00	353.75	2042.50	368.75	2181	-47.25
		2	255.00	5.00	351.88	2071.25	337.50	2206	-46.50
8 ZL	ANTIOZ LVL	1 1 PHR	246.00	5.50	338.12	2057.50	346.88	2169	-46.88
		2 3 PHR	270.00	4.50	367.50	2056.25	359.38	2219	-46.88
9 B	TEST BLOCK	1 1st	342.00	5.25	351.88	2058.75	353.12	2181	-46.88
		2 2nd	174.00	4.75	353.75	2055.00	353.12	2206	-46.88
10 WxWL		1	264.00	4.75	353.12	2050.00	356.25	2194	-46.62
		2	252.00	5.25	352.50	2063.75	350.00	2194	-47.12
11 WL	WAX LEVEL	1 0 PHR	276.00	5.00	361.88	2066.25	356.25	2219	-47.38
		2 4 PHR	240.00	5.00	343.75	2047.50	350.00	2169	-46.38
12 XLxZL		1	261.00	5.00	351.25	2052.50	350.00	2200	-47.12
		2	255.00	5.00	354.38	2061.25	356.25	2188	-46.62
13 XxZ		1	267.00	4.50	352.50	2053.75	350.00	2188	-47.12
		2	249.00	5.50	353.12	2060.00	356.25	2200	-46.62
14 WxZ		1	255.00	5.00	355.00	2042.50	340.62	2200	-46.88
		2	261.00	5.00	350.62	2071.25	365.62	2188	-46.88
15 Z	ANTIOZ.	1 Z1	261.00	4.50	353.75	2043.75	359.38	2175	-47.12
		2 Z2	255.00	5.50	351.88	2070.00	346.88	2212	-46.62

DETERMINING SETTINGS: The original set of priorities indicated that the Dynamic Ozone, High Speed Spin, and Low Temperature Brittleness characteristics were the most important responses. Through use of the Summary Tables, the following formulation was derived:

MATERIAL	AMOUNT	TO IMPROVE:
Neoprene GNA		High Speed Spin, Low Temperature
Antioxidant X1	3 phr	Dynamic Ozone
Antiozonant Z1	3 phr	(Dynamic Ozone ?)
Wax - Poly.	1 phr	(Ozone / Tensile ?)

This combination had not been run as one of the 16 formulations in the experiment. Instead of "Picking the Winner" from among the 16 formulations, the analysis is predicting a different combination. Predictions of average performance with the above combination (conservative estimates using Grand Average + largest Level Effects):

Dynamic Ozone:	Improved, but can't predict hours
DuPont Ozone:	$5.0 - 1.25 - .5 - .5 - .5 = 2.25 = 1B+$ rating
High Speed Spin:	$2194 + 75 + 25 =$ Approx 2300 RPM
Low Temp. Brittleness:	$-46.88 - 1.12 = -48$ deg. F.
Tensile:	$2057 + 57 - 67 = 2047$ PSI
Elongation:	$353 - 31 - 15.5 - 12.5 - 12.5 + 12.5 = 294\%$
Modulus:	$353 + 37 + 15 = 405$ PSI

VERIFYING RESULTS: Results from the experiment trials should always be verified before action is taken. Mixing the above formulation gave results close to the prediction for Tensile, Elongation, Modulus, and High Speed Spin. The dynamic ozone improvements were predicted to be better than previous formulations, but could not be verified in running hours because of the large variability in testing. Samples of CV boots were sent to the customer for functional testing and all parts passed the performance tests as predicted.

Through Statistical Process Control, the Tensile, Elongation, and Modulus tests were monitored to maintain performance. Subsequent process improvements were introduced to improve consistency of the manufactured product.

The methods incorporated in this experiment have been used frequently in the development and production of CV boots and other rubber products. Through the use of Design of Experiments and Statistical Process Control, a better understanding of the chemistry of the formulation and manufacture of the boots was achieved.

Submitted by: Peter L. Knepell
PhD, Operations Research
Logicon/RDA

A Strategic Defense Case Study

Minutes of the First Meeting:

"O.K. folks, here's the architecture we need to study," explained Capt. Kirk, an enthusiastic strategic defense expert. "Don't look at the picture too long because it might change any minute. Some things do remain constant:

- a two-tiered interceptor system (one is space-based and the other is ground-based),
- command centers,
- space-based and ground-based sensors to support the command centers and interceptors, and
- communications links.

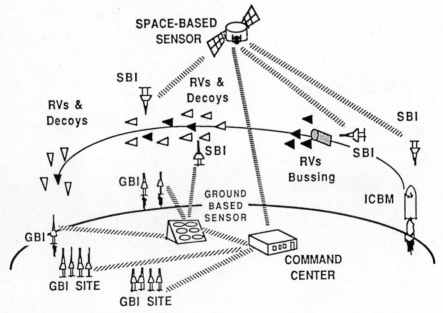

"Look, this study is quite simple." "It's a simple matter of input/output, like knobs on a radio." His knobs were labeled:

Number of SBI	Space-Based Interceptors (SBI)
Number of GBI	Ground-Based Interceptors (GBI)
RVs	Reentry Vehicles (alias: blivits or bombs)
Decoys	The objects used to look like RVs
Pk	The one-shot probability of kill of an interceptor against a target
Trelease	The time it takes someone to give the command to shoot at missiles
Discrimination factor	A measure of how well real missiles are picked over decoys
Tdiscrim	The time it takes to track and discriminate an RV so that an interceptor can shoot at it

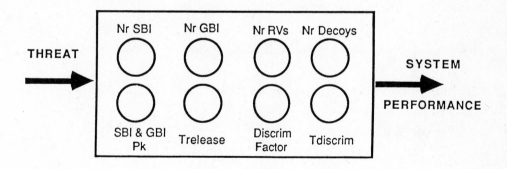

This stimulated a discussion on what measures will be used to gauge 'system performance.' The fundamental measure agreed upon was RVs destroyed. Then things got interesting and everyone got a chance to add knobs to the system.

"Not enough knobs," said Dee Coy, a renowned US expert on Russian threats through the year 2010. "You've got to consider the number of fast burn boosters and fast bussers." (Bussing is a term describing the act of casting several RVs off an ICBM in the proper direction − it's not a rapid transit system.)

"Don't forget the ASAT (anti-satellite) threat," piped in Dr. Larry Leber, a national labs scientist. "A good ASAT attack could 'smear' SBIs all over the galaxy."

"I have a new SBI design that will make all ASATs ineffective," countered Dr. S. B. Iacocca.

"Yeah, but what about SEOs?" remarked the consultant from ERUDITE Corporation.

"Yep, don't forget SEOs."   "SEOs are tricky."   "Oh no, SEOs!"

It got very quiet, especially after I asked, "What are SEOs?"

"Survivability Enhancement Options," someone said disdainfully.

"Whose?" I asked undaunted − after all, I was conducting the 'simple input/output' study.

"The SBI's of course. You've got to consider these in your study," said Kirk.

"What exactly do the SBIs do?"

"THEY haven't decided yet. Anyway, if I knew, I couldn't tell you − it's classified above your level."

"When will THEY decide?"

"Soon, probably after you finish the study."

Minutes of the Second Meeting:

I was much better prepared this time. "Gentlemen, this table shows all factors that you want me to consider.

PARAMETER	LEVELS	PARAMETER	LEVELS
ASAT Attack	4 types	Nr of Decoys	2 levels
SBI Design	2 types	Nr Fast-Burn Boosters	2 levels
Pk	2 levels		
Quantity	4 levels	Trelease	3 levels
GBI Design	2 types	Discrimination Factor	2 levels
Pk	2 levels		
Quantity	4 levels	Tdiscrim	2 levels
FULL FACTORIAL DESIGN = $4 \times 2 \times 2 \times 4 \times 2 \times 2 \times 4 \times 2 \times 2 \times 3 \times 2 \times 2$ = 49152 cases			

I am prepared to conduct a complete factorial analysis (49,152 cases). Since I can simulate an entire war in only 15 minutes, I'll be prepared to brief the results in 512 days (737,280 minutes) − as long as the computer doesn't crash."

"I want an answer in three months," boomed Capt. Kirk.

"O.K., how about a fractional factorial design where I'll tell you which factors were dominant in affecting system performance," I offered.

"No good," he said. "I need to make clear statements about the architecture THEY are designing. All factors are important!"

"All right, please tell me which factors are more likely to remain fixed while we vary everything else." The following list was developed:

SBI & GBI design & Pk	Trelease
Number of decoys	Tdiscrim
Number of fast-burn boosters	Discrimination factor

"I see your problem with the number of factors and I have a suggestion," offered Dee Coy. "If we composed two different threat scenarios, which differ by the number of fast-burn boosters and decoy composition, we could combine two of your factors into one called 'strategic threat.' Then we can measure the effects separately since they occur at different times in the ICBM flyout."

This was an excellent suggestion except for one consideration: "Could there be an interaction effect between the number of fast-burn boosters and decoy composition?" I asked.

Dee Coy was very accommodating: "Given this architecture, there is no reason to believe that there is an interaction. I might also suggest that the effects of Pk, Trelease, and Tdiscrim could be checked in just a few cases to give us a feel for their impact."

"So gentlemen, let's summarize this discussion. It appears that the following factors are foremost in your minds:

ASAT Threat	SBI Design
Number of SBI	GBI Design
Number of GBI	Strategic Threat

Then Dr.Leber had an interesting suggestion: "Why not see how the SBI design performs against the ASAT threats."

"But I want a complete tradeoff analysis for varying the numbers of SBIs and GBIs," said Kirk.

"Gentlemen, if we all agree that our main objective is to conduct in interceptor tradeoff analysis and our secondary objective is to see how changes in other parameters affect these trades, then I think I can have a study design very soon!"

Minutes of the Third Meeting:

I opened the meeting by restating our objective: "If we all agree that our primary mission is to conduct a study on the tradeoffs (i.e., effects) involved with varying the number of SBIs and GBIs in the architecture, I think I have a proposition we all can live with. In particular, I will discuss a study that will support tradeoff analyses for 10 different scenarios and some other limited excursions with less than 200 simulations. With that number of runs, I can meet our schedule."

Here's what I presented: For each of the 10 scenarios, I have a "design space" representing different numbers of interceptors. I would conduct 18 simulations at the lattice points shown.

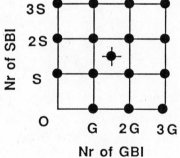

(I don't need to simulate the origin for obvious reasons.)

"There are only 16 lattice points shown. Why 18 runs?"

The reason: "The center point is simulated (replicated) three times to get a measure of variation. This will support the fit of a quadratic response surface. The contours of this surface (level curves) can then be plotted in the design space as shown."

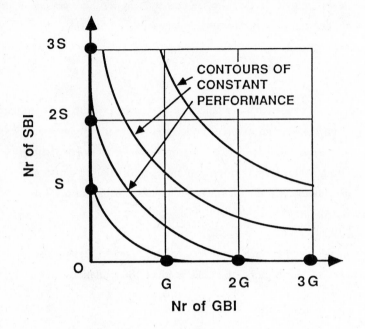

"What do these curves represent?" asked Capt. Kirk.

"One curve shows which combinations of SBIs and GBIs give a constant number of RVs killed. In fact, if you have some baseline number of RV kills that you want to compare the results to, I can generate that curve and highlight it in this display."

"So we could call these curves 'isomorts' for constant mortality," responded Kirk.

At this point, a vocal critic said: "This is statistical voodoo. You must replicate at all points at least 6 times and use the average to counteract variation."

This would increase the number of simulation runs to almost 1000! Discussions of statistical design, analysis of variance, etc. did not calm this persistent nay sayer. Nor did an appeal to the fact that these methods have been accepted by the most respected people in the field. After many unsuccessful attempts we entered into an interesting dialogue:

"You agree that there is a chance for variation at each point in the design space?" I asked.

"Of course," he replied.

"And you agree that with sufficient data, a representative response surface can be created."

"Yes."

"But your fear is that with only one replication for 15 of the points, the response could be too high or too low."

"Correct."

"Could the results at these points <u>all</u> come in too high?"

"Yes, but that would be rare."

"Well then, what would you expect to happen across 16 points in the design space?"

"I guess some would come in too high and some too low."

"Just like we would see with replications at one point?"

"Yes," he agreed.

"Well then, it seems like the 16 points will act like replications where some are too high, some are too low, but together, they should balance each other out to provide a close approximation of the true response surface."

"O.K., but how is this all put together to make a statistical statement?"

"That's voodoo — F statistics!" I answered cryptically.

Capt. Kirk looked at the data and, while he liked the contour plots, he suggested another display which is more appealing to his tastes.

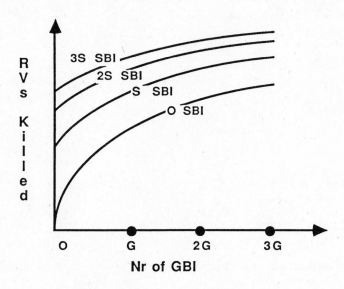

"These curves are essentially generated by cutting the response surface with four planes parallel to the GBI axis, where these planes cut through fixed levels of SBIs," I noted.

"Enough on display of data; what are your scenarios?" asked Kirk.

I suggested the following five cases of SBI type and ASAT threat be considered:

SBI TYPE

ASAT MIX	I	II
NO ASATS	CASE A	
FORCE 1	CASE B	
FORCE 2	CASE C	CASE D
FORCE 3		CASE E

The audience was in wide agreement. I then asked, "Which of these cases would represent a baseline, that we should examine while changing other factors?" For the first time a consensus choice was easy — Case D was chosen as a baseline for all other sensitivity studies. Then using Case D, I suggested that we run complete tradeoff studies while changing settings in:

Discrimination Factor	2 settings
Strategic Threat	2 settings
GBI Design	one change

Thus with the five cases for ASAT-SBI factors and the five excursions noted above, I got agreement on the study design. Each scenario required 18 simulation runs for a total of 180 to provide tradeoff analyses for all scenarios. We then could conduct a handful of simulations to provide a feel for the effects of Pk, Trelease, and Tdiscrim.

Epilog:

This study took six months, primarily due to a massive software development effort to simulate the ASAT threat and the SEOs. Results were widely appreciated for two reasons: people liked our display of results (the tradeoff contours and the tradeoff curves) and we answered a multitude of questions without confounding the results. Specifically, we:

- Characterized the effects and interactions when the number of interceptors was varied (tradeoff analysis)

- Provided a statement on the effects of:
    ASAT Attack
    Interceptor Design
    Strategic Threat
    Discrimination Factor

- Provided a feel for the effects of variations in:
    Pk
    Trelease
    Tdiscrim

What was not appreciated by the customer, but was fundamental to our success, was the importance of a study design to limit the number of simulations. Given time and resource constraints, we had to limit the number of simulation runs early and design our study around that limit. When the limit was discussed with the customer, he had a much greater appreciation of our need to use an efficient design as well as our need to limit the scope of the study. Future studies will now focus on some of the more salient results of our initial study.

Submitted by: Dr. Jack E. Reece, Sr. Principal Process Engineer
David Daniel, Process Engineering Supervisor
Honeywell Solid State Electronics Division

Ms. Robin Bloom, Sr. Process Engineer
United Technologies Microelectronics Center
Colorado Springs, CO

Identifying a Plasma Etch Process Window

1.   The Problem:

Removing a top layer in the presence of an underlayer is a common requirement in the manufacture of integrated circuits. The process must etch the layer uniformly across the wafer and must maintain considerable selectivity to avoid damaging any underlayer. Furthermore, to minimize the risk of damaging other parts of the circuit, operators must have a convenient means for detecting the end point when the process has removed the target layer.

For approximately one year preceding the start of this study, four similar processes run in this equipment had given satisfactory etch rate uniformity and selectivity. In the process a relatively low-powered pulsed plasma activated an etchant gas at low pressure to remove the target layer. Unexpectedly, operators began having difficulty determining end points for the etch process, presumably due to the breakdown of some shielding or circuit that had protected the detector from the pulsing RF in the reactor. In addition while the uniformity of the etch rate across a wafer had been marginally acceptable, design demands for new circuits required better control.

The machine vendor, working cooperatively with factory maintenance personnel, failed to fine an assignable mechanical or electrical cause for the malfunctions. After three weeks with the $200,000 machine offline, the best he could suggest was that the engineers devise new processes using higher-intensity pulsed cathode power to enhance etch rate uniformity. All parties had somewhat agreed that an unstable plasma due to the pulsing at low power contributed to the uniformity problems and that pulsing the plasma in general caused the end point detection interference. But the vendor could not show that he had eliminated the electrical interference with the photocell end point detector nor that the problem would not return.

When the engineers on site suggested using a steady-state, rather than pulsed cathode power, the vendor was adamant that the machine absolutely would not etch the layers uniformly under such conditions — it was designed to use the pulsed power. In fact, preliminary best guess probing experiments suggested that the vendor was probably correct.

Given that they were faced with devising a new process and that they had no real confidence that end point detection problems would cease to plague the machine's operation, the engineers elected to invest the time necessary to characterize its performance using fixed power first.

2. Experimental Approach:

2.1 Factor Settings and Design Matrix:

When restricted to constant power, the machine offered only three process control variables: power, pressure, and flow of etchant gas. Experience gained from the probing experiments and previous use of the equipment identified the logical limits for factor settings, so the engineers elected to use a Box-Wilson central composite design requiring 20 trials. Table 1 shows the factors and their limits for this experiment.

Table 1

Factor Settings

Factor, Units	Low	High
Pressure, mtorr	200	300
Power, watts	75	150
Etchant Flow, sccm	20	40

Simple two-level factorial designs support linear or planar models; to characterize any curvature present in a response as a function of the inputs requires at least 3 levels for each factor. The Box-Wilson Central Composite Design (1, 4) augments the usual 2-level factorial array with *star points* and *center points* such that one explores 5 levels of each factor in relatively few trials.

Among alternatives to this design are a 3-level full factorial which requires 27 trials and the 3-level Box-Behnken, which requires 15 (including replication).

We choose the central composite design as it provides 5 points for estimating curvilinearity and requires only 20 trials (including replication).

Modern experimental design software simplifies the tasks of creating and interpreting designs such as these. The limits for low and high indicated in Table 1 were truly limits, but the software used here allowed us to assign these points to the "star" positions while it constructed the two-level matrix within them, saving the tedium of the arithmetic calculations for factor settings. Table 2 summarizes the design matrix generated.

2.2   Responses:

Responses included the etch rates of layers 1 and 2, the uniformity of each etch, and the selectivity between layers 1 and 2. Measurements at several points on the wafer before and after each trial provided the etch rates. The ratio of etch rates during each trial provided the selectivity.

Etch rate measurements at several points on each wafer gave ranges of values for the rates suggesting that one or more of the factors was contributing a dispersion effect in the process. In addition the range of etch values observed was proportional to the mean etch rate. To describe the dispersion of etch rates we choose a relative of the *coefficient of variation (1)* because it was familiar to those involved in the investigation.

Percent Uniformity = [(Man − Min)/2*Mean]*100   *(1)*

Table 2
Design Matrix of Factor Settings

# Reps	PRESSURE (mtorr)	POWER (watts)	GASFLOW (sccm)
	220	90	24
	280	90	24
	220	135	24
	280	135	24
	220	90	36
	280	90	36
	220	135	36
	280	135	36
	200	113	30
	300	113	30
	250	75	30
	250	150	30
	250	113	20
	250	113	40
6	250	113	30

Calculations are simple and the result provides a concept of a +/− variation about a central value. However, this measurement has a natural limit of 0, and regression models readily predict meaningless <u>negative</u> values. Also, range is sensitive to outliers, so a more robust metric that provides approximately the same answers substitutes $2*\sigma$ for the range.

Taguchi (2) has published a variety of complex equations to model variability and calls them <u>signal to noise</u> ratios. His SN_T for "Target is Best" is also a variant of the coefficient of variation and works well when significant correlation exists between the standard deviation of a series of observations and the mean of those observations. Box, (3) advocates a simpler approach by determining the logarithm of the standard deviations of the observations and basing the analysis on that metric.

The software used in this study supports Box-Cox transformations of responses and performs the arithmetic totally transparent to the user − if the user takes full advantage of the scaling options available, s/he sees only normal response values and appropriate regression coefficients, regardless of the transformation used. Supported transformations include $y^{**}2$, y, $\sqrt{y}$, $\log_e y$, $1/\sqrt{y}$, and $1/y$.

Randomization of experimental trials distributes environmental or equipment bias uniformly among all factor levels; and while the software supports orthogonal blocking, we elected not to use it in this study. The first 3 columns of Table 3 illustrate the random run order software-generated for this investigation. The last 5 columns contain the observations.

3.   Results and Analysis

Each etch rate is the average of 5 measurements on each wafer, and the author used the range of those in equation (1) to model uniformity after transforming the values to the natural logarithm using software utilities. Selectivity is the ratio of the etch rates.

Backwards stepwise regression provided the Taylor expansion series coefficients displayed in Table 4. The author prefers this technique because it efficiently provides the simplest (most parsimonious) model which adequately describes the response as a function of the factors. In this case we set software parameters such that the α risk associated with retaining a coefficient was 0.10. Analysis of Variance (ANOVA) and residual diagnostic

Table 3

Randomized Worksheet and Results
Etch Process Optimization

RUN	PRS	PWR	FLOW	ETCH RATE1	LAYER1 UNIF	ETCH RATE2	LAYER2 UNIF	SELECT
1	280	90	24	827	13.213	193	23.714	4.3
2	250	113	30	1417	8.395	446	9.099	3.2
3	250	113	40	1351	6.295	375	12.912	3.6
4	250	75	30	719	14.289	200	20.701	3.6
5	250	113	30	1365	10.304	372	7.907	3.7
6	220	90	36	978	9.099	279	11.508	3.5
7	280	135	36	1597	9.795	397	6.998	4.0
8	250	113	30	1475	6.607	405	10.399	3.6
9	250	113	20	1376	8.395	316	7.295	4.4
10	250	113	30	1370	8.091	348	9.795	3.9
11	220	90	24	905	12.106	256	11.092	3.5
12	220	135	24	2100	2.000	734	5.794	2.9
13	250	113	30	1400	6.295	367	21.106	3.8
14	220	135	36	1961	2.698	660	5.105	3.0
15	280	90	36	841	11.508	195	21.979	4.3
16	200	113	30	1824	0.800	551	7.096	3.3
17	250	150	30	2065	4.198	615	10.789	3.4
18	300	113	30	1085	11.298	245	25.704	4.4
19	280	135	24	1647	9.099	398	12.388	4.1
20	250	113	30	1449	5.297	462	10.990	3.1

routines within the software help the user decide on the adequacy of a model.

The software coded the predictor variables according to the relationship

$$(X_i - ((MAX\ X_i + MIN\ X_i)/2)) / ((MAX\ X_i - MIN X_i)/2)$$

which results in values of -1 and $+1$ for the low and high predictor values, respectively. Therefore, we may judge the relative importance of each factor on a response by the absolute value of its coefficient. For example, the models for LAYER1 ETCH and LAYER1 UNIF indicate that Pressure and Power are highly coupled or interactive. That is, the effect on the response of changing Pressure from its low to its high setting depends on the setting of Power.

Note that selectivity is a quadratic function of etchant gas flow while the other responses are independent of this factor. In the table above, the coefficients for the uniformity models reflect our choice of a scaling option that displays approximate values for

Table 4

Regression Coefficients

MODEL TERM	ETCH RATE 1	LOG UNIF1	LOG UNIF2	SELECTIVITY LAYER1/LAYER2
CONSTANT	1382.68	8.095	10.251	3.590
PRS	−278.96	7.795	6.232	0.696
PWR	737.89	−5.800	−5.804	−0.237
FLW	-------	------	------	−0.164
PRS**2	-------	−6.320	------	------
PRS*PWR	−213.22	7.385	------	------
PWR**2	-------	------	3.106	------
FLW**2	-------	------	------	0.377
Adj-R**2	0.97	0.826	0.748	0.664
Rel. Press	0.95	0.605	0.581	0.491
RMS Error	71.16	2.503	2.496	0.269

the raw metric regardless of the transformation. This makes interpretation of the model somewhat simpler than using logarithms. In UNIF2, the α risk associated with the PWR**2 coefficient was 0.11.

3.1   Interpretation and Conclusions:

An extremely convenient method for interpreting the Taylor series regression models is contour plotting. Unlike a conventional graph this technique assigns a factor to each axis while holding any others at fixed values. The contour lines are lines of equal response, quite analogous to the contour intervals on a map.

Figures 1 to 6 are computer-generated plots based on the Taylor series models. The author adjusted the maximum value of the Pressure axis to expand the scale of the graphs and focus attention on the area of interest. The '★' on each represent actual data points. The first 4 add contours at etchant flow = 30 sccm -- the center of the investigation. The box in graphs 1 to 4 indicates the best conditions found during probing experiments which seemed to support the vendor's contention that this machine would not operate properly using fixed cathode power. Graphs 1 to 4 show that reducing the pressure and increasing the power improve uniformity and increase potential throughput at the cost of selectivity. Figure 5 shows that increasing the etchant flow to the limit studied improves selectivity, but not enough; Figure 6 shows that decreasing the etchant flow to the minimum investigated improves it further and provides an acceptable result. The box in Figure 6 represents a proposed process window.

An investigator must not forget that the models derived contain experimental error – the values indicated on a contourplot are not absolute, but have confidence intervals. An additional software utility permits estimation of these intervals. The final stage in any investigation requires predicting results and testing them in the process. Table 5 summarizes the predicted values and observed values for each response when the authors tested the modified process.

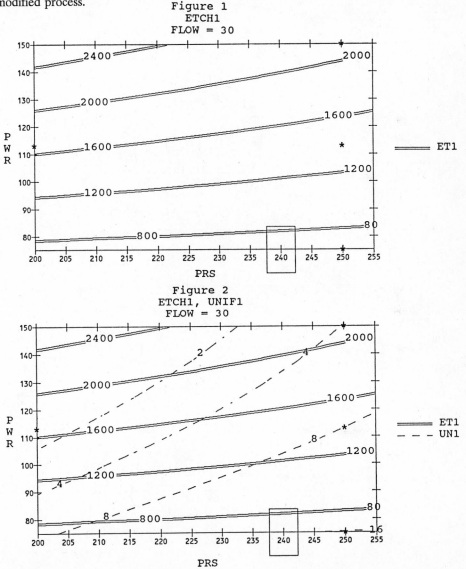

Figure 1
ETCH1
FLOW = 30

Figure 2
ETCH1, UNIF1
FLOW = 30

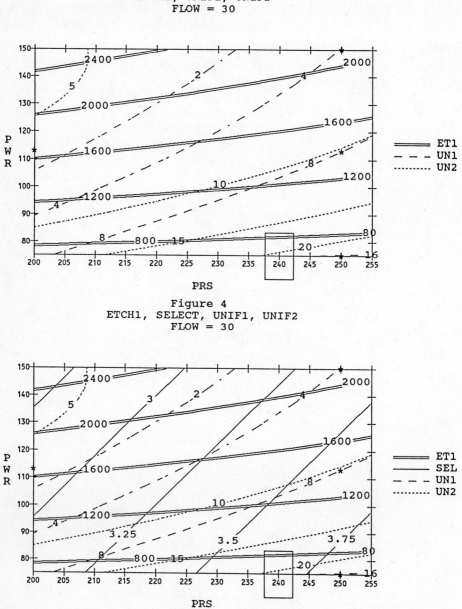

Figure 3
ETCH1, UNIF1, UNIF2
FLOW = 30

Figure 4
ETCH1, SELECT, UNIF1, UNIF2
FLOW = 30

Figure 5
ETCH1, SELECT, UNIF1, UNIF2
FLOW = 40

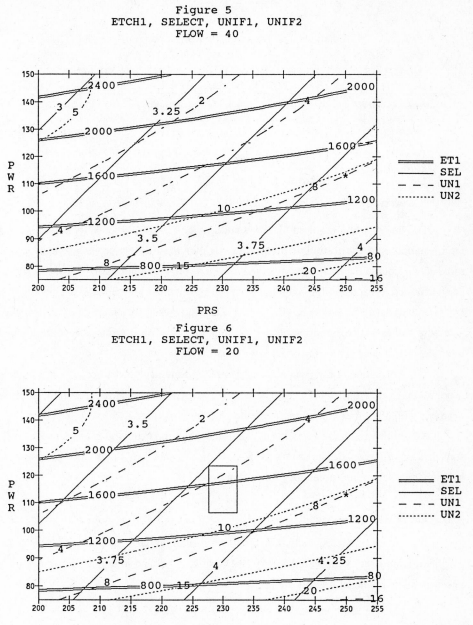

PRS

Figure 6
ETCH1, SELECT, UNIF1, UNIF2
FLOW = 20

PRS

Table 5

Predictions and Observed Values for Responses
in Etch Process (95% Confidence Intervals)

FACTORS	ETCH RATE 1	UNIF LAYER1	UNIF LAYER2	SELECTIVITY LAYER1/LAYER2
PRESSURE	230	230.00	230.00	230.0
POWER	115	115.00	115.00	115.0
GAS FLOW	20	20.00	20.00	20.0
Lower Bound	1477	3.30	6.00	3.0
RESPONSE->	1549	4.60	7.90	3.8
Upper Bound	1620	6.50	10.30	4.6
Observed	1517	4.60	7.70	3.5

This investigation required four days and included 20 experiments, their analysis and interpretation, and the testing of predictions. It identified a new process window which satisfied all process requirements which the vendor swore could not exist. The new process has performed as expected for 30 months, requiring nothing more than regular preventative maintenance and reducing scrap and rework while increasing the throughput at this step. The new process unexpectedly replaced the four previous ones, giving acceptable results regardless of device type or layer.

The application of experimental design and the contourplots derived from the models generated provide a far superior understanding of process windows available to the process engineer than any other experimental method.

Attempting to build these models from other than the orthogonal arrays provided by statistical experimental design usually leads to mathematical correlations among supposedly independent factors and potentially misleading models.

2.   Acknowledgements:

The authors gratefully acknowledge the contributions of Mr. Tom Tjaden, Process Engineer, Honeywell Solid State Electronics Division, in helping to define the study and in interpreting its results and implications.

3.   References:

(1)   Box, G. E. P., Hunter, W. G., and Hunter, J. S.," Statistics for Experimenters," John Wiley & Sons, New York, 1978.

(2)   See for example: Taguchi, G., and Phadke, M. S., "Quality Engineering Through Design Optimization", Conference Record Vol. 3., IEEE Globecom 1984 Conference, Atlanta, GA, 1106-1113; Taguchi, G., and Wu, Y., "Introduction to Off-Line Quality Control," Central Japan Quality Control Association (1986); and Taguchi, G., "Introduction to Quality Engineering: Designing Quality into Products and Processes."   Asian Productivity Organization, UNIPUB, Kraus International Publications, White Plains, NY. See also Phadke, M. S., "Quality Engineering Using Design of Experiments," Proceedings of the American Statistical Association, section on Statistical Education, 11 - 20, (1982).

(3)   Box, G. E. P., "Studies in Quality Improvement: Signal to Noise Ratios, Performance Criteria, and Transformation," Report No. 26, University of Wisconsin Center for Quality and Productivity Improvement, Madison, WI., 1987.

(4)   Box, G. E. P., and Draper, N. R., "Empirical Model-Building and Response Surfaces," John Wiley & Sons, New York, 1987.

Applying Experimental Design Techniques to Operating Systems and Software

Even though there are large numbers of documented designed experiments in the chemical and hardware segments of industry, few documented examples exist in the operating systems and software world. In view of the above, it is now fitting that we share the following case study with you.

A software engineer concerned with overall system performance when running a particularly large computer program decided to run a designed experiment. The experimental objective was to determine the best combination of operating system factors in order to minimize CPU time (regardless of the disk file scheme). Factors of interest to the experimenter were:

FACTOR	LOW	MIDDLE	HIGH
A. Buffered I/O Limit	−1	0	+1
B. "Q" Limit	−1	0	+1
C. Block Size	−1	0	+1
D. File Limit	−1	0	+1
E. Direct I/O Limit	−1	0	+1
F. Working Set Quota	−1	0	+1
G. Byte Limit	−1	0	+1
H. "BF"	A		B

In addition, three different disk file schemes were appropriate to evaluate as a "noise factor" (X_1, X_2, X_3).

For the actual experiment, six different responses were considered worthy of consideration. In our example, however, we will only address one response (CPU time). Utilizing an 18 run computer-generated D-Optimal design, coupled with a "Taguchi style" outer array, the following design matrix with applicable data was:

TABLE I

Design Matrix

Experiment #	A	B	C	D	E	F	G	H	X_1	X_2	X_3
1	-1	-1	-1	-1	-1	-1	-1	A	127	155	269
2	-1	0	0	0	0	0	0	A	118	150	685
3	-1	1	1	1	1	1	1	A	115	169	2321
4	0	-1	-1	0	0	1	1	A	141	163	2237
5	0	0	0	1	1	-1	-1	A	110	134	207
6	0	1	1	-1	-1	0	0	A	131	160	791
7	1	-1	0	-1	1	0	1	A	139	271	1358
8	1	0	1	0	-1	1	-1	A	127	145	216
9	1	1	-1	1	0	-1	0	A	112	139	579
10	-1	-1	1	-1	0	0	-1	B	40	69	144
11	-1	0	-1	0	1	1	0	B	56	125	715
12	-1	1	0	1	-1	-1	1	B	63	92	652
13	0	-1	0	1	-1	1	0	B	41	73	574
14	0	0	1	-1	0	-1	1	B	68	125	889
15	0	1	-1	0	1	0	-1	B	49	82	129
16	1	-1	1	0	1	-1	0	B	51	100	580
17	1	0	-1	1	-1	0	1	B	50	80	786
18	1	1	0	-1	0	1	-1	B	57	87	153

<table>
<tr><td colspan="4">

TABLE II

Average Marginals for Statistic "$\bar{x}$"

	Low	Mid	High
A	337	339	279
B	363	265	327
C	333	276	347
D	315	321	319
E	252	331	373
F	247	290	418
G	128	288	540
H	417	- - -	220

</td><td colspan="4">

TABLE III

Average Marginals for the Statistic LnS *

	Low	Mid	High
A	5.48	5.41	5.26
B	5.55	5.26	5.34
C	5.43	5.27	5.45
D	5.44	5.31	5.39
E	5.27	5.41	5.47
F	5.24	5.31	5.59
G	3.94	5.75	6.45
H	5.56	- - -	5.20

</td></tr>
</table>

* natural log of the sample std dev

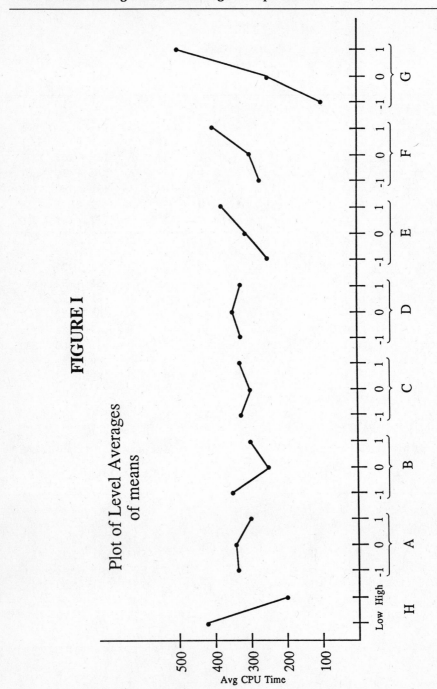

FIGURE I

Plot of Level Averages
of means

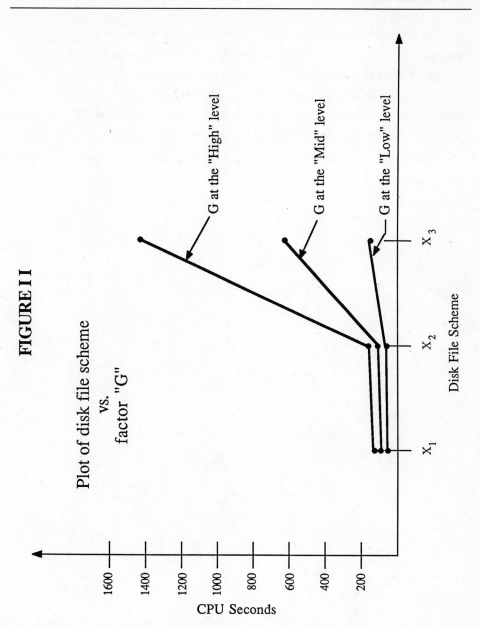

FIGURE I I

Plot of disk file scheme
vs.
factor "G"

G at the "High" level

G at the "Mid" level

G at the "Low" level

X_1 X_2 X_3

Disk File Scheme

1600 1400 1200 1000 800 600 400 200

CPU Seconds

An analysis of the data in Table III indicates that factor G may be a variance reduction factor with the low level being the best setting. Analysis of Table II and Figure I indicate that G, H, and possibly F have an important impact on the mean response. Figure II provides us with a graphical representation of the relationship between disk file scheme and factor G. Of particular interest is the relatively constant low CPU time when G is at the low level, regardless of the disk file scheme being used.

In summary, setting factor G at the low level will allow us to minimize CPU time.

Submitted by: MAXCOR   U.S. AIR FORCE ACADEMY
                 Gary Fenton   C1C James B. Hecker
                 John Hopkinson   C1C Phillip A. Herre
                 William Woodard   C1C Ray L. Plumley
                                        C2C Raymond X. Sagui

"Reducing Variability in a Machining Process: An Industrial Application of Experimental Design"

Background

An industrial bushing has a close tolerance requirement on the outer diameter of 14 mm ± .01 mm (.5512" ± .00039"). The machining process was producing approximately .6% of the product out of the tolerance limits. Some of the variability of the outer diameter was due to an "out of round" condition. A joint effort between MAXCOR and the Air Force Academy was initiated to reduce the amount of runout (as measured by Total Indicator Reading - TIR) by utilizing a designed experiment approach to optimizing the process settings.

Experiment Objective

The objective of the experiment was to see an improvement in the roundness of the bushings. Roundness is defined as the condition on a surface of revolution, such as a cylinder, where all points on the surface intersected by any plane perpendicular to a common axis are equidistant from the center.

Quality Characteristics

The two ends of the bushing may be distinguished by the presence of a dimple near one end (see Figure 1). The manufacturing specifications required that the entire surface (composed of an infinite number of diameters) meet the 14 ± .01 mm condition. To meet this condition, both the taper and the out of roundness must be minimized. Thus, both of these were used as criteria measures (response variables).

Figure 1:   Outer Bushing Top View and Cross Section

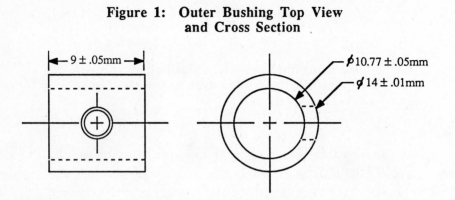

The inspection equipment used a two point contact so that as the part was rotated, both maximum and minimum diameters could be recorded. Both ends of the bushing were inspected, generating four pieces of data for each part. These four data pieces were then used to mathematically calculate the roundness (maximum minus minimum for each end) and the taper (maximum of one end minus maximum of the other end).

Table 1: Quality Characteristics

Response	Type	Anticipated Range
TIR, dimple	quantitative	0 - 0.0110 mm
TIR, non-dimple	quantitative	0 - 0.0110 mm
Taper	quantitative	−0.0075 - 0 mm

Factors

Perhaps the most difficult step of experimental design is deciding which factors should be included in the design. Many times there are so many potential factors that affect the results that the design becomes infeasible because of the size, or the time to run the entire design, or most common, the cost per experiment.

In this case study a brainstorming session with the engineers brought out the possible factors that go into the manufacturing of a single bushing. This initial list consisted of almost twenty different factors shown in Figure 2.

Figure 2:   Fishbone Diagram of Possible Factors

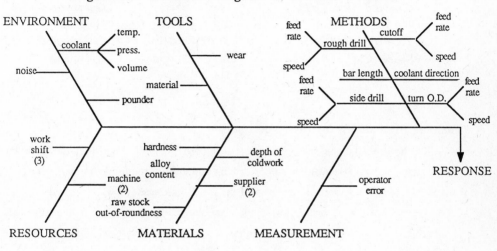

Some factors, such as alloy content, stock roundness, and bar feeding length, were beyond control so they were not included in the experiment. Others, like machine, worker shift, and environment were controlled by randomization or holding them constant. After two weeks of weighing the priority and appropriateness of each factor, ten two-level factors were chosen for the design.

Table 2: Experimental Factors

Factor	Levels	Type
Hardness (HD)	Lo - Hi	Qualitative
Rgh Drill Feed Rate (RDFR)	.0015 - .0025	Quantitative
Rgh Drill Speed (RDS)	2000 - 3000 rpm	Quantitative
Sd Hole Drill Feed Rate (SHFR)	.07 - .17 ipm	Quantitative
Sd Hole Drill Speed (SHDS)	1800 - 2800 rpm	Quantitative
Turn O.D. Feed Rate (TFR)	.002 - .003	Quantitative
Turn O.D. Speed (TODS)	3000 - 5000 rpm	Quantitative
Cutoff Feed Rate (CFR)	.0002 - .0005	Quantitative
Cutoff Speed (CS)	2500 - 4000 rpm	Quantitative
Coolant Flow (CF)	Top - Bottom	Qualitative

Choosing the Factors for the Experiment

The hardness of the raw stock was considered the most important factor to the MAXCOR engineers. Because of all the drilling and cutting required to make a bushing, the drill bit can severely deform the metal if the metal hardness is low. The feed rate and speed factors were considered important because they cause movement of the raw stock during drilling and cutting. The place of coolant flow was also considered important because of affect on lubrication and cooling.

Run Limitations

The parts manufactured during the experiment were eventually to be sent to a customer, so it was important to select factor level settings that would not intentionally produce non-conforming products. The appropriate sample size for each run was selected by calculations based upon the historical standard deviation and the amount of shift that was to be detected, along with the alpha and beta risks that were reasonable. (See Chapter 2).

Appropriate Experimental Designs

Due to MAXCOR's familiarity with the Taguchi design, an L_{16} Taguchi sequence was used. Except for the factor of material hardness, which could not be easily changed due to equipment set-ups, the runs were randomized by making use of a random number table.

Experimental Implementation and Data Collection

MAXCOR's technicians were instructed to set up the machinery in accordance with the randomized Taguchi L_{16} design. After each run, the quality characteristics were measured and recorded for each bushing. Throughout this entire process, management was present to make certain there was no deviation from the experimental design.

Analysis and Results

The analytical method used was a regression analysis on the three-response mean for each replicated run to determine main effects. For each of the three regressions, there was no significance at the 95% confidence level. Due to the small differences in factor settings, the resulting regression weights were not statistically significant. A set of "vital few" factors were suspected to exist despite the regression results. Therefore, it was necessary to rank order the regression weights to determine which factors appeared to be more significant than others. The coefficients of the orthogonally coded factors were used to set up the prediction equations; these coefficients are listed in Table 3.

A regression on the variance of the three responses for each replicated run was used to determine dispersion effects. Again, the regressions were not statistically significant. Likewise, these regression weights were used to determine the most likely significant factors. The coefficients of the orthogonally coded factors were used to set up the prediction equation. (See Table 3.)

Using the coefficients listed in Table 3, each factor was set to minimize the mean and variance of the response variables. This involved setting factors with a positive coefficient at the low level, and factors with a negative coefficient at the high level. This procedure was accomplished for each of the response variables. In some instances, the factor coefficients for the mean and variance responses were opposite in sign. Since MAXCOR desired to emphasize decreasing the variance in the responses, it was decided to give precedence to the coefficients of the variance response. The best factor setting for optimizing the mean and variance can be seen in Table 4.

Table 3: Coefficients

Factor	Y1	Y2	Y3	S_1^2	S_2^2	S_3^2
HD					0.00505	
RDFR			0.024		-0.00462	
RDS						
SHFR	0.042	0.046	0.030	0.00641		
SHDS		0.031	0.037		-0.00385	
TFR				0.00459	0.00466	
TODS			-0.036		0.00444	
CFR	-0.038			-0.00453	-0.00571	
CS				0.00498		
CF				-0.00526	0.00539	-0.00478

Table 4: Optimal Settings

Factor	Setting
HD	low
RDFR	high
RDS	high or low
SHFR	low
SHDS	high
TFR	low
TODS	low
CFR	high
CS	low
CF	high

Confirmation Runs

MAXCOR ran 200 confirmation runs at the recommended levels. To obtain an overall picture of the bushing, the TIR responses for both the dimpled and non-dimpled ends were combined. Although there was no significant change in the mean TIR response, the variation was decreased by 50% (see Appendices 1 and 2). Although the confirmation runs produced parts with reasonable taper, there is not way to discern any improvement in this response due to the lack of historical data.

APPENDIX 1

MAXCOR MFG. - HISTORICAL DATA
TIR ON OUTER BUSHINGS

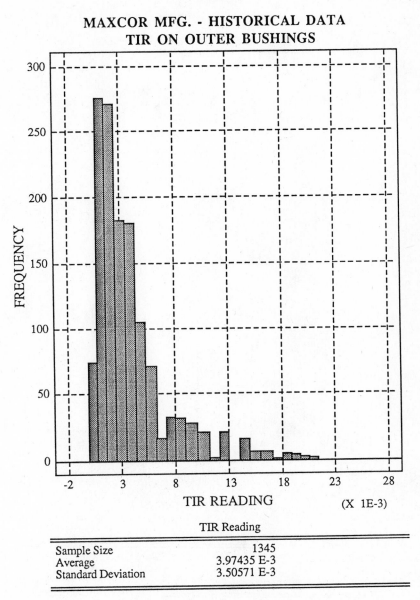

TIR READING (X 1E-3)

TIR Reading	
Sample Size	1345
Average	3.97435 E-3
Standard Deviation	3.50571 E-3

APPENDIX 2

MAXCOR MFG. - CONFIRMATION DATA
TIR BOTH ENDS COMBINED

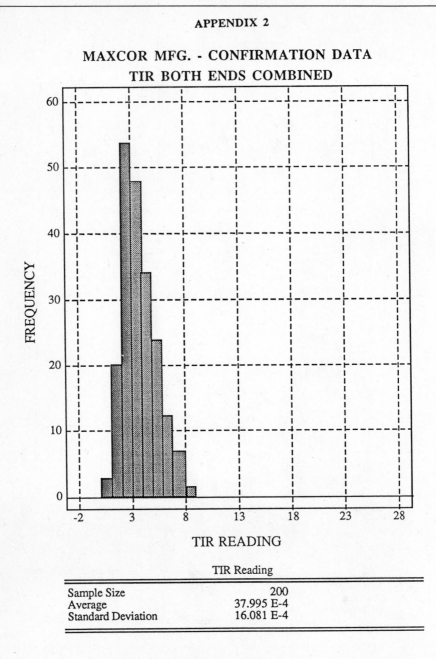

TIR READING

	TIR Reading
Sample Size	200
Average	37.995 E-4
Standard Deviation	16.081 E-4

Submitted by: Tom Gardner
James D. Riggs

The Effect of the CE Rivet H Parameter on Head Protrusion

Introduction

Problem Statement: Determine the effect of the CE rivet h parameter (see Figure 1) on head protrusion in skin panel rivet installations.

Objective: Use a designed experiment and knowledge gained from previous experimentation to quantify the effect of the h parameter on rivet protrusion. This information will be applied to determine the required countersink dimension for each value of h.

Background: The variation in head diameter of CE rivets has been found to be a major contributor to the variation in head protrusion in CE rivet skin installations. A request to rivet vendors to better control head diameter was rejected in favor of controlling the h parameter. Apparently, the head diameter is controlled by the h parameter. Examination of Figure 1 shows the geometric reasoning for this statement. A correlation analysis between h and head diameter showed marginal correlation, thus challenging the geometric foundation. This experiment is designed to establish the role of h on head protrusion.

Results:

The h parameter does effect rivet protrusion, but the magnitude of the effect (0.0011") is much smaller than that of the combination of countersink and head diameter (0.0093"). See Figures 2 and 3. Figure 2 shows the expected increase in protrusion with an increase in h. Minimal correlation exists between h and the head diameter. The countersink required to control protrusion to 0.0020" can be determined from h only marginally. Head diameter is the factor from which the countersink should be determined.

This experiment shows that the countersink ranges over which the protrusion can be controlled to 0.0020" are different from the existing ranges shown below. This may be due to a change in rivet vendor — now Precision Form and formerly Allfast.

	CE 5	CE 6
Existing range	0.240" − 0.245"	0.290" − 0.295"
Experiment range	0.245" − 0.252"	0.292" − 0.294"

Figure 3 shows two effects. First, as the countersink increases, protrusion decreases as expected. Secondly, as the head diameter increases, the protrusion decreases when the countersink is small. When the countersink is larger, the protrusion increases as the head diameter increases. This effect is due to the mechanics compensating for small countersink diameter with increased pressure on the rivet gun.

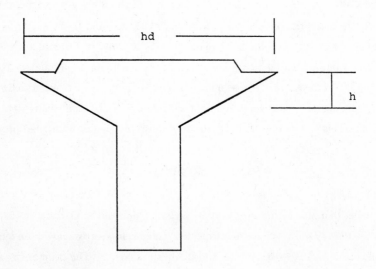

Figure 1.   The CE Rivet H Parameter

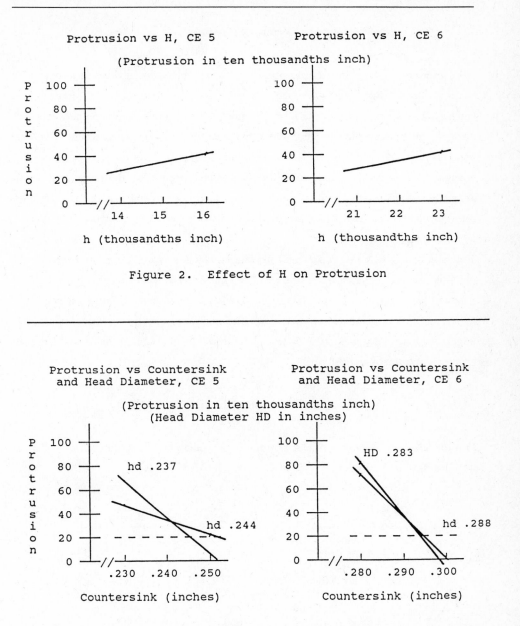

Figure 2. Effect of H on Protrusion

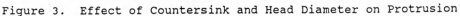

Figure 3. Effect of Countersink and Head Diameter on Protrusion

Recommendations

This experiment used rivets from a variety of batches which gives no clue as to the within batch variation. Both head diameter and h need to be measured from within several batches. The within variation can be determined, as well as the correlation between head diameter and h. Also, the number of rivets that must be measured to find accurate averages for these factors can be found. If the correlation between head diameter an h is good enough, h can be used to characterize the batch and to determine the target countersink diameter. Otherwise, the head diameter must be used. In addition, if h can be correlated, recommendations to vendors for target h values and variances can be made.

The graphs in Figure 3 give countersink ranges which will produce protrusions for the different head diameters for both the CE 5 and CE 6 rivets. Batches (boxes) on the floor need to have average head diameters and variation measured to allow the graph to give the target countersink diameter for each box. The goal is to find two or three countersink targets for each rivet type (CE 5 or 6). This will permit utilization of two or three countersink cutters per rivet type.

Experiment Design

Previous experimentation in similar skin riveting installations shows the following factors, with levels, to be important.

Factor	Levels	Symbol
panel	0	
complexity	0	
shimming	0	
rivet head diameter	2	hd
h	3	csk
countersink diameter	3	h
rivet head protrusion	random	PROT

All samples are taken from a single panel in Acc 315 thus eliminating the panel-to-panel effect.

The complexity factor was eliminated as all the combinations of h and csk were available in the open (easy) areas of installation on the selected panel.

The shimming effect was eliminated by working only in areas where no shims are present.

The rivet head diameter, i.e., CE 5 and CE 6 is a fixed factor over which the experiment is blocked. The actual measurements are expected to be 0.228" to 0.248" and 0.279" to 0.299", respectively. The deviation from nominal for both the CE 5 (0.238") and CE 6 (0.288") rivets is used in the analysis.

The h parameter has the following levels within hd. The deviation from nominal (level 2 for each type) is used.

	Level	Actual
CE 5	1	0.0140"
	2	0.0148"
	3	0.0162"
CE 6	1	0.0209"
	2	0.0216"
	3	0.0230"

The countersink diameter is considered fixed within hd as follows. The deviation from nominal is used in the analysis and the nominal are the same as those of the head diameter.

	Level	Actual
CE 5	1	0.228"
	2	0.238"
	3	0.248"
CE 6	1	0.279"
	2	0.288"
	3	0.298"

The rivet head protrusion is the response variable. It has been shown that controlling this factor clearly reduces defects associated with skin panel installation. These include both rivet installation defects (such as exposed countersinks, gaps, shanked rivets, cracked butts) and skin defects (including scratches, tool marks, pillowing, canning).

Model

The informal model to be tested is:

PROTRUSION = mean + hd + h + csk + hd*h + hd*csk + h*csk + h*h + csk*csk + hd*h*csk + hd*h*h + hd*csk*csk + RESIDUALS

The anticipated Anova table is:

AA TABLE

Source	df	FC	RC
hd	1	$18n\phi(hd)$	σ_R^2
h	2	$18n\phi(h)$	σ_R^2
csk	2	$18n\phi(hd)$	σ_R^2
hd*h	2	$9n\phi(hd*h)$	σ_R^2
hd*csk	2	$9n\phi(hd*csk)$	σ_R^2
h*csk	4	$3n\phi(h*csk)$	σ_R^2
hd*h*csk	4	$n\phi(hd*h*csk)$	σ_R^2
REP	$18(n-1)$	0	σ_R^2
Total	$18n - 1$		

Sample Size

To compute the number of samples needed per cell (h and csk in full factorial combinations) we first assume the df to be infinity, then recompute the sample size based on the previous estimate. The difference (in ten thousandths) we need to detect is 5, and σ_E is equal to 3 (determined from past experience). This gives us:

$$t_{0.025}(\infty) * \sigma_E (2/n)^{1/2} = 5$$
$$=> 1.96 * 3 (2/n)^{1/2} = 5$$

$$=> \quad n = 2 * ((1.96 * 3) / 5)^2$$
$$n = 2.77 \quad => \quad 3$$

If n = 3, we will take a total of 42 samples. Let df for t = 41.

$$=> \quad n = 2 * ((2.02 * 3) / 5)^2$$
$$n = 2.94 \quad => \quad 3$$

The sample size needed is 3 per cell, yielding a total of 42 samples.

Randomization

Randomization is by runs within a block. Random numbers are generated and assigned to the factorial runs within a block. See Figure 4 for a diagram of the normal order of installation-first CE 6's, CE 5's, CE 6's, and then CE 5's.

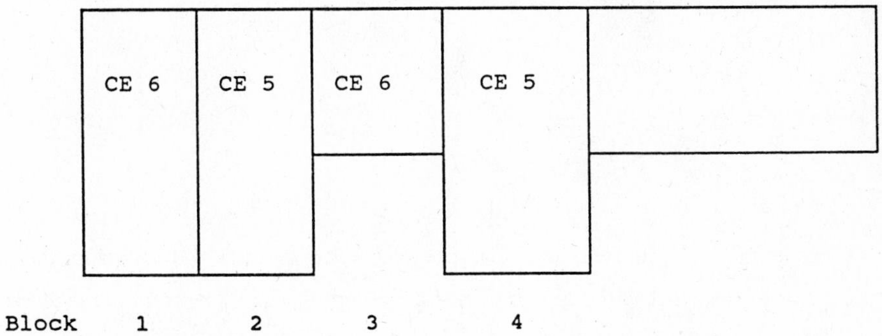

Block   1   2   3   4

Four blocks containing either CE 5 or CE 6 rivets are laid out on the 65-45800-139 skin panel in ACC 315. The order of installation is from left to right in the diagram.

Figure 4. Test Panel Layout

Within each block, the randomized run combinations are as follows:

	Block 1				Block 2	
Run	h	c		Run	h	c
1	1	2		1	2	1
2	2	1		2	3	2
3	2	3		3	3	1
4	3	3		4	2	3
5	2	2		5	1	2
6	3	2		6	2	2
7	3	1		7	3	3
8	1	3		8	1	1
9	1	1		9	1	3

	Block 3				Block 4	
Run	h	c		Run	h	c
1	2	2		1	2	3
2	1	1		2	1	2
3	2	3		3	2	1
4	1	2		4	1	1
5	3	1		5	3	3
6	2	1		6	2	2
7	1	3		7	3	1
8	3	2		8	1	3
9	3	3		9	2	1

Data Analysis

The experimental data collected, listed in Appendix A, are severely reduced from the plan above. None of the blocks are balanced. Block 1 is the only block with no duplicate runs.

The data for csk, h, and hd are the differences of the specification nominal from the actual readings. The length (len) of the rivet is taken from the part number rather than from actual measurements as previous experimentation shows high correlation between the measured length and the length obtained from the part number.

The mechanics suggested two factors that could (but did not) show a sequence effect. Tape is applied to the surface of the rivet gun which interfaces with the rivet head. As rivets are installed, this tape wears thus diminishing a cushioning effect. The other factor is the firmness of the structure into which the rivets are installed. As the installation proceeds from left to right (see Figure 4), the structure becomes progressively less firm.

Multiple regression was used to analyze the data. The factors which contribute to the variability of the rivet head protrusion are, countersink, head diameter, and the h parameter, as shown in Figures 2 and 3 above. Figure 5 shows a plot of the residuals versus the model output. No clear patterns are revealed. The first row of the data was removed from the analysis after viewing standardized residuals showed this row to be an outlier.

Least clear differences are calculated to make block to block comparisons due to the unbalanced data set. LSD_1 is used for comparing blocks 1 and 2, LSD_2 for blocks 1 and 3, LSD_3 for 1 and 4, LSD_4 for 2 and 3, LSD_5 for 2 and 4, and LSD_6 for comparing 3 and 4.

$$t_{0.025}(27) = 2.052$$

$$s = \sqrt{\frac{4 \times 16.82^2 + 2 \times 20.60^2 + 9 \times 37.12^2 + 10 \times 20.33^2}{4 + 2 + 9 + 10}} = 27.21$$

$$LSD1 = 2.052 \times 27.21 \times \text{sqrt} (1/5 + 1/3) = 40.78$$
$$LSD2 = 2.052 \times 27.21 \times \text{sqrt} (1/5 + 1/9) = 31.14$$
$$LSD3 = 2.052 \times 27.21 \times \text{sqrt} (1/5 + 1/10) = 30.58$$
$$LSD4 = 2.052 \times 27.21 \times \text{sqrt} (1/3 + 1/9) = 37.22$$
$$LSD5 = 2.052 \times 27.21 \times \text{sqrt} (1/3 + 1/10) = 36.76$$
$$LSD6 = 2.052 \times 27.21 \times \text{sqrt} (1/9 + 1/10) = 28.83$$

The Bonferroni Limits are calculated such that BSD1 compares block 1 with block 2, etc., as is done with LSD above. The calculations use $t_{0.025/6}(27) = 2.848$.

$$BSD = 2.848 \times 27.21 \times \text{sqrt} (1/5 + 1/3) = 56.59$$
$$BSD = 2.848 \times 27.21 \times \text{sqrt} (1/5 + 1/9) = 43.22$$
$$BSD = 2.848 \times 27.21 \times \text{sqrt} (1/5 + 1/10) = 42.44$$
$$BSD = 2.848 \times 27.21 \times \text{sqrt} (1/3 + 1/9) = 51.66$$
$$BSD = 2.848 \times 27.21 \times \text{sqrt} (1/3 + 1/10) = 51.01$$
$$BSD = 2.848 \times 27.21 \times \text{sqrt} (1/9 + 1/10) = 35.61$$

The following table shows the results of both of the comparisons by block. Note that if zero is contained in an interval then no clear difference exists between the represented blocks.

	Block 2	Block 3	Block 4
Block 1	−2.73	−21.96	−14.90
LSD	(−43.51, 38.05)	(−53.10, 9.18)	(−45.48, 15.68)
BSD	(−59.32, 53.86)	(−65.18, 21.26)	(−57.34, 27.54)
Block 2		−19.23	−12.17
LSD		(−56.45, 17.99)	(−48.92, 23.59)
BSD		(−70.89, 32.43)	(−63.18, 38.84)
Block 3			7.06
LSD			(−21.77, 35.89)
BSD			(−28.55, 42.67)

It's clear that no differences exist between any of the blocks.

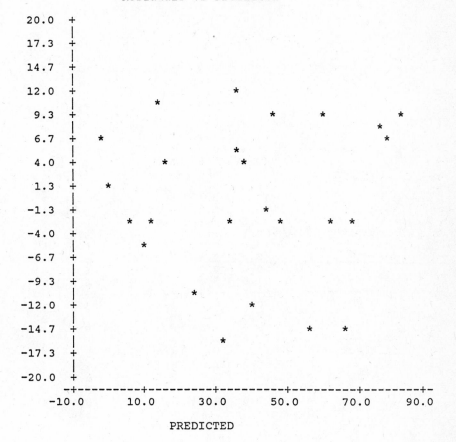

H EXPERIMENT - 26 JAN 89

Residuals vs Predicted

Figure 5.   Residuals Plot

APPENDIX A.   EXPERIMENT DATA

	Prot	csk	h	Len	hd	ce	Run
	35.0	-8.0	0.0	5.0	-1.5	6.0	2
	2.0	11.0	0.0	5.0	-2.5	6.0	3
Block 1	18.0	10.0	1.5	5.0	0.0	6.0	4
	14.0	2.0	0.2	5.0	-4.5	6.0	5
	44.0	0.0	1.4	6.0	-1.0	6.0	6
	47.0	0.0	-0.7	5.0	1.5	5.0	2
Block 2	23.0	10.0	-0.4	5.0	1.5	5.0	7
	6.0	10.0	-0.6	5.0	0.5	5.0	7
	39.0	0.0	-0.5	5.0	-0.5	6.0	1
	82.0	-10.0	-0.7	5.0	-3.0	6.0	2
	1.0	9.0	-0.5	5.0	-1.0	6.0	3
	0.0	10.0	0.1	5.0	-3.0	6.0	3
Block 3	90.0	-10.0	1.0	5.0	-1.5	6.0	5
	63.0	-7.0	0.1	5.0	-2.0	6.0	6
	83.0	-8.0	0.8	7.0	-4.0	6.0	6
	41.0	0.0	0.8	7.0	-4.5	6.0	8
	2.0	11.0	1.0	7.0	-4.5	6.0	9
	25.0	0.0	-0.2	5.0	2.0	5.0	2
	40.0	0.0	-0.7	5.0	1.0	5.0	2
	66.0	-8.0	0.4	5.0	2.5	5.0	3
	40.0	-7.0	-0.8	5.0	-0.5	5.0	4
Block 4	28.0	10.0	1.4	6.0	5.5	5.0	5
	11.0	12.0	1.0	6.0	4.0	5.0	5
	54.0	0.0	1.4	6.0	5.5	5.0	7
	49.0	-11.0	1.3	6.0	4.0	5.0	7
	4.0	11.0	-0.7	5.0	1.5	5.0	8
	58.0	-10.0	-0.1	5.0	2.0	5.0	9

APPENDIX B. RESIDUALS

Row	Observed	Predicted	Residual
2	2.00000	-4.70511	6.70511
3	18.00000	14.00258	3.99742
4	14.00000	30.57593	-16.57593
5	44.00000	46.29594	-2.29594
6	47.00000	34.83098	12.16902
7	23.00000	12.56487	10.43513
8	6.00000	9.14704	-3.14704
9	39.00000	35.31270	3.68730
10	82.00000	74.39116	7.60884
11	1.00000	3.51947	-2.51947
12	.00000	-1.45349	1.45349
13	90.00000	80.53669	9.46331
14	63.00000	65.22723	-2.22723
15	83.00000	76.97873	6.02127
16	41.00000	42.82755	-1.82755
17	2.00000	-4.32410	6.32410
18	25.00000	37.23959	-12.23959
19	40.00000	34.83098	5.16902
20	66.00000	57.13543	8.86457
21	40.00000	54.38330	-14.38330
22	28.00000	31.05339	-3.05339
23	11.00000	21.92986	-10.92986
24	54.00000	44.94712	9.05288
25	49.00000	63.79825	-14.79825
26	4.00000	8.74858	-4.74858
27	58.00000	60.20539	-2.20539

DURBIN WATSON SERIAL CORRELATION = 2.29

CATAPULTING STATISTICS INTO ENGINEERING CURRICULA

S. E. Jones, PhD, University of Alabama
Stephen R. Schmidt, PhD, USAF Academy
Bruce Johnson, MS, HQ USAF/LE-RD

Today's rapidly changing technologies often result in products and processes that are not easily modelled using engineering theory alone. Other tools, such as statistics, can be used to supplement engineering knowlege and thereby speed up product and process development times. The use of a mechanical device such as the catapult (see Figure 1) discussed in this paper has been shown to be a very effective tool for demonstrating the power of blending engineering theory with statistically designed experiments.

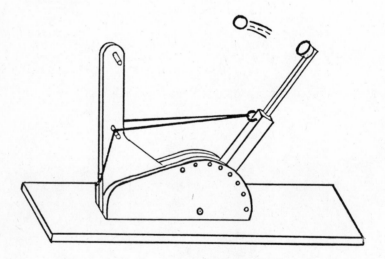

Figure 1. Catapult

Analysis of the Catapult Using only Engineering Theory

The catapult is a device for launching a small projectile (ball) toward a predetermined, downrange impact site R (see Figure 2). The force F (see Figure 3), which rotates the extensible moment arm of the catapult, is provided by a stretched rubber band. The rubber band may be attached to several points on the superstructure of the device and may assume several stretch lengths for a given configuration. The force exerted by the rubber band on the moment arm of the catapult can be a complicated function of the stretch length. However, we assume that the stretched length is a function of the current position of the moment arm, $F = F(\theta)$.

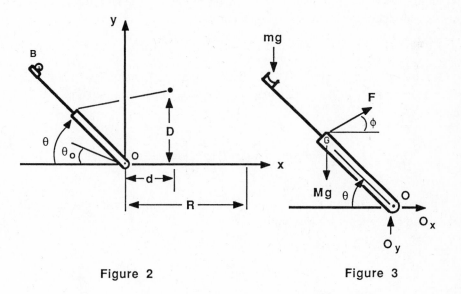

Figure 2   Figure 3

Taking moments about 0, and assuming that there are no applied moments at 0, we find

$$I_0\ddot{\theta} = r_F F(\theta)\sin\theta\cos\phi - (Mg\, r_G + mg\, r_B)\sin\theta \qquad (1)$$

where I_0 is the mass moment of inertia of the compound rod and ball assembly, r_F is the distance from 0 to the rubber band connection, r_G is the distance from 0 to the center of mass of the rod, r_B is the distance from 0 to the center of mass of the ball, M is the mass of the rod, m is the mass of the ball, and g is the local gravitational constant.

Since $F = F(\theta)$ and

$$\tan\phi = \frac{D - r_F \sin\theta}{d + r_F \cos\theta} \, , \tag{2}$$

equation (1) can be integrated once to give

$$\frac{1}{2}I_0\dot\theta^2 = r_F\int_{\theta_0}^{\theta} F(\theta)\sin\theta\cos\phi\, d\theta - (Mg\, r_G + mg\, r_B)(\sin\theta - \sin\theta_0) \tag{3}$$

where we have assumed that the device is released from rest at $\theta = \theta_0$. This equation expresses the current angular rate $\dot\theta$ in terms of the current angular position θ (see Figure 2). When a stop position at $\theta = \theta_1$ is reached (see Figure 3), the angular speed is $\dot\theta_1$ and the ball will be launched with speed

$$v_B = r_B\dot\theta_1 \tag{4}$$

toward the downrange target at R. We assume that the launch angle is normal to the rod position at $\theta = \theta_1$. From (4), $\dot\theta_1$ can be found, and the expression for it takes the form

$$\frac{1}{2}I_0\dot\theta_1^2 = r_F\int_{\theta_0}^{\theta_1} F(\theta)\sin\theta\cos\phi\, d\theta - (Mg\, r_G + mg\, r_B)(\sin\theta_1 - \sin\theta_0). \tag{5}$$

Combining this equation with (4), we can find the initial velocity for the ball when the prescribed stop angle θ_1 is reached.

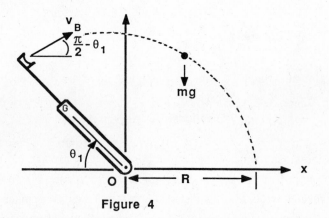

Figure 4

The small ball is launched with initial speed v_B in a direction that is normal to the catapult, as shown in Figure 4. The stop in the mechanism at $\theta = \theta_1$ provides for the initial launch angle, which is $\pi/2 - \theta_1$ radians. The initial launch position has coordinates $(-r_B\cos\theta_1, r_B\sin\theta_1)$ in the rectangular, cartesian coordinate system with origin at 0 (Figure 4).

Neglecting the influence of atmospheric drag, the equations of motion for the ball are

$$\ddot{x} = 0 \quad, \quad \ddot{y} = -g \tag{6}$$

where g is the local gravitational constant. These equations can be easily integrated, and the solution is

$$x = v_B \cos\left(\frac{\pi}{2} - \theta_1\right) t - \frac{1}{2} r_B \cos\theta_1 \tag{7}$$

and

$$y = r_B \sin\theta_1 + v_B \sin\left(\frac{\pi}{2} - \theta_1\right) t - \frac{1}{2} gt^2. \tag{8}$$

At impact, $x = R$ and $y = 0$ at the terminal time $t = \bar{t}$. Inserting these conditions into (7) and (8) and eliminating the terminal time between the resulting equations gives one relation involving v_B and θ_1.

$$r_B\sin\theta_1 + (R+r_B\cos\theta_1)\tan\left(\frac{\pi}{2} - \theta_1\right) - \frac{g}{2\,v_B^2}\frac{(R+r_B\cos\theta_1)^2}{\cos^2\left(\frac{\pi}{2} - \theta_1\right)} = 0. \tag{9}$$

This equation can be combined with (4) and (5) to form a nonlinear system for the unknowns θ_1, $\dot{\theta}_1$, and v_B when the initial launch angle θ_0 is given. We may also view this same system as one in which the unknowns are θ_0, $\dot{\theta}_1$, and v_B when the stop angle θ_1 is prescribed. When the constitutive properties of the rubber band are supplied, this system can be used to determine the launch conditions for a given, downrange site R.

The algebraic structure of the system of equations (4), (5) and (9) is such that they may be easily combined into one equation:

$$\frac{gl_0}{4r_B} \frac{(R + r_B\cos\theta_1)^2}{\cos^2\left(\frac{\pi}{2} - \theta_1\right)\left[r_B\sin\theta_1 + (R + r_B\cos\theta_1)\tan\left(\frac{\pi}{2} - \theta_1\right)\right]} \tag{10}$$

$$= r_F\int_{\theta_0}^{\theta_1} F(\theta)\sin\theta\cos\phi\,d\theta - (Mgr_G + mg\,r_B)(\sin\theta_1 - \sin\theta_0)$$

This equation is a complicated transcendental involving an integral. The integrand contains a force function $F = F(\theta)$ which is nonlinear for large stretches. The integrand also contains the cosine of the angle ϕ which is itself a complicated function of θ (see equation (2)). There is no reasonable prospect for evaluation of this integral in terms of elementary functions. However, it can be dealt with numerically, and this will be the approach in subsequent work.

One comment must be made at this juncture. For a given value of R and a given force law $F = F(\theta)$, there may be no solution to (10) for any values of θ_0 and θ_1. On the other hand, when the force in the rubber band is sufficiently high, there will be solutions for a wide range of initial and stop angles θ_0 and θ_1.

There are several points of concern in the catapult analysis just presented. First, the force law for the rubber band has not been specified. Each rubber band that is used with the device must be tested and the force as a function of stretch must be determined. Rubber is generally a nonlinear elastic material. So, the force-extension law may be nonlinear.

We will assume that the stretch in the rubber band is uniform. However, for some configurations, the rubber band passes over pegs which can offer enough friction to make the stretch nonuniform. When these pegs are lubricated, this effect can be minimized.

A second concern is that the moment of inertia I_0 and the center of mass of the extensible moment arm r_G have not been specified. However, both of these quantities are easy to determine experimentally as functions of the rod extension length r_B. The moment of inertia can be computed for any given rod length by treating the compound moment arm as a pendulum and measuring the period of oscillation for various arm lengths. This will compute the moment of inertia, I_0, very conveniently and accurately. Similarly, the position of the mass center as a function of rod length can be determined from an elementary balance experiment.

Finally, the amount of negligible friction which is present in any device of this type is always a point of concern. The analysis does not account for any applied moments due to friction. We assume that the pin surface at 0 can be lubricated to eliminate any frictional moments.

However, an even greater source of difficulty is presented by the interaction between the ball (projectile) and the cup surface. If the ball is pressed into the cup, then friction is induced on the ball surface which may change the direction of the initial velocity vector. The ball will tend to roll out of the cup, as opposed to being launched cleanly at an angle of 90° to the moment arm. We can remedy this situation by either lubricating the cup or by taking care prior to launch not to press the ball into the cup.

In conclusion, a deterministic, engineering analysis for the catapult has been presented. The result of this analysis is a single equation (10) which relates the initial and terminal angles, θ_0 and θ_1, to the applied force $F(\theta)$, the terminal distance R, and the various distances and weights in the system. In order to implement the analysis to select appropriate launch conditions, we must specify all of the input parameters from the problem and write a computer program to extract the information from equation (10). The input parameters consist of all of the distances r_B, r_F, and r_G, the moment of inertia I_0, the rubber band force characteristics, and the range R. Working with students, this project would take the better part of a semester to complete.

Blending Engineering with Statistics

Since it is going to be difficult at best to model the catapult using only an engineering approach, it is advisable to seek an alternative approximation technique to speed up the modeling process. Assume that we know the true functional relationship of distance with each of the variables contained in the catapult. This function could be approximated with a Taylor series model through the use of calculus. For the catapult, as well as today's complex products and processes, the problem is that this functional relationship is typically unknown or difficult to determine. A Taylor series model, however, can still be constructed using empirical data. For the catapult problem, a two level experimental design matrix can be used to generate the appropriate empirical data required to build a first order Taylor series approximation which can include desired 2–way linear interactions. Less than four hours of instruction are normally required to explain a simple approach to accomplish this type of modeling. Air Force Academy faculty and cadets attending an experimental design class recently used this approach for catapult modeling. They used engineering knowledge to brainstorm the four most likely variables that affect distance. These four variables were then included in an eight shot design matrix of highs and lows for each variable. The distance for these eight shots and the design matrix highs and lows were used to construct a first–order Taylor series model. Confirmation tests of the model indicated that ±2–inch accuracy can be achieved over the 200 inch operational range of the device. After initial training, the entire experimental approach was accomplished in two hours.

The obvious conclusions are: (1) The Taylor series modeling approach is fast and accurate; (2) In today's competitive market environment, engineers can potentially increase their ability to reduce product and process development times through the use of statistically designed experiments blended with existing engineering theory; and (3) The use of training devices such as the catapult not only provides hands–on training, but also makes the learning process enjoyable.

Submitted by: Majors James Brickell and Kenneth Knox
Department of Civil Engineering
United States Air Force Academy

Designed Experiments Case History - Environmental Engineering
Operation of an Activated Sludge System

Purpose: Demonstrate how experimental design can be used in environmental engineering. Specifically, show how operation of an activated sludge reactor can be better understood and optimized.

The Activated Sludge System

The activated sludge process is a biological treatment system used at many wastewater treatment plants. Briefly, biological organisms (termed sludge) within the reactor are used to aerobically convert incoming waste (influent) into additional biomass or innocuous carbon dioxide and water. The activated sludge reactor is followed immediately by a settling tank, called a secondary clarifier, where the liquid and solids (sludge) are separated. The sludge from the clarifier is then either wasted, or recycled back to the reactor to maintain an acceptable biomass population. Figure 1 shows a schematic of a typical completely-mixed activated sludge system.

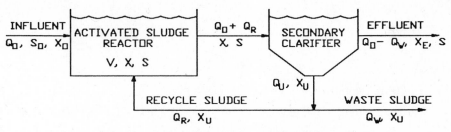

Figure 1. Schematic of Activated Sludge System

The parameters shown in Figure 1 are defined as follows:

Q_o, Q_u, Q_w, Q_r = flow rate in influent, clarifier underflow, waste, and recycled flows, respectively (m^3 / d)

S_o, S = soluble food concentration (measured as BOD_5) in the fluent and reactor, respectively (mg / L)

X_o, X, X_e, X_u = biomass concentration in influent, reactor, effluent, and clarifier underflow, respectively (mg / L)

V = reactor volume (m^3)

Performance of an activated sludge system is measured using the biochemical oxygen demand, BOD. The BOD is the amount of oxygen required to stabilize the decomposition matter in a water using aerobic biochemical action. In general, a lower BOD is indicative of a higher quality water. A typical municipal wastewater would have a BOD_5 concentration (BOD measured after 5 days at $20°C$) of 110 - 400 mg/L. The EPA effluent standards for wastewater, on an average monthly basis, are 30 mg/L BOD_5 (Peavy et al., 1985).

A mass balance analysis of the system described in Figure 1, using the Monod equation to describe the rate of bacterial growth, yields the following analytical relationship:

$$(S_o - S) = \frac{(1 + k_d VX / Q_w X_u)}{(Q_o Y / Q_w X_u)}$$

where:

$S_o - S$ = amount of BOD removed by the system

k_d = biological growth rate constant (d^{-1})

Y = fraction of food (S) converted to biomass (X), or

$$Y = \frac{kg\ X\ produced\ per\ day}{kg\ S\ consumed\ per\ day}$$

Experimental Design

To demonstrate application of experimental design, a simulated activated sludge system was conceived. Key parameters of the system were then varied between expected ranges, and the results noted.

The parameters which affect the treatment efficiency for removing BOD ($S_o - S$) from wastewater can be gleaned from the analytical equation describing the process (equation 1). Metcalf and Eddy, 1979, provide typical ranges for these parameters for an average municipal activated sludge system:

X, reactor biomass concentration $\quad$ 3000 - 6000 mg / L
X_u, clarifier biomass concentration $\quad$ 8000 - 12000 mg / L
k_d, biological growth rate constant $\quad$ 0.040 - 0.075 d^{-1}
Y, fraction of food to biomass $\quad$ 0.40 - 0.80 kg / kg

The system that will be used for this analysis originated from an example problem (10-1) presented by Metcalf and Eddy, 1979, where the reactor volume, V, is 4700 m^3, and the influent flow rate, Q_o, is 21,600 m^3 / d. Based on typical wasting rates, the waste sludge flow rate for this system was calculated to be 78.5 - 940.0 m^3 / d.

Initially, a series of experiments was designed using the five variables (X, X_u, k_d, Y, and Q_w). A two-level design was used, with the low ($-$) level at the lower limit of the typical range, and the high ($+$) level at the higher limit for each variable. A full-factorial design would have required 32 runs (2^5), which may be excessive for an actual plant to perform. Therefore, a quarter-factorial design was used, requiring 8 runs (2^3). Aliasing and the results of each run (based on equation 1) are shown in the design matrix below.

Quarter-Factorial Design Matrix

Run	X	X_u	Q_w	$k_d{=}XX_u$	XQ_w	X_uQ_w	$Y{=}XX_uQ_w$	Response (mg/L)
1	+	+	+	+	+	+	+	775
2	+	+	−	+	−	−	−	354
3	+	−	+	−	+	−	−	1001
4	+	−	−	−	−	+	+	102
5	−	+	+	−	−	+	−	1371
6	−	+	−	−	+	−	+	87
7	−	−	+	+	−	−	+	496
8	−	−	−	+	+	+	−	195
Avg(+)	558	647	911	455	515	611	365	$\bar{y}{=}548$
Avg(−)	537	448	184	640	581	485	730	
Δ	21	198	726	−185	−66	126	−365	
Δ/2	10	99	363	−93	−33	63	−183	

Prediction equation (First-order Taylor series approximation):

$$S_o - S = 548 + 10\bar{X} + 99\bar{X}_u + 363\bar{Q}_w - 93\bar{k}_d - 33\overline{XQ}_w + 63\overline{X_uQ}_w - 183\bar{Y}$$

The bars over the variables indicate these are linearly "coded" variables, with the value of -1 at the lower limit, $+1$ at the higher limit, and 0 at the midrange value. To evaluate the usefulness of the derived Taylor approximation of equation 1, a series of 25 runs was made with the five variables allowed to randomly vary between their typical upper and lower limits. The predicted response was then compared with the "correct" response from equation 1, the results of which are shown in Figure 2. The least squares regression line of the results has a respectable coefficient of determination (R_2) of 0.905.

An inspection of the prediction equation shows that the most important variables for controlling removal of BOD from wastewater seem to be Q_w and Y. A Pareto Diagram, Figure 3, more clearly demonstrates the contribution to BOD removal of each variable. In this figure, X_u and k_d also seem to contribute significantly to the observed responses, while X and the tested interactions (XQ_w and X_uQ_w) appear less consequential.

Untested interactions and aliasing could materially impact the prediction. Aliasing cannot be evaluated without further testing. The untested interactions, however, were analyzed using two-factor interaction graphs (Figures 4a-e) between the most significant main effects variables (Q_w, Y, X_u, and k_d). From Figure 4 the following observations can be made:

* Since the interaction graphs are not parallel, significant interaction exists between Q_w and Y (Figure 4a), Q_w, and k_d (Figure 4b), and between Y and k_d (Figure 4d).

* Little to no interaction exists between Y and X_u (Figure 4c), and between X_u and k_d (Figure 4e).

Even though it may be impractical using a "real" activated sludge system, a two-level full factorial model was tested on the computer using all five variables (32 runs) in order to more rigorously evaluate the desired experiments approach. Figure 5 shows that the prediction equation does a good job of modeling the actual BOD removal equation, with an R^2 over 0.99. The corresponding Pareto Diagram, Figure 6, shows that with minor exceptions the activated sludge system accurately identified the key variables using the designed experiments approach with only eight runs.

Operational Evaluation of Results

Based on the results of the quarter-factorial design, to maximize BOD removal from wastewater one should maximize Q_w and X_u, while minimizing k_d and Y. X and the tested

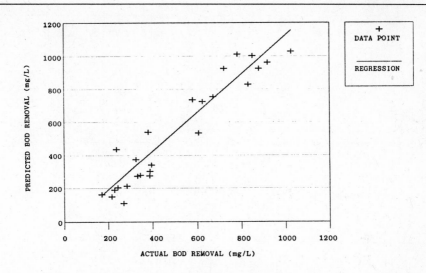

Figure 2. Comparison of predicted versus actual BOD removal rates from the quarter-factorial design model.

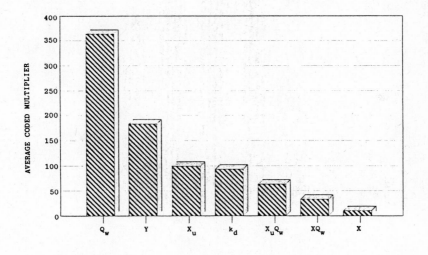

Figure 3. Pareto Diagram for quarter-factorial design.

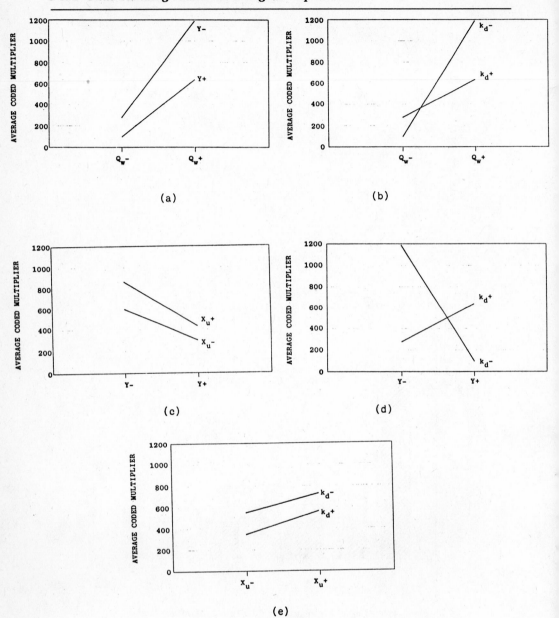

Figure 4. Two-Factor Interaction Graphs.

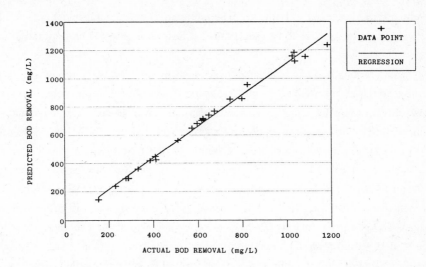

Figure 5.   Comparison of predicted versus actual BOD removal rates from the full-factorial design model.

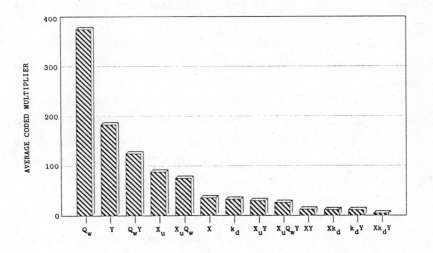

Figure 6.   Pareto Diagram for full-factorial design.

interactions are of lesser importance. The interactions between Q_w and Y, Q_w and k_d, and between Y and k_d appear to be significant, but further tests would be required to quantify these relationships.

Operationally, the only variables over which an operator has direct control (by turning a valve, for instance) are X, X_u, and Q_w. k_d and Y are biological rate constants which can only be controlled by relatively extreme measures such as changing the reactor water temperature or changing the type of microorganism used to stabilize the waste products. Therefore, a reasonable operational recommendation based on the results of this study would be to maximize the flow rate of waste sludge (Q_w) and the clarifier biomass concentration (X_u) in order to maximize removal of BOD from wastewater. If the biological growth rate constant, k_d, proved to be unstable for some reason, Figure 4b suggests that adjusting Q_w to approximately 275 m^3 / d (the intersection of the two lines, accounting for the "coded" variable) would make the activated sludge system robust to k_d. The power of the designed experiments approach is demonstrated here by noting that these conclusions are certainly not obvious from a simple inspection of equation 1.

An actual activated sludge process is extremely complex, however, and other considerations would come into play before undertaking changes to the system. One consideration is cost. For example, increasing Q_w will naturally increase the costs of disposing the waste sludge.

Operational problems may also arise, such as the sludge in the clarifier may form nitrogen gas and float if the residence time in the settler becomes excessive while attempting to increase X_u.

This case study has attempted to demonstrate the power of the designed experiments approach. Using some simple statistical techniques coupled with engineering judgement, complex systems can be economically modeled, and their operation greatly simplified. As a result, adjustments and modifications to the system can be made, and the consequences of these changes predicted with some degree of confidence.

References

Metcalf & Eddy, Inc., *Wastewater Engineering: Treatment, Disposal, Reuse*, 2nd ed., McGraw-Hill, New York, 1979.

Peavy, Howard S., Donald R. Rowe, and George Tchobanoglous, *Environmental Engineering*, McGraw-Hill, New York, 1985.

Submitted by   Dr. Thomas F. Curry
LOGICON/RDA Colorado Springs

Captain John J. Tomick
Department of Mathematical Sciences
United States Air Force Academy

Captain Barbara A. Yost
Department of Mathematical Sciences
United States Air Force Academy

Dr. Stephen T. Dziuban
LOGICON/RDA Colorado Springs

Minimizing the Number of Large Scale Computer Simulation Runs:

The Testing, Validation, and Implementation of the

Interagency Working Group (IWG)

Acquired Immune Deficiency Syndrome (AIDS) Model

Abstract

The IWG AIDS model requires over 600 lines of input data, most of which is uncertain. To determine the effect of the uncertain data, a sensitivity analysis is performed on 134 input variables. A Plackett-Burman experimental design is used to minimize the number of computer runs necessary to estimate the first-order approximation of the gradient of the local response surface of five dependent variables. For the input variables analyzed, this intermediate version of the model indicates that the six most important variables driving the gradient are the rate of progression from HIV to AIDS for adults , the percent of females age 15-44 that are married, the transmission rate of cocirculating sexually transmitted diseases, the rate of progression from AIDS to death for adults, the percent of men age 15-44 that are married, the percent of low risk married females that are HIV positive, and the number of contacts per month of high risk single females age 15-44. This suggests that intervention strategies should first consider methods to reduce risky behavior and control cocirculating sexually transmitted diseases.

Introduction and Background

In December 1988, researchers from the Department of Mathematical Sciences were asked to consult on the testing and validation of a simulation model designed to forecast, for most any nation, the effects of the AIDS epidemic. The request came from the State Department Interagency Working Group (IWG) which is chaired by the Deputy Assistant Secretary for the Environment. The IWG is a consortium of several agencies including the Subcommittee for AIDS Models and Methods, the US Agency for International Development, Health and Human Services, the National Institute of Health, the Center for Disease Control, the Department of Energy, and the Department of Defense. The AIDS model was developed during 1989, and this paper contains the experimental design methodology and the results of the initial testing and validation.

The AIDS model is a set of approximately 360 deterministic differential equations developed and programmed by scientists at Los Alamos National Laboratory (LANL) and the Merriam Laboratory at the University of Illinois at Urbana. The model uses risk-based methodology similar to models described in [3]. Data for the model is provided by the US Bureau of Census, Center for International Research (CIR).

The Problem: Uncertain Input Data and Lots of It

In spite of the excellent data collection procedures used by CIR, the AIDS data from Africa and the rest of the world are uncertain. Thus, the question arises as to the value of forecasts that are based on input parameters estimated from such data. The answer depends upon what one is asking the model to compute. Uncertainty in the model's output depends upon uncertainty in the input data as well as upon uncertainties in the model assumptions and formulation. We discuss the techniques used to analyze the sensitivity of the model to the input data, and how this sensitivity analysis can be used to help form decisions based on model results. The question of uncertainty due to incorrect model formulation is left to future studies.

The IWG AIDS model requires a large amount of input data. The input data file is over 600 lines long and contains demographic and epidemiological information including population distributions, fertility rates, migration rates, sexual risk behavior categories, sexual contact rates, cocirculating sexually transmitted diseases, condom use, percent of the population that is seropositive (has the HIV virus), risk of infection per sexual contact, rate of progression from infection to AIDS, and blood transfusion and blood screening. Most of these data are enumerated by sex, age, and marital status resulting in over 350 separate input variables that could be analyzed for their effect on the various response surfaces. Using "engineering knowledge," the team selected 134 input variables that were of interest.

Each run of the AIDS model requires approximately 15 minutes on a 386 based microcomputer. If "one-at-a-time" input data perturbation techniques were used, the required 2^{134} runs would take 10^{29} years. Clearly, an efficient design was needed to evaluate the model. In conferences at LANL, we determined that Plackett-Burman designs (using Hadamard matrices) were well suited for the initial first-order approximation of the gradient of the "local response surfaces."

Experimental Design

To study the relative sensitivity of the results to each of the 134 parameters in the input file, we investigated the response surfaces by varying the input data by small amounts around a baseline set of values for a "generic" country. Using the techniques described in [1,2,6,7], we developed a 136 by 136 Hadamard matrix to set up 136 runs. All parameters were varied simultaneously, and the effects of each were ranked. Since the parameters each take on their high and low values for half the runs, there is enough information to do t-tests to statistically compare the means of the outputs from the high value and low value runs.

Results

In our study of a recent version of the model, most inputs were varied about their baseline values by approximately 10% of their range. Using 10 year forecasts, the epidemic response surfaces were most sensitive to the rate of progression from HIV to AIDS for adults (expressed as the percent of the HIV+ population that converts to AIDS per year), and the percent of females age 15-44 that are married. Obtaining better estimates of these parameters appears to be the key to narrowing confidence bounds on the model results.

In all, five epidemic response surfaces were examined. The nine most important input variables driving each response are listed in the following table:

Frequently appearing variables key:

%MarFem15-44	=	Percent of females age 15-44 that are married
%MarMen15-44	=	Percent of males age 15-44 that are married
Inf Rate Men	=	Risk of infection per sexual contact for men
STD Trans	=	Unprotected transmission rate of cocirculating sexually transmitted diseases (STD's)
Cntct/mth HRSF 15-44	=	# of sexual contacts per month by high risk single females age 15-44
Rate AIDS to Death 5+	=	Rate of conversion from AIDS to Death for age 5 and older
Rate HIV to AIDS 5+	=	Rate of conversion from HIV to AIDS for age 5 and older
Months Dur STD Fem	=	Duration in months of and STD episode for females
%HIV+ LRMF 15-44	=	Percent of low risk married females that are HIV+

Response variable key:

1	=	Percent of population infected
2	=	Cumulative AIDS deaths
3	=	Percent of deaths due to AIDS
4	=	Number of infected females age 15-44
5	=	Percent of females age 15-44 infected

Frequently appearing variables	Response Variable					
	1	2	3	4	5	Avg
Rate HIVtoAIDS5+	1	1	1	1	1	1.0
%MarFem15-44	2	2	2	2	2	2.0
STD Trans	3	10	9	4	3	5.8
Rate AIDStoDeath5+	4	11	11	5	4	7.0
%MarMen15-44	7	7	8	10	7	7.8
%HIV+ LRMF15-44	14	5	5	12	9	9.0
Cntct/mth HRSF 15-44	8	9	12	11	10	10.0
MonthsDur STD Fem	6	16	15	7	6	10.0
Inf Rate Men	5	18	23	6	5	11.4

These results suggest the use of intervention strategies that reduce high risk behavior and control sexually transmitted diseases. When Ugandan President Yoweri K. Museveni was shown these model results, he immediately reversed a long-held position and urged Ugandans to use condoms [5].

The Value of the Experimental Design Results

The above table lists only the top nine variables controlling the response surfaces. When these results were briefed to the IWG, several health officials stated that this was the first time they had seen any information on what variables were important in driving the AIDS epidemic. They noted that the complete ranking of all 134 input parameters can be used to determine the "best" way to minimize the number of deaths due to AIDS. These results can also be used to guide data collection. Indeed, it is evident that accurate values for these variables must be determined before approximate confidence bounds on the forecasts can be calculated.

The experimental design runs also helped us determine model deficiencies. In other testing, by examining the pattern of the parameter settings in failed runs, we determined that when the number of men was increased by 10%, and the number of women was decreased by 10%, the model failed because there were insufficient numbers of women for the men to marry. This resulted in the programming of a variable marriage rate that can handle depletion of certain age groups in the population. This also suggests that important changes in societal structure and behavior may occur due to large segments of a population dying from AIDS.

The model was run with both 5% and 10% perturbations and the rank ordering of the independent variables remained the same indicating that the response surfaces are fairly smooth in the region of interest.

Lastly, the results of the main effects study have been used to screen the input parameters for analysis of two-way interactions. Specifically, using a "two-to-the-seven-minus-one" experimental design, we examined seven input variables for pairwise interactions. From the marginal mean plots, we determined that (at 96% confidence) there was a statistically significant interaction between the number of men in the high risk group and the risk of infection per sexual contact for males. This supplies additional support for the need to accurately determine the risk of infection per sexual contact.

In summary, we found the use of experimental design techniques to be very valuable. They reduced the time required for model testing from years to weeks, provided us with valuable rankings of main effects, and gave direction to future data collection, and future model analyses.

References

[1]   J. Hadamard, "Resolution d'une Question Relative aux Determinants," *Bulletin des Sciences Mathematiques*, No.2, Vol. 17, Part 1,pp 240-246 (1893).

[2]   A. Hedayat and W.D. Wallis, "Hadamard Matrices and Their Applications," *The Annals of Statistics*, Vol. 6, No. 6, pp 1184-1238 (1978).

[3]   J.M. Hyman, and E.A. Stanley, "Using Mathematical Models to Understand the AIDS Epidemic," *Mathematical Biosciences*, Vol 90, pp415-473 (1988).

[4]   R.E.A.C. Paley, "On Orthogonal Matrices," *Journal of Mathematics and Physics*, Vol. 12, pp 311-320 (1933).

[5]   J. Perlez, "Spread of AIDS is Worrying Uganda," *The New York Times*, (January 30, 1991).

[6]   R.L. Plackett, and J.P. Burman, "The Design of Optimum Multifactorial Experiments," *Biometrika*, Vol. 33, pp 305-325 (1946).

[7]   S. R. Schmidt, and R.G. Launsby, *Understanding Industrial Designed Experiments*, CQG Ltd Printing, Longmont, CO (1989).

Submitted by: Rita Whiteley and Robert Lawson
                          Cell Technology, Inc., Boulder, CO

Karl Fischer Moisture Designed Experiment

Introduction:

The Karl Fischer method of water determination, titrates water with iodine in the presence of sulfur dioxide, methanol, and a suitable base. See the following chemical reaction:

$H_2O + I_2 + SO_2 + CH_3OH + 3RN \longrightarrow [RNH]SO_4CH_3 + 2[RNH]I$ where RN = Base.

Use of the Mettler DL18 titrator allows the titration to be followed exactly through the use of a two-pin platinum electrode which has a current source applied to its poles. The voltage measured at the polarized electrode pins is used by the controls as an input signal. When the last traces of water have been titrated, voltage drops to virtually zero, the electrodes are then depolarized by the iodine now present.

The Karl Fischer titration permits determination of available water only. This means water must be brought into solution by suitable means before the titration can be carried out. Sample preparation for freeze-dried ImuVert is very critical. Since a portion of the moisture found in lyophilized ImuVert is tightly bound to the crystalline structure and difficult to make available for titration, an external extraction step must be performed prior to titration.

Objective:

Determine how external extraction variables, such as solvents and shake time, effect the measured percent moisture titrated from lyophilized ImuVert.

Factors and Levels: Factors and levels chosen are as follows:

Factors	Levels
Methanol	0, 0.5 ml
Chloroform	0, 0.5 ml
Formamide	0.5, 1.0 ml
Shake Time	30 sec, 3630 sec

Response: The response looked at was percent moisture with hopes to maximize the measured amount of moisture determined by the Karl Fischer titration.

Experimental Design:

Defining Contrast: I = ABCD
2**K−P Fractional Factorial
Yates Order

Run #	TC	Methanol	Chloroform	Formamide	Shake Time	Response y_1	Response y_2
4	(1)	0	0	0.5 ml	30 sec	1.2402	1.1005
2	a	0.5 ml	0	0.5 ml	3630 sec	0.8562	0.9602
3	b	0	0.5 ml	0.5 ml	3630 sec	0.9883	0.9907
1	ab	0.5 ml	0.5 ml	0.5 ml	30 sec	1.0212	1.0845
7	c	0	0	1.0 ml	3630 sec	0.9993	1.0068
8	ac	0.5 ml	0	1.0 ml	30 sec	0.9625	1.0568
5	bc	0	0.5 ml	1.0 ml	30 sec	0.8733	1.0253
6	abc	0.5 ml	0.5 ml	1.0 ml	3630 sec	1.0151	0.9992

The experiment was performed in a randomized order.

Procedure:

Prior to sample analysis, the titrator must be calibrated to ensure accurate mechanical operation. Calibration is completed by checking titrant concentration and instrument drift. Both parameters were within statistically acceptable ranges.

A blank vial determination (to determine moisture in the solvents as well as residual moisture attributed to an empty vial) is the next step. This is accomplished by using an empty vial which was processed exactly like lyophilized product; then, testing the blank vial with the same factor levels as will be done for the sample. For example, if run #2 required 0.5 ml formamide, 0.5 ml methanol, no chloroform, and a 3630 second shake time − the blank for that run would be done in the same manner.

Sample analysis for the designed experiment is the next stage. For each experimental run, two freeze-dried samples were tested with three to four repeat samplings per vial. Run #3 was an exception − one sampling/vial was done due to solvent volume. Solvents were added in a specific order as follows: formamide, methanol, and lastly chloroform. Swirling of the sample occurred after the formamide addition to expediate resolution of the freeze-dried product. The experimental design runs that received only the 30 second vigorous vortexing were tested immediately; while, other design runs after vortexing were allowed to shake continuously for one hour on a platform shaker before being tested.

In order for the titrator to determine the percentage moisture, several parameter values need to be entered by the operator. Sample volume to be tested was held constant at 0.4 ml. The blank vial's moisture content was entered based on the blank determination for that given run. Sample weight, which is based on the product's dry cake weight, solvent volume, and sample size varied due to the solvent volume used for the external extraction. The following sample weights were used: 2.0 ml solvent--0.0073 g, 1.5 ml solvent--0.0097 g, 1.0 ml solvent--0.0146 g, and 0.5 ml solvent--0.0292 g.

Data/Analysis:

Following is an ANOVA table showing the results obtained from the designed experiment.

$$I = ABCD$$

Index #	Total Observation	Sum of Square	Half Effect	Measures Average	DF	Mean Square	F Ratio
1	2.1057	16.3622	1.011				
2	1.8164	0.0025	0.013	A^1=BCD	1	0.0025	0.608
3	1.9790	0.0163	0.032	B^2=ACD	1	0.0163	3.904
4	2.3407	0.0188	0.034	AB=CD	1	0.0188	4.513
5	1.8986	0.0058	−0.019	C^3=ABD	1	0.0058	1.382
6	2.0143	0.0002	0.004	AC=BD	1	0.0002	0.048
7	2.0061	0.0051	−0.018	BC=AD	1	0.0051	1.219
8	2.0193	0.0355	−0.047	ABC=D^4	1	0.0355	8.518
	Error	0.0333			8	0.0042	
	Total	0.1174			15		

[1] Methanol
[2] Chloroform
[3] Formamide
[4] Shake Time

Using a $F_{T0.05}$ = 5.31, the only significant factor is D (shake time) or the three-way interaction ABC (methanol, chloroform, and formamide). However, at the 90% confidence level, using a $F_{T0.10} = 3.46$, the B factor (chloroform) and the AB or CD interaction become significant.

See the following graphical display of the single factor effects generated by using low and high points for each factor.

Since the largest single effects (shake time and chloroform) are confounded with the ABC and ACD interactions, another experiment was needed to determine which was causing the significant effect.

Second designed experiment contained the following factors and levels:

Factors	Levels
Chloroform	0.5, 1.0 ml
Shake Time	30, 3630 sec

Percent moisture was the response examined. Constants in the experiment were: methanol at 0.5 ml and formamide at 0.5 ml.

Experimental design:

2**K Factorial
Yates Order

Run #	TC	Chloroform	Shake Time	Response y_1	y_2
2	(1)	0.5 ml	30 sec	1.2264	1.2141
3	a	1.0 ml	30 sec	1.1330	1.1467
1	b	0.5 ml	3630 sec	1.0419	1.1271
4	ab	1.0 ml	3630 sec	1.1718	1.1533

The experiment was run in a random order.

The data can be summarized in the following ANOVA table and graphs.

I = AB

Index #	Total Observations	Sum of Squares	Half Effect	Measures Average	DF	Mean Square	F Ratio
1	2.1690	10.6129	1.152				
2	2.4405	0.0126	0.040	A(CHCL$_3$)	1	0.0126	12.651
3	2.2797	0.0000	−0.001	B(Shake)	1	0.0000	0.003
4	2.3251	0.0064	−0.028	AB	1	0.0064	6.440
	Error	0.0040			4	0.0010	
	Total	0.0229			7		

At the 95% confidence level only chloroform is an important effect; whereas, at the 90% confidence level the AB interaction becomes significant. In either case shake time alone has no significance, which means that the important effect noted in experiment one was a three factor interaction.

Conclusions:

The most significant effect on the external extraction process performed on ImuVert is the three solvent interaction. Based on the first experiments single effect plots, 0.5 ml of formamide, 0.5 ml of methanol, and 1.0 ml of chloroform are used for the external extraction.

Since shaking the product versus just vortexing the product was shown to have no effect in the designed experiment, for future testing the 30 second vortex and 60 minute shake will be used. This method did produce smaller coefficients of variation within sample testing per vial.

Attempts were made to remove the methanol to simplify the external extraction since it did not contribute significantly to the external extraction. However, when methanol was removed, two phases formed and measurements were harder to make. Therefore, all three external extraction solvent components were kept in the system.

Further optimization of this external extraction process could be done, including trying even less formamide. Also, the two factor interaction AB or CD which is significant at the 90% level may be important and a deconfounding experiment should be completed.

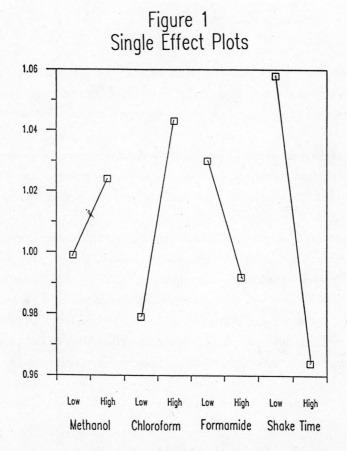

Figure 1
Single Effect Plots

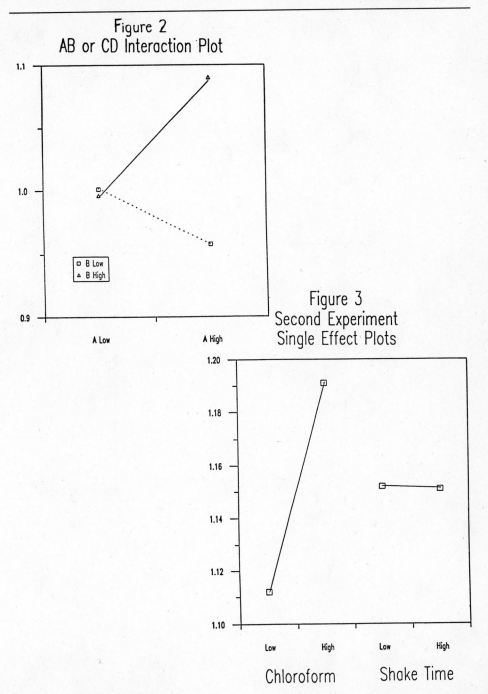

Figure 2
AB or CD Interaction Plot

□ B Low
▲ B High

A Low   A High

Figure 3
Second Experiment
Single Effect Plots

Chloroform   Shake Time

Figure 4
Second Experiment
AB Interaction

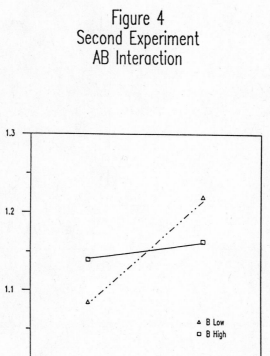

Submitted by: C. B. (Kip) Rogers
Digital Equipment Corporation, Marlboro, MA

THE END COUNT METHOD
POWER WITH SMALL SAMPLES

In your company as in mine, we deal with IC's, SCR's, bearings, chemicals and a range of equipment in making a reliable and cost effective products. To keep things from crashing in our world we also need to make the right decisions often with limited samples and/or time. In this context End Count is a wonderful tool to add to your data analysis tool kit. Easy to learn, simple to apply, exact in its answers; it is a most useful supplement to one's existing statistical techniques. Now let's see what it is and how it works.

END COUNT

In the following paragraphs we will explore an increasingly popular technique for making sound decisions with very small samples. It is a technique which is of special use to those for whom:

    (a)   Samples are very expensive,

    (b)   Time is critical or,

    (c)   Both.

First let's look at how we make decisions in cases like these, then we'll do a couple of brief examples using the End Count method. Finally, for the statistically curious we'll show why the method works.

In decision making, we rarely can get all the data. Instead we rely on samples drawn from the entire data set (population). We then compare the results of these samples with other data and make a relative judgment on which we base our actions. This process involves some risk, which we hope to minimize, in return for the benefit of not having to measure the entire data set.

With End Count we are comparing two alternative solutions to a problem. Our goal is to decide whether the two alternatives are significantly different from each other. End Count is a simple technique for making this decision with very few assumptions or samples.

It provides the exact probability of the observed outcome happening by chance (i.e., there is no difference between the two items or methods being tested). Then you, as a decision maker, can decide if the risk is worth taking.

EXAMPLE 1

Two different styles of heat sinks are being evaluated. A standard test setup is designed and the heat sink results (on-chip temp) are as follows:

> Style A: 110   113   108   111   (Results in degrees Centigrade)
> Style B: 107   105   103   106

Looking at the data, your 'gut' reaction is that Style B, which yields a lower temperature, is better, but you wonder 'could this be a fluke?' – 'Should I do more testing?'. In this case the answers are No and No! Let's see why.

The first step is to rank order the data (keeping the data identified with its Style I.D.). From low to high we have:

<pre>
103   105   106   107   108   110   111   113
 B   B   B   B   A   A   A   A
| ⇐ 4 ⇒ | ⇐ 4 ⇒ |
</pre>

By counting in the number of consecutive letters from each end until a letter change occurs and combining these "End Counts" we get 4 plus 4 equals 8. This is a measure of NON-homogeneity signifying that the two designs are really different. How likely is this degree of stratification to occur by chance if the designs were equal (which they aren't)? Table 1 has the answer. Letting style B = STD and style A = EXP we locate the 4/4 row and read across to the End Count = 8 column. The answer 2/70 or 2.86% is the likelihood of getting this degree of stratification from equal populations. In other words we can be 97% certain that the two designs are really different.

Example 2

An adhesion test is performed to see if two flux types influence the bonding strength achieved with a new solder process. Noting that Flux Y had a much longer processing time, we take advantage of the fact that the End Count method does not require equal sample sizes. Results are:

Flux X:　435　　440　　428　　432　　437　　(Data in grams)

Flux Y:　422　　431　　426　　424

Rank ordering we get:

```
422   424   426   428   431   432   435   437   440
 Y     Y     Y     X     Y     X     X     X     X
|  ⇐    3    ⇒  |         |  ⇐    4    ⇒  |
```

Once again the same nagging questions occur. 'Gut' says X gives stronger bonds but logic asks 'Could this be an accident?' and 'Shouldn't I test more?'. Referring to the End Count table we see that for 4 EXP and 5 STD samples with an End Count of $3 + 4 = 7$ we have only a 4/126 (3.17%) chance of this happening by accident and we conclude that type X really does give stronger bonds.

Table 1 — End Count Probabilities

PROBABILITY OF OBTAINING AN END COUNT THIS HIGH (OR HIGHER) WHEN STD = EXP

EXP Sample	STD Sample	END COUNT								
		10	9	8	7	6	5	4	3	2
5	5	2/252	N/A	4/252	8/252	16/252	32/252	60/252	100/252	140/252
5	4		2/126	N/A	4/126	8/126	16/126	30/126	50/126	70/126
4	5		2/126	N/A	4/126	8/126	16/126	30/126	50/126	70/126
4	4			2/70	N/A	4/70	8/70	16/70	28/70	40/70
4	3				2/35	N/A	4/35	8/35	14/35	20/35
3	4				2/35	N/A	4/35	8/35	14/35	20/35
3	3					2/20	N/A	4/20	8/20	12/20

INSTRUCTIONS FOR USING THE END COUNT METHOD

1. Rank order data in a single row.

2. Write an E under each experimental value and an S under each standard value.[*]

3. Start at the left and count the number of consecutive letters which are the same. Go to the right end and repeat using the opposite letter.

4. Add the two counts. The total is your End Count. (If the same letter occurs at both ends then the End Count = 0 which simply means that the End Count method cannot detect a difference.)

5. Refer to the above table for the likelihood that an end count this high could have happened by chance.

6. Decide whether this likelihood is low enough for you to be comfortable stating that the observed difference is real.

[*] If any data values are both E and S then flip a coin or put the letters in a hat to determine the assignment for those values.

FOR THE TECHNICALLY CURIOUS

The End Count technique belongs to a class of statistics known as non-parametric methods. In these cases we don't assume a normal (or any other) distribution. The first published appearance of the techniques goes back to 1959, when John Tukey published his paper on what is referred to as the Tukey Quick Test. This procedure has been modified and popularized by a well known industrial consultant, Dorian Shainin.

Our End Count examples can be explained using a simple coin shuffling methodology. Starting with the combination of 4 pennies followed by 4 dimes we can by systematically moving the D's to the right generate all 70 possible combinations of coin flips (Figure 2). Just as flips of an unbiased coin represent a 50/50 proposition so should test results from two identical processes fall equally above and below the median. In this case if EXP = STD, then each of the 70 combinations is equally likely! Given that the combinations are equally likely then let us first check to see if half the combinations begin with D and half with P. Thirty-five each—right!

Now let us consider the highly stratified cases with an End Count of 8. In this case there are only two such possible combinations out of all 70 orderings. Stated another way, the drawing 4 pennies followed by 4 dimes is very unlikely: 1/70 + 1/70 or 2.9%! Putting on our statistician's hat we conclude that this occurrence, though possible, is soooo unlikely that we reject the hypothesis that this ordering happened by chance. Transferring this concept to the real world, and our tests, we conclude that tests which produce End Counts with low probabilities in all likelihood come from populations having different averages.

Note: The preceding examples were directed towards answering the question "Are STD and EXP *different* from each other?" In statistical terms we were performing a two-tailed test. Also worth considering is the (one-tailed) test case where we will only take action if the stratification is in one direction i.e., PPPPDDDD but *not* DDDDPPPP. This is often the case when a new process will be considered only if it is better (not just different from the standard). In this case all the End Count table values are halved (2/20 becomes 1/20) effectively doubling the sensitivity of the End Count method.

Table 2 — Generation of 4/4 Combinations by shifting "P's"

N	COMB	END COUNT	N	COMB	END COUNT	N	COMB	END COUNT
1	PPPPDDDD	8	25	PDPDDDPP	0	49	DPDPDPPD	0
2	PPDPDDDD	6	26	PDDPPPDD	3	50	DPDPDPDP	2
3	PPPDDPDD	5	27	PDDPPDPD	2	51	DPDPDDPP	3
4	PPPDDDPD	4	28	PDDPPDDP	0	52	DPDDPPPD	0
5	PPPDDDDP	0	29	PDDPDPPD	2	53	DPDDPPDP	2
6	PPDPPDDD	5	30	PDDPDPDP	0	54	DPDDPDPP	3
7	PPDPDPDD	4	31	PDDPDDPP	0	55	DPDDDPPP	4
8	PPDPDDPD	3	32	PDDDPPPD	2	56	DDPPPPDD	0
9	PPDPDDDP	0	33	PDDDPPDP	0	57	DDPPPDPD	0
10	PPDDPPDD	4	34	PDDDPDPP	0	58	DDPPPDDP	3
11	PPDDPDPD	3	35	PDDDDPPP	0	59	DDPPDPPD	0
12	PPDDPDDP	0	36	DPPPPDDD	0	60	DDPPDPDP	3
13	PPDDDPPD	3	37	DPPPDPDD	0	61	DDPPDDPP	4
14	PPDDDPDP	0	38	DPPPDDPD	0	62	DDPDPPPD	0
15	PPDDDDPP	0	39	DPPPDDDP	2	63	DDPDPPDP	3
16	PDPPPDDD	4	40	DPPDPPDD	0	64	DDPDPDPP	4
17	PDPPDPDD	3	41	DPPDPDPD	0	65	DDPDDPPP	5
18	PDPPDDPD	2	42	DPPDPDDP	2	66	DDDPPPPD	0
19	PDPPDDDP	0	43	DPPDDPPD	0	67	DDDPPPDP	4
20	PDPDPPDD	3	44	DPPDDPDP	2	68	DDDPPDPP	5
21	PDPDPDPD	2	45	DPPDDDPP	3	69	DDDPDPPP	6
22	PDPDPDDP	0	46	DPDPPPDD	0	70	DDDDPPPP	8
23	PDPDDPPD	2	47	DPDPPDPD	0			
24	PDPDDPDP	0	48	DPDPPDDP	2			

Note: Using Table 2 as a guide, a simple software program can be developed which covers all practical sample sizes. The technique involves moving the rightmost P sequentially to the right; then popping it back as the next P moves one space to the right...etc.

Appendixes

			Column			
Row	1–5	6–10	11–15	16–20	21–25	26–30	31–35
1	50527	38063	89093	22648	96867	34685	62753
2	40483	25946	46848	63507	86984	05724	73793
3	80462	82084	21308	94040	77165	29466	52079
4	90545	19515	78767	05365	88790	57080	41826
5	19169	99595	73714	19238	72251	05922	87552
6	52665	26336	40764	26694	19988	67596	36786
7	90799	50016	47988	52706	16346	31800	00971
8	45285	63834	42650	90117	15536	72206	44917
9	19441	42344	28192	99346	23158	19639	48645
10	37222	08805	23852	40256	85211	32672	64952
11	37065	26209	66719	27944	16944	81393	35310
12	48899	17553	47175	31199	26672	19188	38099
13	44622	87968	72342	43257	54973	60454	12025
14	73638	18241	52230	89038	67734	44774	55787
15	42221	90634	01822	75261	08195	60196	28252
16	89921	11815	52797	74868	34720	20249	95911
17	38555	79182	87003	42410	09189	29731	44611
18	26213	44234	84459	21895	01186	21979	70360
19	68455	81564	04625	69299	58598	41757	58267
20	91866	70002	70036	66806	00693	89570	22355
21	99347	68962	57338	30529	02209	82909	71087
22	82928	44796	04134	13362	69602	24334	86744
23	16242	87679	08615	30668	50641	63391	00152
24	71385	29319	42870	79158	67826	32664	21638
25	51420	85638	02017	31406	16928	66618	74314
26	07700	56663	07642	72029	15147	43861	48231
27	38514	82498	84649	38830	02403	07300	42212
28	39756	23408	97266	33946	58018	78874	25099
29	12796	42649	09878	25483	30689	39905	99100
30	93589	01789	63970	76873	33782	38460	87506
31	11213	10296	36181	92910	14471	21381	04219
32	61233	66409	94223	11695	36021	92241	85479
33	98421	75296	41506	71427	57443	64683	94253
34	04977	06872	34764	00911	95711	00664	08983
35	59298	13735	95985	56378	94783	54725	05701
36	15017	53836	00873	37602	85937	05872	74263
37	29311	95124	47110	83427	44799	68966	49907
38	82609	13408	68318	84181	36411	20749	83699
39	39952	93967	91401	50899	75843	47751	55027
40	09731	99395	53883	53855	88758	14556	83042
41	45821	84519	07493	12282	80953	13641	67947
42	84309	87119	70436	73462	98107	82617	35553
43	02929	44362	01024	67928	43415	79721	67653
44	42989	49536	64420	38132	58927	32576	51847
45	63063	49898	06019	98905	37156	84088	64254
46	57588	27694	01911	79714	87368	96074	90567
47	26093	88111	84279	85312	51340	23309	98691
48	73300	95180	55945	73516	31579	81353	15724
49	54085	91495	91575	46931	54054	18622	64911
50	76838	77175	69384	09011	69902	88684	81569

Binomial Probabilities calculated from: $P(x) = \binom{n}{x} p^x (1-p)^{n-x}$

n	x	.01	.02	.03	.04	.05	.06	.07	.08	.09		x	n
2	0	.9801	.9604	.9409	.9216	.9025	.8836	.8649	.8464	.8281		2	
	1	.0198	.0392	.0582	.0768	.0950	.1128	.1302	.1472	.1630		1	
	2	.0001	.0004	.0009	.0016	.0025	.0036	.0049	.0064	.0081		0	2
3	0	.9703	.9412	.9127	.8847	.8574	.8306	.8044	.7787	.7536		3	
	1	.0294	.0576	.0847	.1106	.1354	.1590	.1816	.2031	.2236		2	
	2	.0003	.0012	.0026	.0046	.0071	.0102	.0137	.0177	.0221		1	
	3	.0000	.0000	.0000	.0001	.0001	.0002	.0003	.0005	.0007		0	3
4	0	.9606	.9224	.8853	.8493	.8145	.7807	.7481	.7164	.6857		4	
	1	.0388	.0753	.1095	.1416	.1715	.1993	.2252	.2492	.2713		3	
	2	.0006	.0023	.0051	.0088	.0135	.0191	.0254	.0325	.0402		2	
	3	.0000	.0000	.0001	.0002	.0005	.0008	.0013	.0019	.0027		1	
	4	.0000	.0000	.0000	.0000	.0000	.0000	.0000	.0001	.0001		0	4
5	0	.9510	.9039	.8587	.8154	.7738	.7339	.6957	.6591	.6240		5	
	1	.0480	.0922	.1328	.1699	.2036	.2342	.2618	.2866	.3086		4	
	2	.0010	.0038	.0082	.0142	.0214	.0299	.0394	.0498	.0610		3	
	3	.0000	.0001	.0003	.0006	.0011	.0019	.0030	.0043	.0060		2	
	4	.0000	.0000	.0000	.0000	.0000	.0001	.0001	.0002	.0003		1	
	5	.0000	.0000	.0000	.0000	.0000	.0000	.0000	.0000	.0000		0	5
6	0	.9415	.8858	.8330	.7828	.7351	.6899	.6470	.6064	.5679		6	
	1	.0571	.1085	.1546	.1957	.2321	.2642	.2922	.3164	.3370		5	
	2	.0014	.0055	.0120	.0204	.0305	.0422	.0550	.0688	.0833		4	
	3	.0000	.0002	.0005	.0011	.0021	.0036	.0055	.0080	.0110		3	
	4	.0000	.0000	.0000	.0000	.0001	.0002	.0003	.0005	.0008		2	
	5	.0000	.0000	.0000	.0000	.0000	.0000	.0000	.0000	.0000		1	
	6	.0000	.0000	.0000	.0000	.0000	.0000	.0000	.0000	.0000		0	6
7	0	.9321	.8681	.8080	.7514	.6983	.6485	.6017	.5578	.5168		7	
	1	.0659	.1240	.1749	.2192	.2573	.2897	.3170	.3396	.3578		6	
	2	.0020	.0076	.0162	.0274	.0406	.0555	.0716	.0886	.1061		5	
	3	.0000	.0003	.0008	.0019	.0036	.0059	.0090	.0128	.0175		4	
	4	.0000	.0000	.0000	.0001	.0002	.0004	.0007	.0011	.0017		3	
	5	.0000	.0000	.0000	.0000	.0000	.0000	.0000	.0001	.0001		2	
	6	.0000	.0000	.0000	.0000	.0000	.0000	.0000	.0000	.0000		1	
	7	.0000	.0000	.0000	.0000	.0000	.0000	.0000	.0000	.0000		0	7
8	0	.9227	.8508	.7837	.7214	.6634	.6096	.5596	.5132	.4703		8	
	1	.0746	.1389	.1939	.2405	.2793	.3113	.3370	.3570	.3721		7	
	2	.0026	.0099	.0210	.0351	.0515	.0695	.0888	.1067	.1288		6	
	3	.0001	.0004	.0013	.0029	.0054	.0089	.0134	.0189	.0255		5	
	4	.0000	.0000	.0001	.0002	.0004	.0007	.0013	.0021	.0031		4	
	5	.0000	.0000	.0000	.0000	.0000	.0000	.0001	.0001	.0002		3	
	6	.0000	.0000	.0000	.0000	.0000	.0000	.0000	.0000	.0000		2	
	7	.0000	.0000	.0000	.0000	.0000	.0000	.0000	.0000	.0000		1	
	8	.0000	.0000	.0000	.0000	.0000	.0000	.0000	.0000	.0000		0	8
9	0	.9135	.8337	.7602	.6925	.6302	.5730	.5204	.4722	.4279		9	
	1	.0830	.1531	.2116	.2597	.2985	.3292	.3525	.3695	.3809		8	
	2	.0034	.0125	.0262	.0433	.0629	.0840	.1061	.1285	.1507		7	
	3	.0001	.0006	.0019	.0042	.0077	.0125	.0186	.0261	.0348		6	
	4	.0000	.0000	.0001	.0003	.0006	.0012	.0021	.0034	.0052		5	
	5	.0000	.0000	.0000	.0000	.0000	.0001	.0002	.0003	.0005		4	
	6	.0000	.0000	.0000	.0000	.0000	.0000	.0000	.0000	.0000		3	
	7	.0000	.0000	.0000	.0000	.0000	.0000	.0000	.0000	.0000		2	
	8	.0000	.0000	.0000	.0000	.0000	.0000	.0000	.0000	.0000		1	
	9	.0000	.0000	.0000	.0000	.0000	.0000	.0000	.0000	.0000		0	9
		.99	.98	.97	.96	.95	.94	.93	.92	.91		x	n

n	x	.10	.15	.20	.25	p .30	.35	.40	.45	.50		
2	0	.8100	.7225	.6400	.5625	.4900	.4225	.3600	.3025	.2500	2	
	1	.1800	.2550	.3200	.3750	.4200	.4550	.4800	.4950	.5000	1	
	2	.0100	.0225	.0400	.0625	.0900	.1225	.1600	.2025	.2500	0	2
3	0	.7290	.6141	.5120	.4219	.3430	.2746	.2160	.1664	.1250	3	
	1	.2430	.3251	.3840	.4219	.4410	.4436	.4320	.4084	.3750	2	
	2	.0270	.0574	.0960	.1406	.1890	.2389	.2880	.3341	.3750	1	
	3	.0010	.0034	.0080	.0156	.0270	.0429	.0640	.0911	.1250	0	3
4	0	.6561	.5220	.4096	.3164	.2401	.1785	.1296	.0915	.0625	4	
	1	.2916	.3685	.4096	.4219	.4116	.3845	.3456	.2995	.2500	3	
	2	.0486	.0975	.1536	.2109	.2646	.3105	.3456	.3675	.3750	2	
	3	.0036	.0115	.0256	.0469	.0756	.1115	.1536	.2005	.2500	1	
	4	.0001	.0005	.0016	.0039	.0081	.0150	.0256	.0410	.0625	0	4
5	0	.5905	.4437	.3277	.2373	.1681	.1160	.0778	.0503	.0312	5	
	1	.3280	.3915	.4096	.3955	.3601	.3124	.2592	.2059	.1562	4	
	2	.0729	.1382	.2048	.2637	.3087	.3364	.3456	.3369	.3125	3	
	3	.0081	.0244	.0512	.0879	.1323	.1811	.2304	.2757	.3125	2	
	4	.0004	.0022	.0064	.0146	.0283	.0488	.0768	.1128	.1562	1	
	5	.0000	.0001	.0003	.0010	.0024	.0053	.0102	.0185	.0312	0	5
6	0	.5314	.3771	.2621	.1780	.1176	.0754	.0467	.0277	.0156	6	
	1	.3543	.3993	.3932	.3560	.3025	.2437	.1866	.1359	.0938	5	
	2	.0984	.1762	.2458	.2966	.3241	.3280	.3110	.2780	.2344	4	
	3	.0146	.0415	.0819	.1318	.1852	.2355	.2765	.3032	.3125	3	
	4	.0012	.0055	.0154	.0330	.0595	.0951	.1382	.1861	.2344	2	
	5	.0001	.0004	.0015	.0044	.0102	.0205	.0369	.0609	.0938	1	
	6	.0000	.0000	.0001	.0002	.0007	.0018	.0041	.0083	.0156	0	6
7	0	.4783	.3206	.2097	.1335	.0824	.0490	.0280	.0152	.0078	7	
	1	.3720	.3960	.3670	.3115	.2471	.1848	.1306	.0872	.0547	6	
	2	.1240	.2097	.2753	.3115	.3177	.2985	.2613	.2140	.1641	5	
	3	.0230	.0617	.1147	.1730	.2269	.2679	.2903	.2918	.2734	4	
	4	.0026	.0109	.0287	.0577	.0972	.1442	.1935	.2388	.2734	3	
	5	.0002	.0012	.0043	.0115	.0250	.0466	.0774	.1172	.1641	2	
	6	.0000	.0001	.0004	.0013	.0036	.0084	.0172	.0320	.0547	1	
	7	.0000	.0000	.0000	.0001	.0002	.0006	.0016	.0037	.0078	0	7
8	0	.4305	.2725	.1678	.1001	.0576	.0319	.0168	.0084	.0039	8	
	1	.3826	.3847	.3355	.2670	.1977	.1373	.0896	.0548	.0312	7	
	2	.1488	.2376	.2936	.3115	.2965	.2587	.2090	.1569	.1094	6	
	3	.0331	.0839	.1468	.2076	.2541	.2786	.2787	.2568	.2188	5	
	4	.0046	.0185	.0459	.0865	.1361	.1875	.2322	.2627	.2734	4	
	5	.0004	.0026	.0092	.0231	.0467	.0808	.1239	.1719	.2188	3	
	6	.0000	.0002	.0011	.0038	.0100	.0217	.0413	.0703	.1094	2	
	7	.0000	.0000	.0001	.0004	.0012	.0033	.0079	.0164	.0312	1	
	8	.0000	.0000	.0000	.0000	.0001	.0002	.0007	.0017	.0039	0	8
9	0	.3874	.2316	.1342	.0751	.0404	.0207	.0101	.0046	.0020	9	
	1	.3874	.3679	.3020	.2253	.1556	.1004	.0605	.0339	.0176	8	
	2	.1722	.2597	.3020	.3003	.2668	.2162	.1612	.1110	.0703	7	
	3	.0446	.1069	.1762	.2336	.2668	.2716	.2508	.2119	.1641	6	
	4	.0074	.0283	.0661	.1168	.1715	.2194	.2508	.2600	.2461	5	
	5	.0008	.0050	.0165	.0389	.0735	.1181	.1672	.2128	.2461	4	
	6	.0001	.0006	.0028	.0087	.0210	.0424	.0743	.1160	.1641	3	
	7	.0000	.0000	.0003	.0012	.0039	.0098	.0212	.0407	.0703	2	
	8	.0000	.0000	.0000	.0001	.0004	.0013	.0035	.0083	.0176	1	
	9	.0000	.0000	.0000	.0000	.0000	.0001	.0003	.0008	.0020	0	9
		.90	.85	.80	.75	.70	.65	.60	.55	.50	x	n

p

n	x	.01	.02	.03	.04	.05	.06	.07	.08	.09		
10	0	.9044	.8171	.7374	.6648	.5987	.5386	.4840	.4344	.3894	10	
	1	.0914	.1667	.2281	.2770	.3151	.3438	.3643	.3777	.3851	9	
	2	.0042	.0153	.0317	.0519	.0746	.0988	.1234	.1478	.1714	8	
	3	.0001	.0008	.0026	.0058	.0105	.0168	.0248	.0343	.0452	7	
	4	.0000	.0000	.0001	.0004	.0010	.0019	.0033	.0052	.0078	6	
	5	.0000	.0000	.0000	.0000	.0001	.0001	.0003	.0005	.0009	5	
	6	.0000	.0000	.0000	.0000	.0000	.0000	.0000	.0000	.0001	4	
	7	.0000	.0000	.0000	.0000	.0000	.0000	.0000	.0000	.0000	3	
	8	.0000	.0000	.0000	.0000	.0000	.0000	.0000	.0000	.0000	2	
	9	.0000	.0000	.0000	.0000	.0000	.0000	.0000	.0000	.0000	1	
	10	.0000	.0000	.0000	.0000	.0000	.0000	.0000	.0000	.0000	0	10
12	0	.8864	.7847	.6938	.6127	.5404	.4759	.4186	.3677	.3225	12	
	1	.1074	.1922	.2575	.3064	.3413	.3645	.3781	.3837	.3827	11	
	2	.0060	.0216	.0438	.0702	.0988	.1280	.1565	.1835	.2082	10	
	3	.0002	.0015	.0045	.0098	.0173	.0272	.0393	.0532	.0686	9	
	4	.0000	.0001	.0003	.0009	.0021	.0039	.0067	.0104	.0153	8	
	5	.0000	.0000	.0000	.0001	.0002	.0004	.0008	.0014	.0024	7	
	6	.0000	.0000	.0000	.0000	.0000	.0000	.0001	.0001	.0003	6	
	7	.0000	.0000	.0000	.0000	.0000	.0000	.0000	.0000	.0000	5	
	8	.0000	.0000	.0000	.0000	.0000	.0000	.0000	.0000	.0000	4	
	9	.0000	.0000	.0000	.0000	.0000	.0000	.0000	.0000	.0000	3	
	10	.0000	.0000	.0000	.0000	.0000	.0000	.0000	.0000	.0000	2	
	11	.0000	.0000	.0000	.0000	.0000	.0000	.0000	.0000	.0000	1	
	12	.0000	.0000	.0000	.0000	.0000	.0000	.0000	.0000	.0000	0	12
15	0	.8601	.7386	.6333	.5421	.4633	.3953	.3367	.2863	.2430	15	
	1	.1303	.2261	.2938	.3388	.3658	.3785	.3801	.3734	.3605	14	
	2	.0092	.0323	.0636	.0988	.1348	.1691	.2003	.2273	.2496	13	
	3	.0004	.0029	.0085	.0178	.0307	.0468	.0653	.0857	.1070	12	
	4	.0000	.0002	.0008	.0022	.0049	.0090	.0148	.0223	.0317	11	
	5	.0000	.0000	.0001	.0002	.0006	.0013	.0024	.0043	.0069	10	
	6	.0000	.0000	.0000	.0000	.0000	.0001	.0003	.0006	.0011	9	
	7	.0000	.0000	.0000	.0000	.0000	.0000	.0000	.0001	.0001	8	
	8	.0000	.0000	.0000	.0000	.0000	.0000	.0000	.0000	.0000	7	
	9	.0000	.0000	.0000	.0000	.0000	.0000	.0000	.0000	.0000	6	
	10	.0000	.0000	.0000	.0000	.0000	.0000	.0000	.0000	.0000	5	
	11	.0000	.0000	.0000	.0000	.0000	.0000	.0000	.0000	.0000	4	
	12	.0000	.0000	.0000	.0000	.0000	.0000	.0000	.0000	.0000	3	
	13	.0000	.0000	.0000	.0000	.0000	.0000	.0000	.0000	.0000	2	
	14	.0000	.0000	.0000	.0000	.0000	.0000	.0000	.0000	.0000	1	
	15	.0000	.0000	.0000	.0000	.0000	.0000	.0000	.0000	.0000	0	15
20	0	.8179	.6676	.5438	.4420	.3585	.2901	.2342	.1887	.1516	20	
	1	.1652	.2725	.3364	.3683	.3774	.3703	.3526	.3282	.3000	19	
	2	.0159	.0528	.0988	.1458	.1887	.2246	.2521	.2711	.2818	18	
	3	.0010	.0065	.0183	.0364	.0596	.0860	.1139	.1414	.1672	17	
	4	.0000	.0006	.0024	.0065	.0133	.0233	.0364	.0523	.0703	16	
	5	.0000	.0000	.0002	.0009	.0022	.0048	.0088	.0145	.0222	15	
	6	.0000	.0000	.0000	.0001	.0003	.0008	.0017	.0032	.0055	14	
	7	.0000	.0000	.0000	.0000	.0000	.0001	.0002	.0005	.0011	13	
	8	.0000	.0000	.0000	.0000	.0000	.0000	.0000	.0001	.0002	12	
	9	.0000	.0000	.0000	.0000	.0000	.0000	.0000	.0000	.0000	11	
	10	.0000	.0000	.0000	.0000	.0000	.0000	.0000	.0000	.0000	10	
	11	.0000	.0000	.0000	.0000	.0000	.0000	.0000	.0000	.0000	9	
	12	.0000	.0000	.0000	.0000	.0000	.0000	.0000	.0000	.0000	8	
	13	.0000	.0000	.0000	.0000	.0000	.0000	.0000	.0000	.0000	7	
	14	.0000	.0000	.0000	.0000	.0000	.0000	.0000	.0000	.0000	6	
	15	.0000	.0000	.0000	.0000	.0000	.0000	.0000	.0000	.0000	5	
	16	.0000	.0000	.0000	.0000	.0000	.0000	.0000	.0000	.0000	4	
	17	.0000	.0000	.0000	.0000	.0000	.0000	.0000	.0000	.0000	3	
	18	.0000	.0000	.0000	.0000	.0000	.0000	.0000	.0000	.0000	2	
	19	.0000	.0000	.0000	.0000	.0000	.0000	.0000	.0000	.0000	1	
	20	.0000	.0000	.0000	.0000	.0000	.0000	.0000	.0000	.0000	0	20
		.99	.98	.97	.96	.95	.94	.93	.92	.91	x	n

p

n	x	.10	.15	.20	.25	.30	.35	.40	.45	.50	x	n
10	0	.3487	.1969	.1074	.0563	.0282	.0135	.0060	.0025	.0010	10	10
	1	.3874	.3474	.2684	.1877	.1211	.0725	.0403	.0207	.0098	9	
	2	.1937	.2759	.3020	.2816	.2335	.1757	.1209	.0763	.0439	8	
	3	.0574	.1298	.2013	.2503	.2668	.2522	.2150	.1665	.1172	7	
	4	.0112	.0401	.0881	.1460	.2001	.2377	.2508	.2384	.2051	6	
	5	.0015	.0085	.0264	.0584	.1029	.1536	.2007	.2340	.2461	5	
	6	.0001	.0012	.0055	.0162	.0368	.0689	.1115	.1596	.2051	4	
	7	.0000	.0001	.0008	.0031	.0090	.0212	.0425	.0746	.1172	3	
	8	.0000	.0000	.0001	.0004	.0014	.0043	.0106	.0229	.0439	2	
	9	.0000	.0000	.0000	.0000	.0001	.0005	.0016	.0042	.0098	1	
	10	.0000	.0000	.0000	.0000	.0000	.0000	.0001	.0003	.0010	0	10
12	0	.2824	.1422	.0687	.0317	.0138	.0057	.0022	.0008	.0002	12	12
	1	.3766	.3012	.2062	.1267	.0712	.0368	.0174	.0075	.0029	11	
	2	.2301	.2924	.2835	.2323	.1678	.1088	.0639	.0339	.0161	10	
	3	.0852	.1720	.2362	.2581	.2397	.1954	.1419	.0923	.0537	9	
	4	.0213	.0683	.1329	.1936	.2311	.2367	.2128	.1700	.1208	8	
	5	.0038	.0193	.0532	.1032	.1585	.2039	.2270	.2225	.1934	7	
	6	.0005	.0040	.0155	.0401	.0792	.1281	.1766	.2124	.2256	6	
	7	.0000	.0006	.0033	.0115	.0291	.0591	.1009	.1489	.1934	5	
	8	.0000	.0001	.0005	.0024	.0078	.0199	.0420	.0762	.1208	4	
	9	.0000	.0000	.0001	.0004	.0015	.0048	.0125	.0277	.0537	3	
	10	.0000	.0000	.0000	.0000	.0002	.0008	.0025	.0068	.0161	2	
	11	.0000	.0000	.0000	.0000	.0000	.0001	.0003	.0010	.0029	1	
	12	.0000	.0000	.0000	.0000	.0000	.0000	.0000	.0001	.0002	0	12
15	0	.2059	.0874	.0352	.0134	.0047	.0016	.0005	.0001	.0000	15	15
	1	.3432	.2312	.1319	.0668	.0305	.0126	.0047	.0016	.0005	14	
	2	.2669	.2856	.2309	.1559	.0916	.0476	.0219	.0090	.0032	13	
	3	.1285	.2184	.2501	.2252	.1700	.1110	.0634	.0318	.0139	12	
	4	.0428	.1156	.1876	.2252	.2186	.1792	.1268	.0780	.0417	11	
	5	.0105	.0449	.1032	.1651	.2061	.2123	.1859	.1404	.0916	10	
	6	.0019	.0132	.0430	.0917	.1472	.1906	.2066	.1914	.1527	9	
	7	.0003	.0030	.0138	.0393	.0811	.1319	.1771	.2013	.1964	8	
	8	.0000	.0005	.0035	.0131	.0348	.0710	.1181	.1647	.1964	7	
	9	.0000	.0001	.0007	.0034	.0116	.0298	.0612	.1048	.1527	6	
	10	.0000	.0000	.0001	.0007	.0030	.0096	.0245	.0515	.0916	5	
	11	.0000	.0000	.0000	.0001	.0006	.0024	.0074	.0191	.0417	4	
	12	.0000	.0000	.0000	.0000	.0001	.0004	.0016	.0052	.0139	3	
	13	.0000	.0000	.0000	.0000	.0000	.0001	.0003	.0010	.0032	2	
	14	.0000	.0000	.0000	.0000	.0000	.0000	.0000	.0001	.0005	1	
	15	.0000	.0000	.0000	.0000	.0000	.0000	.0000	.0000	.0000	0	15
20	0	.1216	.0388	.0115	.0032	.0008	.0002	.0000	.0000	.0000	20	20
	1	.2702	.1368	.0576	.0211	.0068	.0020	.0005	.0001	.0000	19	
	2	.2852	.2293	.1369	.0669	.0278	.0100	.0031	.0008	.0002	18	
	3	.1901	.2428	.2054	.1339	.0716	.0323	.0123	.0040	.0011	17	
	4	.0898	.1821	.2182	.1897	.1304	.0738	.0350	.0139	.0046	16	
	5	.0319	.1028	.1746	.2023	.1789	.1272	.0746	.0365	.0148	15	
	6	.0089	.0454	.1091	.1686	.1916	.1712	.1244	.0746	.0370	14	
	7	.0020	.0160	.0545	.1124	.1643	.1844	.1659	.1221	.0739	13	
	8	.0004	.0046	.0222	.0609	.1144	.1614	.1797	.1623	.1201	12	
	9	.0001	.0011	.0074	.0271	.0654	.1158	.1597	.1771	.1602	11	
	10	.0000	.0002	.0020	.0099	.0308	.0686	.1171	.1593	.1762	10	
	11	.0000	.0000	.0005	.0030	.0120	.0336	.0710	.1185	.1602	9	
	12	.0000	.0000	.0001	.0008	.0039	.0136	.0355	.0727	.1201	8	
	13	.0000	.0000	.0000	.0002	.0010	.0045	.0146	.0366	.0739	7	
	14	.0000	.0000	.0000	.0000	.0002	.0012	.0049	.0150	.0370	6	
	15	.0000	.0000	.0000	.0000	.0000	.0003	.0013	.0049	.0148	5	
	16	.0000	.0000	.0000	.0000	.0000	.0000	.0003	.0013	.0046	4	
	17	.0000	.0000	.0000	.0000	.0000	.0000	.0000	.0002	.0011	3	
	18	.0000	.0000	.0000	.0000	.0000	.0000	.0000	.0000	.0002	2	
	19	.0000	.0000	.0000	.0000	.0000	.0000	.0000	.0000	.0000	1	
	20	.0000	.0000	.0000	.0000	.0000	.0000	.0000	.0000	.0000	0	20
		.90	.85	.80	.75	.70	.65	.60	.55	.50	x	n

Poisson Probabilities calculated from: $P(x) = \dfrac{\lambda^x e^{-\lambda}}{x!}$

x	λ .1	.2	.3	.4	.5	.6	.7	.8	.9
0	.9048	.8187	.7408	.6703	.6065	.5488	.4966	.4493	.4066
1	.0905	.1637	.2222	.2681	.3033	.3293	.3476	.3595	.3659
2	.0045	.0164	.0333	.0536	.0758	.0988	.1217	.1438	.1647
3	.0002	.0011	.0033	.0072	.0126	.0198	.0284	.0383	.0494
4	.0000	.0001	.0003	.0007	.0016	.0030	.0050	.0077	.0111
5	.0000	.0000	.0000	.0001	.0002	.0004	.0007	.0012	.0020
6	.0000	.0000	.0000	.0000	.0000	.0000	.0001	.0002	.0003

x	λ 1.0	1.5	2.0	2.5	3.0	3.5	4.0	4.5	5.0
0	.3679	.2231	.1353	.0821	.0498	.0302	.0183	.0111	.0067
1	.3679	.3347	.2707	.2052	.1494	.1057	.0733	.0500	.0337
2	.1839	.2510	.2707	.2565	.2240	.1850	.1465	.1125	.0842
3	.0613	.1255	.1804	.2138	.2240	.2158	.1954	.1687	.1404
4	.0153	.0471	.0902	.1336	.1680	.1888	.1954	.1898	.1755
5	.0031	.0141	.0361	.0668	.1008	.1322	.1563	.1708	.1755
6	.0005	.0035	.0120	.0278	.0504	.0771	.1042	.1281	.1462
7	.0001	.0008	.0034	.0099	.0216	.0385	.0595	.0824	.1044
8	.0000	.0001	.0009	.0031	.0081	.0169	.0298	.0463	.0653
9	.0000	.0000	.0002	.0009	.0027	.0066	.0132	.0232	.0363
10	.0000	.0000	.0000	.0002	.0008	.0023	.0053	.0104	.0181
11	.0000	.0000	.0000	.0000	.0002	.0007	.0019	.0043	.0082
12	.0000	.0000	.0000	.0000	.0001	.0002	.0006	.0016	.0034
13	.0000	.0000	.0000	.0000	.0000	.0001	.0002	.0006	.0013
14	.0000	.0000	.0000	.0000	.0000	.0000	.0001	.0002	.0005
15	.0000	.0000	.0000	.0000	.0000	.0000	.0000	.0001	.0002

x	λ 5.5	6.0	6.5	7.0	7.5	8.0	9.0	10.0	11.0
0	.0041	.0025	.0015	.0009	.0006	.0003	.0001	.0000	.0000
1	.0225	.0149	.0098	.0064	.0041	.0027	.0011	.0005	.0002
2	.0618	.0446	.0318	.0223	.0156	.0107	.0050	.0023	.0010
3	.1133	.0892	.0688	.0521	.0389	.0286	.0150	.0076	.0037
4	.1558	.1339	.1118	.0912	.0729	.0573	.0337	.0189	.0102
5	.1714	.1606	.1454	.1277	.1094	.0916	.0607	.0378	.0224
6	.1571	.1606	.1575	.1490	.1367	.1221	.0911	.0631	.0411
7	.1234	.1377	.1462	.1490	.1465	.1396	.1171	.0901	.0646
8	.0849	.1033	.1188	.1304	.1373	.1396	.1318	.1126	.0888
9	.0519	.0688	.0858	.1014	.1144	.1241	.1318	.1251	.1085
10	.0285	.0413	.0558	.0710	.0858	.0993	.1186	.1251	.1194
11	.0143	.0225	.0330	.0452	.0585	.0722	.0970	.1137	.1194
12	.0065	.0113	.0179	.0263	.0366	.0481	.0728	.0948	.1094
13	.0028	.0052	.0089	.0142	.0211	.0296	.0504	.0729	.0926
14	.0011	.0022	.0041	.0071	.0113	.0169	.0324	.0521	.0728
15	.0004	.0009	.0018	.0033	.0057	.0090	.0194	.0347	.0534
16	.0001	.0003	.0007	.0014	.0026	.0045	.0109	.0217	.0367
17	.0000	.0001	.0003	.0006	.0012	.0021	.0058	.0128	.0237
18	.0000	.0000	.0001	.0002	.0005	.0009	.0029	.0071	.0145
19	.0000	.0000	.0000	.0001	.0002	.0004	.0014	.0037	.0084
20	.0000	.0000	.0000	.0000	.0001	.0002	.0006	.0019	.0046
21	.0000	.0000	.0000	.0000	.0000	.0001	.0003	.0009	.0024
22	.0000	.0000	.0000	.0000	.0000	.0000	.0001	.0004	.0012
23	.0000	.0000	.0000	.0000	.0000	.0000	.0000	.0002	.0006
24	.0000	.0000	.0000	.0000	.0000	.0000	.0000	.0001	.0003
25	.0000	.0000	.0000	.0000	.0000	.0000	.0000	.0000	.0001

x	\(\lambda\)								
	12	13	14	15	16	17	18	19	20
0	.0000	.0000	.0000	.0000	.0000	.0000	.0000	.0000	.0000
1	.0001	.0000	.0000	.0000	.0000	.0000	.0000	.0000	.0000
2	.0004	.0002	.0001	.0000	.0000	.0000	.0000	.0000	.0000
3	.0018	.0008	.0004	.0002	.0001	.0000	.0000	.0000	.0000
4	.0053	.0027	.0013	.0006	.0003	.0001	.0001	.0000	.0000
5	.0127	.0070	.0037	.0019	.0010	.0005	.0002	.0001	.0001
6	.0255	.0152	.0087	.0048	.0026	.0014	.0007	.0004	.0002
7	.0437	.0281	.0174	.0104	.0060	.0034	.0019	.0010	.0005
8	.0655	.0457	.0304	.0194	.0120	.0072	.0042	.0024	.0013
9	.0874	.0661	.0473	.0324	.0213	.0135	.0083	.0050	.0029
10	.1048	.0859	.0663	.0486	.0341	.0230	.0150	.0095	.0058
11	.1144	.1015	.0844	.0663	.0496	.0355	.0245	.0164	.0106
12	.1144	.1099	.0984	.0829	.0661	.0504	.0368	.0259	.0176
13	.1056	.1099	.1060	.0956	.0814	.0658	.0509	.0378	.0271
14	.0905	.1021	.1060	.1024	.0930	.0800	.0655	.0514	.0387
15	.0724	.0885	.0989	.1024	.0992	.0906	.0786	.0650	.0516
16	.0543	.0719	.0866	.0960	.0992	.0963	.0884	.0772	.0646
17	.0383	.0550	.0713	.0847	.0934	.0963	.0936	.0863	.0760
18	.0255	.0397	.0554	.0706	.0830	.0909	.0936	.0911	.0844
19	.0161	.0272	.0409	.0557	.0699	.0814	.0887	.0911	.0888
20	.0097	.0177	.0286	.0418	.0559	.0692	.0798	.0866	.0888
21	.0055	.0109	.0191	.0299	.0426	.0560	.0684	.0783	.0846
22	.0030	.0065	.0121	.0204	.0310	.0433	.0560	.0676	.0769
23	.0016	.0037	.0074	.0133	.0216	.0320	.0438	.0559	.0669
24	.0008	.0020	.0043	.0083	.0144	.0226	.0328	.0442	.0557
25	.0004	.0010	.0024	.0050	.0092	.0154	.0237	.0336	.0446
26	.0002	.0005	.0013	.0029	.0057	.0101	.0164	.0246	.0343
27	.0001	.0002	.0007	.0016	.0034	.0063	.0109	.0173	.0254
28	.0000	.0001	.0003	.0009	.0019	.0038	.0070	.0117	.0181
29	.0000	.0001	.0002	.0005	.0011	.0023	.0044	.0077	.0125
30	.0000	.0000	.0001	.0002	.0006	.0013	.0026	.0049	.0083
31	.0000	.0000	.0001	.0001	.0003	.0007	.0015	.0030	.0054
32	.0000	.0000	.0000	.0001	.0001	.0004	.0009	.0018	.0034
33	.0000	.0000	.0000	.0000	.0001	.0002	.0005	.0010	.0020
34	.0000	.0000	.0000	.0000	.0000	.0001	.0002	.0006	.0012
35	.0000	.0000	.0000	.0000	.0000	.0001	.0001	.0003	.0007
36	.0000	.0000	.0000	.0000	.0000	.0000	.0001	.0002	.0004
37	.0000	.0000	.0000	.0000	.0000	.0000	.0000	.0001	.0002
38	.0000	.0000	.0000	.0000	.0000	.0000	.0000	.0000	.0001
39	.0000	.0000	.0000	.0000	.0000	.0000	.0000	.0000	.0001

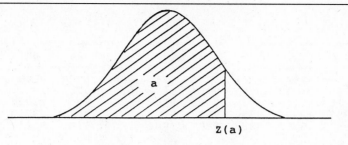

Z(a)

Cumulative Probabilities

Each 4-digit entry is the area "a" under the standard normal curve
from −∞ to Z(a).   Reference the sketch above.

z	.00	.01	.02	.03	.04	.05	.06	.07	.08	.09
.0	.5000	.5040	.5080	.5120	.5160	.5199	.5239	.5279	.5319	.5359
.1	.5398	.5438	.5478	.5517	.5557	.5596	.5636	.5675	.5714	.5753
.2	.5793	.5832	.5871	.5910	.5948	.5987	.6026	.6064	.6103	.6141
.3	.6179	.6217	.6255	.6293	.6331	.6368	.6406	.6443	.6480	.6517
.4	.6554	.6591	.6628	.6664	.6700	.6736	.6772	.6808	.6844	.6879
.5	.6915	.6950	.6985	.7019	.7054	.7088	.7123	.7157	.7190	.7224
.6	.7257	.7291	.7324	.7357	.7389	.7422	.7454	.7486	.7517	.7549
.7	.7580	.7611	.7642	.7673	.7704	.7734	.7764	.7794	.7823	.7852
.8	.7881	.7910	.7939	.7967	.7995	.8023	.8051	.8078	.8106	.8133
.9	.8159	.8186	.8212	.8238	.8264	.8289	.8315	.8340	.8365	.8389
1.0	.8413	.8438	.8461	.8485	.8508	.8531	.8554	.8577	.8599	.8621
1.1	.8643	.8665	.8686	.8708	.8729	.8749	.8770	.8790	.8810	.8830
1.2	.8849	.8869	.8888	.8907	.8925	.8944	.8962	.8980	.8997	.9015
1.3	.9032	.9049	.9066	.9082	.9099	.9115	.9131	.9147	.9162	.9177
1.4	.9192	.9207	.9222	.9236	.9251	.9265	.9279	.9292	.9306	.9319
1.5	.9332	.9345	.9357	.9370	.9382	.9394	.9406	.9418	.9429	.9441
1.6	.9452	.9463	.9474	.9484	.9495	.9505	.9515	.9525	.9535	.9545
1.7	.9554	.9564	.9573	.9582	.9591	.9599	.9608	.9616	.9625	.9633
1.8	.9641	.9649	.9656	.9664	.9671	.9678	.9686	.9693	.9699	.9706
1.9	.9713	.9719	.9726	.9732	.9738	.9744	.9750	.9756	.9761	.9767
2.0	.9772	.9778	.9783	.9788	.9793	.9798	.9803	.9808	.9812	.9817
2.1	.9821	.9826	.9830	.9834	.9838	.9842	.9846	.9850	.9854	.9857
2.2	.9861	.9864	.9868	.9871	.9875	.9878	.9881	.9884	.9887	.9890
2.3	.9893	.9896	.9898	.9901	.9904	.9906	.9909	.9911	.9913	.9916
2.4	.9918	.9920	.9922	.9925	.9927	.9929	.9931	.9932	.9934	.9936

z	.00	.01	.02	.03	.04	.05	.06	.07	.08	.09
2.5	.9938	.9940	.9941	.9943	.9945	.9946	.9948	.9949	.9951	.9952
2.6	.9953	.9955	.9956	.9957	.9959	.9960	.9961	.9962	.9963	.9964
2.7	.9965	.9966	.9967	.9968	.9969	.9970	.9971	.9972	.9973	.9974
2.8	.9974	.9975	.9976	.9977	.9977	.9978	.9979	.9979	.9980	.9981
2.9	.9981	.9982	.9982	.9983	.9984	.9984	.9985	.9985	.9986	.9986
3.0	.9987	.9987	.9987	.9988	.9988	.9989	.9989	.9989	.9990	.9990
3.1	.9990	.9991	.9991	.9991	.9992	.9992	.9992	.9992	.9993	.9993
3.2	.9993	.9993	.9994	.9994	.9994	.9994	.9994	.9995	.9995	.9995
3.3	.9995	.9995	.9995	.9996	.9996	.9996	.9996	.9996	.9996	.9997
3.4	.9997	.9997	.9997	.9997	.9997	.9997	.9997	.9997	.9997	.9998

Selected Percentiles

Each entry is $Z(a)$ where $P[\ Z \le Z(a)\ ] = a$. For example,
$P(Z \le 1.645) = .95$ so $Z(.95) = 1.645$.

a	.10	.05	.025	.02	.01	.005	.001
Z(a)	-1.282	-1.645	-1.960	-2.054	-2.326	-2.576	-3.090

a	.90	.95	.975	.98	.99	.995	.999
Z(a)	1.282	1.645	1.960	2.054	2.326	2.576	3.090

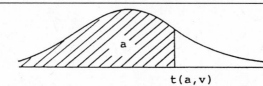

$$t(a,v)$$

df				a			
v	.75	.90	.95	.975	.99	.995	.9995
1	1.000	3.078	6.314	12.706	31.821	63.657	636.619
2	0.816	1.886	2.920	4.303	6.965	9.925	31.599
3	0.765	1.638	2.353	3.182	4.541	5.841	12.924
4	0.741	1.533	2.132	2.776	3.747	4.604	8.610
5	0.727	1.476	2.015	2.571	3.365	4.032	6.869
6	0.718	1.440	1.943	2.447	3.143	3.707	5.959
7	0.711	1.415	1.895	2.365	2.998	3.499	5.408
8	0.706	1.397	1.860	2.306	2.896	3.355	5.041
9	0.703	1.383	1.833	2.262	2.821	3.250	4.781
10	0.700	1.372	1.812	2.228	2.764	3.169	4.587
11	0.697	1.363	1.796	2.201	2.718	3.106	4.437
12	0.695	1.356	1.782	2.179	2.681	3.055	4.318
13	0.694	1.350	1.771	2.160	2.650	3.012	4.221
14	0.692	1.345	1.761	2.145	2.624	2.977	4.140
15	0.691	1.341	1.753	2.131	2.602	2.947	4.073
16	0.690	1.337	1.746	2.120	2.583	2.921	4.015
17	0.689	1.333	1.740	2.110	2.567	2.898	3.965
18	0.688	1.330	1.734	2.101	2.552	2.878	3.922
19	0.688	1.328	1.729	2.093	2.539	2.861	3.883
20	0.687	1.325	1.725	2.086	2.528	2.845	3.850
21	0.686	1.323	1.721	2.080	2.518	2.831	3.819
22	0.686	1.321	1.717	2.074	2.508	2.819	3.792
23	0.685	1.319	1.714	2.069	2.500	2.807	3.768
24	0.685	1.318	1.711	2.064	2.492	2.797	3.745
25	0.684	1.316	1.708	2.060	2.485	2.787	3.725
26	0.684	1.315	1.706	2.056	2.479	2.779	3.707
27	0.684	1.314	1.703	2.052	2.473	2.771	3.690
28	0.683	1.313	1.701	2.048	2.467	2.763	3.674
29	0.683	1.311	1.699	2.045	2.462	2.756	3.659
30	0.683	1.310	1.697	2.042	2.457	2.750	3.646
40	0.681	1.303	1.684	2.021	2.423	2.704	3.551
60	0.679	1.296	1.671	2.000	2.390	2.660	3.460
120	0.677	1.289	1.658	1.980	2.358	2.617	3.373
∞	0.674	1.282	1.645	1.960	2.326	2.576	3.291

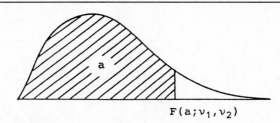

$$F(a;v_1,v_2)$$

The F distribution is skewed to the right and has two parameters:

   v_1 = numerator degrees of freedom, and

   v_2 = denominator degrees of freedom.

The random variable F is denoted by $F(v_1,v_2)$.   Entries in the tables denote the positive values $F(a;v_1,v_2)$ such that

   $P[F(v_1,v_2) \leq F(a;v_1,v_2)] = a$ (reference the above sketch).

The superscripted entries, like 4053^2, should be read as 405300.

Because the F distribution has two parameters, any set of tables containing even a few percentiles will be extensive.   We have included the following percentiles in the pages that follow: 50th, 75th, 90th, 95th, 97.5^{th}, 99th, and 99.9^{th}.   This set of seven percentiles is given for each of 20 different values that v_1 and v_2 can assume (i.e., 400 possible combinations of v_1 and v_2).

Percentiles less than the 50th can be found by using the following relationship:

$$F(a;v_1,v_2) \quad = \quad \frac{1}{F(1-a;v_2,v_1)}$$

Please note that the order of the parameters is reversed in the two F expressions.   For example, if a = .05, v_1 = 8, and v_2 = 30, then

$$F(.05;8,30) \quad = \quad \frac{1}{F(.95;30,8)} \quad = \quad \frac{1}{3.08} \quad = \quad .325$$

ν_2	a	ν_1									
		1	2	3	4	5	6	7	8	9	10
1	.50	1.00	1.50	1.71	1.82	1.89	1.94	1.98	2.00	2.03	2.04
	.75	5.83	7.50	8.20	8.58	8.82	8.98	9.10	9.19	9.26	9.32
	.90	39.9	49.5	53.6	55.8	57.2	58.2	58.9	59.4	59.9	60.2
	.95	161	200	216	225	230	234	237	239	241	242
	.975	648	800	864	900	922	937	948	957	963	969
	.99	4052	5000	5403	5625	5764	5859	5928	5982	6022	6056
	.999	4053^2	5000^2	5404^2	5625^2	5764^2	5859^2	5929^2	5981^2	6023^2	6056^2
2	.50	.667	1.00	1.13	1.21	1.25	1.28	1.30	1.32	1.33	1.34
	.75	2.57	3.00	3.15	3.23	3.28	3.31	3.34	3.35	3.37	3.38
	.90	8.53	9.00	9.16	9.24	9.29	9.33	9.35	9.37	9.38	9.39
	.95	18.5	19.0	19.2	19.2	19.3	19.3	19.4	19.4	19.4	19.4
	.975	38.5	39.0	39.2	39.2	39.3	39.3	39.4	39.4	39.4	39.4
	.99	98.5	99.0	99.2	99.2	99.3	99.3	99.4	99.4	99.4	99.4
	.999	998	999	999	999	999	999	999	999	999	999
3	.50	.585	.881	1.00	1.06	1.10	1.13	1.15	1.16	1.17	1.18
	.75	2.02	2.28	2.36	2.39	2.41	2.42	2.43	2.44	2.44	2.44
	.90	5.54	5.46	5.39	5.34	5.31	5.28	5.27	5.25	5.24	5.23
	.95	10.1	9.55	9.28	9.12	9.01	8.94	8.89	8.85	8.81	8.79
	.975	17.4	16.0	15.4	15.1	14.9	14.7	14.6	14.5	14.5	14.4
	.99	34.1	30.8	29.5	28.7	28.2	27.9	27.7	27.5	27.3	27.2
	.999	167	149	141	137	135	133	132	131	130	129
4	.50	.549	.828	.941	1.00	1.04	1.06	1.08	1.09	1.10	1.11
	.75	1.81	2.00	2.05	2.06	2.07	2.08	2.08	2.08	2.08	2.08
	.90	4.54	4.32	4.19	4.11	4.05	4.01	3.98	3.95	3.94	3.92
	.95	7.71	6.94	6.59	6.39	6.26	6.16	6.09	6.04	6.00	5.96
	.975	12.2	10.6	9.98	9.60	9.36	9.20	9.07	8.98	8.90	8.84
	.99	21.2	18.0	16.7	16.0	15.5	15.2	15.0	14.8	14.7	14.5
	.999	74.1	61.2	56.2	53.4	51.7	50.5	49.7	49.0	48.5	48.0
5	.50	.528	.799	.907	.965	1.00	1.02	1.04	1.05	1.06	1.07
	.75	1.69	1.85	1.88	1.89	1.89	1.89	1.89	1.89	1.89	1.89
	.90	4.06	3.78	3.62	3.52	3.45	3.40	3.37	3.34	3.32	3.30
	.95	6.61	5.79	5.41	5.19	5.05	4.95	4.88	4.82	4.77	4.74
	.975	10.0	8.43	7.76	7.39	7.15	6.98	6.85	6.76	6.68	6.62
	.99	16.3	13.3	12.1	11.4	11.0	10.7	10.5	10.3	10.2	10.1
	.999	47.2	37.1	33.2	31.1	29.7	28.8	28.2	27.6	27.2	26.9
6	.50	.515	.780	.886	.942	.977	1.00	1.02	1.03	1.04	1.05
	.75	1.62	1.76	1.78	1.79	1.79	1.78	1.78	1.78	1.77	1.77
	.90	3.78	3.46	3.29	3.18	3.11	3.05	3.01	2.98	2.96	2.94
	.95	5.99	5.14	4.76	4.53	4.39	4.28	4.21	4.15	4.10	4.06
	.975	8.81	7.26	6.60	6.23	5.99	5.82	5.70	5.60	5.52	5.46
	.99	13.7	10.9	9.78	9.15	8.75	8.47	8.26	8.10	7.98	7.87
	.999	35.5	27.0	23.7	21.9	20.8	20.0	19.5	19.0	18.7	18.4
7	.50	.506	.767	.871	.926	.960	.983	1.00	1.01	1.02	1.03
	.75	1.57	1.70	1.72	1.72	1.71	1.71	1.70	1.70	1.69	1.69
	.90	3.59	3.26	3.07	2.96	2.88	2.83	2.78	2.75	2.72	2.70
	.95	5.59	4.74	4.35	4.12	3.97	3.87	3.79	3.73	3.68	3.64
	.975	8.07	6.54	5.89	5.52	5.29	5.12	4.99	4.90	4.82	4.76
	.99	12.2	9.55	8.45	7.85	7.46	7.19	6.99	6.84	6.72	6.62
	.999	29.2	21.7	18.8	17.2	16.2	15.5	15.0	14.6	14.3	14.1

ν_2	a	ν_1									
		11	12	15	20	24	30	40	60	120	∞
1	.50	2.05	2.07	2.09	2.12	2.13	2.15	2.16	2.17	2.18	2.20
	.75	9.36	9.41	9.49	9.58	9.63	9.67	9.71	9.76	9.80	9.85
	.90	60.5	60.7	61.2	61.7	62.0	62.3	62.5	62.8	63.1	63.3
	.95	243	244	246	248	249	250	251	252	253	254
	.975	973	977	985	993	997	1000	1010	1010	1010	1020
	.99	6080	6110	6160	6210	6230	6260	6290	6310	6340	6370
	.999	6090^2	6110^2	6160^2	6210^2	6230^2	6260^2	6290^2	6310^2	6340^2	6370^2
2	.50	1.35	1.36	1.38	1.39	1.40	1.41	1.42	1.43	1.43	1.44
	.75	3.39	3.39	3.41	3.43	3.43	3.44	3.45	3.46	3.47	3.48
	.90	9.40	9.41	9.42	9.44	9.45	9.46	9.47	9.47	9.48	9.49
	.95	19.4	19.4	19.4	19.4	19.5	19.5	19.5	19.5	19.5	19.5
	.975	39.4	39.4	39.4	39.4	39.5	39.5	39.5	39.5	39.5	39.5
	.99	99.4	99.4	99.4	99.4	99.5	99.5	99.5	99.5	99.5	99.5
	.999	999	999	999	999	999	999	999	999	999	999
3	.50	1.19	1.20	1.21	1.23	1.23	1.24	1.25	1.25	1.26	1.27
	.75	2.45	2.45	2.46	2.46	2.46	2.47	2.47	2.47	2.47	2.47
	.90	5.22	5.22	5.20	5.18	5.18	5.17	5.16	5.15	5.14	5.13
	.95	8.76	8.74	8.70	8.66	8.63	8.62	8.59	8.57	8.55	8.53
	.975	14.4	14.3	14.3	14.2	14.1	14.1	14.0	14.0	13.9	13.9
	.99	27.1	27.1	26.9	26.7	26.6	26.5	26.4	26.3	26.2	26.1
	.999	129	128	127	126	126	125	125	124	124	123
4	.50	1.12	1.13	1.14	1.15	1.16	1.16	1.17	1.18	1.18	1.19
	.75	2.08	2.08	2.08	2.08	2.08	2.08	2.08	2.08	2.08	2.08
	.90	3.91	3.90	3.87	3.84	3.83	3.82	3.80	3.79	3.78	3.76
	.95	5.94	5.91	5.86	5.80	5.77	5.75	5.72	5.69	5.66	5.63
	.975	8.79	8.75	8.66	8.56	8.51	8.46	8.41	8.36	8.31	8.26
	.99	14.4	14.4	14.2	14.0	13.9	13.8	13.7	13.7	13.6	13.5
	.999	47.7	47.4	46.8	46.1	45.8	45.4	45.1	44.7	44.4	44.0
5	.50	1.08	1.09	1.10	1.11	1.12	1.12	1.13	1.14	1.14	1.15
	.75	1.89	1.89	1.89	1.88	1.88	1.88	1.88	1.87	1.87	1.87
	.90	3.28	3.27	3.24	3.21	3.19	3.17	3.16	3.14	3.12	3.10
	.95	4.71	4.68	4.62	4.56	4.53	4.50	4.46	4.43	4.40	4.36
	.975	6.57	6.52	6.43	6.33	6.28	6.23	6.18	6.12	6.07	6.02
	.99	9.96	9.89	9.72	9.55	9.47	9.38	9.29	9.20	9.11	9.02
	.999	26.6	26.4	25.9	25.4	25.1	24.9	24.6	24.3	24.1	23.8
6	.50	1.05	1.06	1.07	1.08	1.09	1.10	1.10	1.11	1.12	1.12
	.75	1.77	1.77	1.76	1.76	1.75	1.75	1.75	1.74	1.74	1.74
	.90	2.92	2.90	2.87	2.84	2.82	2.80	2.78	2.76	2.74	2.72
	.95	4.03	4.00	3.94	3.87	3.84	3.81	3.77	3.74	3.70	3.67
	.975	5.41	5.37	5.27	5.17	5.12	5.07	5.01	4.96	4.90	4.85
	.99	7.79	7.72	7.56	7.40	7.31	7.23	7.14	7.06	6.97	6.88
	.999	18.2	18.0	17.6	17.1	16.9	16.7	16.4	16.2	16.0	15.7
7	.50	1.04	1.04	1.05	1.07	1.07	1.08	1.08	1.09	1.10	1.10
	.75	1.69	1.68	1.68	1.67	1.67	1.66	1.66	1.65	1.65	1.65
	.90	2.68	2.67	2.63	2.59	2.58	2.56	2.54	2.51	2.49	2.47
	.95	3.60	3.57	3.51	3.44	3.41	3.38	3.34	3.30	3.27	3.23
	.975	4.71	4.67	4.57	4.47	4.42	4.36	4.31	4.25	4.20	4.14
	.99	6.54	6.47	6.31	6.16	6.07	5.99	5.91	5.82	5.74	5.65
	.999	13.9	13.7	13.3	12.9	12.7	12.5	12.3	12.1	11.9	11.7

ν_2	a	ν_1 1	2	3	4	5	6	7	8	9	10
8	.50	.499	.757	.860	.915	.948	.971	.988	1.00	1.01	1.02
	.75	1.54	1.66	1.67	1.66	1.66	1.65	1.64	1.64	1.64	1.63
	.90	3.46	3.11	2.92	2.81	2.73	2.67	2.62	2.59	2.56	2.54
	.95	5.32	4.46	4.07	3.84	3.69	3.58	3.50	3.44	3.39	3.35
	.975	7.57	6.06	5.42	5.05	4.82	4.65	4.53	4.43	4.36	4.30
	.99	11.3	8.65	7.59	7.01	6.63	6.37	6.18	6.03	5.91	5.81
	.999	25.4	18.5	15.8	14.4	13.5	12.9	12.4	12.0	11.8	11.5
9	.50	.494	.749	.852	.906	.939	.962	.978	.990	1.00	1.01
	.75	1.51	1.62	1.63	1.63	1.62	1.61	1.60	1.60	1.59	1.59
	.90	3.36	3.01	2.81	2.69	2.61	2.55	2.51	2.47	2.44	2.42
	.95	5.12	4.26	3.86	3.63	3.48	3.37	3.29	3.23	3.18	3.14
	.975	7.21	5.71	5.08	4.72	4.48	4.32	4.20	4.10	4.03	3.96
	.99	10.6	8.02	6.99	6.42	6.06	5.80	5.61	5.47	5.35	5.26
	.999	22.9	16.4	13.9	12.6	11.7	11.1	10.7	10.4	10.1	9.89
10	.50	.490	.743	.845	.899	.932	.954	.971	.983	.992	1.00
	.75	1.49	1.60	1.60	1.59	1.59	1.58	1.57	1.56	1.56	1.55
	.90	3.28	2.92	2.73	2.61	2.52	2.46	2.41	2.38	2.35	2.32
	.95	4.96	4.10	3.71	3.48	3.33	3.22	3.14	3.07	3.02	2.98
	.975	6.94	5.46	4.83	4.47	4.24	4.07	3.95	3.85	3.78	3.72
	.99	10.0	7.56	6.55	5.99	5.64	5.39	5.20	5.06	4.94	4.85
	.999	21.0	14.9	12.6	11.3	10.5	9.92	9.52	9.20	8.96	8.75
11	.50	.486	.739	.840	.893	.926	.948	.964	.977	.986	.994
	.75	1.47	1.58	1.58	1.57	1.56	1.55	1.54	1.53	1.53	1.52
	.90	3.23	2.86	2.66	2.54	2.45	2.39	2.34	2.39	2.27	2.25
	.95	4.84	3.98	3.59	3.36	3.20	3.09	3.01	2.95	2.90	2.85
	.975	6.72	5.26	4.63	4.28	4.04	3.88	3.76	3.66	3.59	3.53
	.99	9.65	7.21	6.22	5.67	5.32	5.07	4.89	4.74	4.63	4.54
	.999	19.7	13.8	11.6	10.3	9.58	9.05	8.66	8.35	8.12	7.92
12	.50	.484	.735	.835	.888	.921	.943	.959	.972	.981	.989
	.75	1.46	1.56	1.56	1.55	1.54	1.53	1.52	1.51	1.51	1.50
	.90	3.18	2.81	2.61	2.48	2.39	2.33	2.28	2.24	2.21	2.19
	.95	4.75	3.89	3.49	3.26	3.11	3.00	2.91	2.85	2.80	2.75
	.975	6.55	5.10	4.47	4.12	3.89	3.73	3.61	3.51	3.44	3.37
	.99	9.33	6.93	5.95	5.41	5.06	4.82	4.64	4.50	4.39	4.30
	.999	18.6	13.0	10.8	9.63	8.89	8.38	8.00	7.71	7.48	7.29
15	.50	.478	.726	.826	.878	.911	.933	.948	.960	.970	.977
	.75	1.43	1.52	1.52	1.51	1.49	1.48	1.47	1.46	1.46	1.45
	.90	3.07	2.70	2.49	2.36	2.27	2.21	2.16	2.12	2.09	2.06
	.95	4.54	3.68	3.29	3.06	2.90	2.79	2.71	2.64	2.59	2.54
	.975	6.20	4.76	4.15	3.80	3.58	3.41	3.29	3.20	3.12	3.06
	.99	8.68	6.36	5.42	4.89	4.56	4.32	4.14	4.00	3.89	3.80
	.999	16.6	11.3	9.34	8.25	7.57	7.09	6.74	6.47	6.26	6.08
20	.50	.472	.718	.816	.868	.900	.922	.938	.950	.959	.966
	.75	1.40	1.49	1.48	1.47	1.45	1.44	1.43	1.42	1.41	1.40
	.90	2.97	2.59	2.38	2.25	2.16	2.09	2.04	2.00	1.96	1.94
	.95	4.35	3.49	3.10	2.87	2.71	2.60	2.51	2.45	2.39	2.35
	.975	5.87	4.46	3.86	3.51	3.29	3.13	3.01	2.91	2.84	2.77
	.99	8.10	5.85	4.94	4.43	4.10	3.87	3.70	3.56	3.46	3.37
	.999	14.8	9.95	8.10	7.10	6.46	6.02	5.69	5.44	5.24	5.08

ν_2	a	ν_1 11	12	15	20	24	30	40	60	120	∞
8	.50	1.02	1.03	1.04	1.05	1.06	1.07	1.07	1.08	1.08	1.09
	.75	1.63	1.62	1.62	1.61	1.60	1.60	1.59	1.59	1.58	1.58
	.90	2.52	2.50	2.46	2.42	2.40	2.38	2.36	2.34	2.32	2.29
	.95	3.31	3.28	3.22	3.15	3.12	3.08	3.04	3.01	2.97	2.93
	.975	4.24	4.20	4.10	4.00	3.95	3.89	3.84	3.78	3.73	3.67
	.99	5.73	5.67	5.52	5.36	5.28	5.20	5.12	5.03	4.95	4.86
	.999	11.4	11.2	10.8	10.5	10.3	10.1	9.92	9.73	9.54	9.34
9	.50	1.01	1.02	1.03	1.04	1.05	1.05	1.06	1.07	1.07	1.08
	.75	1.58	1.58	1.57	1.56	1.56	1.55	1.55	1.54	1.53	1.53
	.90	2.40	2.38	2.34	2.30	2.28	2.25	2.23	2.21	2.18	2.16
	.95	3.10	3.07	3.01	2.94	2.90	2.86	2.83	2.79	2.75	2.71
	.975	3.91	3.87	3.77	3.67	3.61	3.56	3.51	3.45	3.39	3.33
	.99	5.18	5.11	4.96	4.81	4.73	4.65	4.57	4.48	4.40	4.31
	.999	9.71	9.57	9.24	8.90	8.72	8.55	8.37	8.19	8.00	7.81
10	.50	1.01	1.01	1.02	1.03	1.04	1.05	1.05	1.06	1.06	1.07
	.75	1.55	1.54	1.53	1.52	1.52	1.51	1.51	1.50	1.49	1.48
	.90	2.30	2.28	2.24	2.20	2.18	2.16	2.13	2.11	2.08	2.06
	.95	2.94	2.91	2.85	2.77	2.74	2.70	2.66	2.62	2.58	2.54
	.975	3.66	3.62	3.52	3.42	3.37	3.31	3.26	3.20	3.14	3.08
	.99	4.77	4.71	4.56	4.41	4.33	4.25	4.17	4.08	4.00	3.91
	.999	8.58	8.44	8.13	7.80	7.64	7.47	7.30	7.12	6.94	6.76
11	.50	1.00	1.01	1.02	1.03	1.03	1.04	1.05	1.05	1.06	1.06
	.75	1.52	1.51	1.50	1.49	1.49	1.48	1.47	1.47	1.46	1.45
	.90	2.23	2.21	2.17	2.12	2.10	2.08	2.05	2.03	2.00	1.97
	.95	2.82	2.79	2.72	2.65	2.61	2.57	2.53	2.49	2.45	2.40
	.975	3.47	3.43	3.33	3.23	3.17	3.12	3.06	3.00	2.94	2.88
	.99	4.46	4.40	4.25	4.10	4.02	3.94	3.86	3.78	3.69	3.60
	.999	7.76	7.62	7.32	7.01	6.85	6.68	6.52	6.35	6.17	6.00
12	.50	.995	1.00	1.01	1.02	1.03	1.03	1.04	1.05	1.05	1.06
	.75	1.50	1.49	1.48	1.47	1.46	1.45	1.45	1.44	1.43	1.42
	.90	2.17	2.15	2.11	2.06	2.04	2.01	1.99	1.96	1.93	1.90
	.95	2.72	2.69	2.62	2.54	2.51	2.47	2.43	2.38	2.34	2.30
	.975	3.32	3.28	3.18	3.07	3.02	2.96	2.91	2.85	2.79	2.72
	.99	4.22	4.16	4.01	3.86	3.78	3.70	3.62	3.54	3.45	3.36
	.999	7.14	7.01	6.71	6.40	6.25	6.09	5.93	5.76	5.59	5.42
15	.50	.984	.989	1.00	1.01	1.02	1.02	1.03	1.03	1.04	1.05
	.75	1.44	1.44	1.43	1.41	1.41	1.40	1.39	1.38	1.37	1.36
	.90	2.04	2.02	1.97	1.92	1.90	1.87	1.85	1.82	1.79	1.76
	.95	2.51	2.48	2.40	2.33	2.39	2.25	2.20	2.16	2.11	2.07
	.975	3.01	2.96	2.86	2.76	2.70	2.64	2.59	2.52	2.46	2.40
	.99	3.73	3.67	3.52	3.37	3.29	3.21	3.13	3.05	2.96	2.87
	.999	5.93	5.81	5.54	5.25	5.10	4.95	4.80	4.64	4.47	4.31
20	.50	.972	.977	.989	1.00	1.01	1.01	1.02	1.02	1.03	1.03
	.75	1.39	1.39	1.37	1.36	1.35	1.34	1.33	1.32	1.31	1.29
	.90	1.91	1.89	1.84	1.79	1.77	1.74	1.71	1.68	1.64	1.61
	.95	2.31	2.28	2.20	2.12	2.08	2.04	1.99	1.95	1.90	1.84
	.975	2.72	2.68	2.57	2.46	2.41	2.35	2.29	2.22	2.16	2.09
	.99	3.29	3.23	3.09	2.94	2.86	2.78	2.69	2.61	2.52	2.42
	.999	4.94	4.82	4.56	4.29	4.15	4.01	3.86	3.70	3.54	3.38

ν_2	a	ν_1									
		1	2	3	4	5	6	7	8	9	10
24	.50	.469	.714	.812	.863	.895	.917	.932	.944	.953	.961
	.75	1.39	1.47	1.46	1.44	1.43	1.41	1.40	1.39	1.38	1.38
	.90	2.93	2.54	2.33	2.19	2.10	2.04	1.98	1.94	1.91	1.88
	.95	4.26	3.40	3.01	2.78	2.62	2.51	2.42	2.36	2.30	2.25
	.975	5.72	4.32	3.72	3.38	3.15	2.99	2.87	2.78	2.70	2.64
	.99	7.82	5.61	4.72	4.22	3.90	3.67	3.50	3.36	3.26	3.17
	.999	14.0	9.34	7.55	6.59	5.98	5.55	5.23	4.99	4.80	4.64
30	.50	.466	.709	.807	.858	.890	.912	.927	.939	.948	.955
	.75	1.38	1.45	1.44	1.42	1.41	1.39	1.38	1.37	1.36	1.35
	.90	2.88	2.49	2.28	2.14	2.05	1.98	1.93	1.88	1.85	1.82
	.95	4.17	3.32	2.92	2.69	2.53	2.42	2.33	2.27	2.21	2.16
	.975	5.57	4.18	3.59	3.25	3.03	2.87	2.75	2.65	2.57	2.51
	.99	7.56	5.39	4.51	4.02	3.70	3.47	3.30	3.17	3.07	2.98
	.999	13.3	8.77	7.05	6.12	5.53	5.12	4.82	4.58	4.39	4.24
40	.50	.463	.705	.802	.854	.885	.907	.922	.934	.943	.950
	.75	1.36	1.44	1.42	1.40	1.39	1.37	1.36	1.35	1.34	1.33
	.90	2.84	2.44	2.23	2.09	2.00	1.93	1.87	1.83	1.79	1.76
	.95	4.08	3.23	2.84	2.61	2.45	2.34	2.25	2.18	2.12	2.08
	.975	5.42	4.05	3.46	3.13	2.90	2.74	2.62	2.53	2.45	2.39
	.99	7.31	5.18	4.31	3.83	3.51	3.29	3.12	2.99	2.89	2.80
	.999	12.6	8.25	6.60	5.70	5.13	4.73	4.44	4.21	4.02	3.87
60	.50	.461	.701	.798	.849	.880	.901	.917	.928	.937	.945
	.75	1.35	1.42	1.41	1.38	1.37	1.35	1.33	1.32	1.31	1.30
	.90	2.79	2.39	2.18	2.04	1.95	1.87	1.82	1.77	1.74	1.71
	.95	4.00	3.15	2.76	2.53	2.37	2.25	2.17	2.10	2.04	1.99
	.975	5.29	3.93	3.34	3.01	2.79	2.63	2.51	2.41	2.33	2.27
	.99	7.08	4.98	4.13	3.65	3.34	3.12	2.95	2.82	2.72	2.63
	.999	12.0	7.76	6.17	5.31	4.76	4.37	4.09	3.87	3.69	3.54
120	.50	.458	.697	.793	.844	.875	.896	.912	.923	.932	.939
	.75	1.34	1.40	1.39	1.37	1.35	1.33	1.31	1.30	1.29	1.28
	.90	2.75	2.35	2.13	1.99	1.90	1.82	1.77	1.72	1.68	1.65
	.95	3.92	3.07	2.68	2.45	2.29	2.18	2.09	2.02	1.96	1.91
	.975	5.15	3.80	3.23	2.89	2.67	2.52	2.39	2.30	2.22	2.16
	.99	6.85	4.79	3.95	3.48	3.17	2.96	2.79	2.66	2.56	2.47
	.999	11.4	7.32	5.79	4.95	4.42	4.04	3.77	3.55	3.38	3.24
∞	.50	.455	.693	.789	.839	.870	.891	.907	.918	.927	.934
	.75	1.32	1.39	1.37	1.35	1.33	1.31	1.29	1.28	1.27	1.25
	.90	2.71	2.30	2.08	1.94	1.85	1.77	1.72	1.67	1.63	1.60
	.95	3.84	3.00	2.60	2.37	2.21	2.10	2.01	1.94	1.88	1.83
	.975	5.02	3.69	3.12	2.79	2.57	2.41	2.29	2.19	2.11	2.05
	.99	6.63	4.61	3.78	3.32	3.02	2.80	2.64	2.51	2.41	2.32
	.999	10.8	6.91	5.42	4.62	4.10	3.74	3.47	3.27	3.10	2.96

v_2	a	v_1									
		11	12	15	20	24	30	40	60	120	∞
24	.50	.967	.972	.983	.994	1.00	1.01	1.01	1.02	1.02	1.03
	.75	1.37	1.36	1.35	1.33	1.32	1.31	1.30	1.29	1.28	1.26
	.90	1.85	1.83	1.78	1.73	1.70	1.67	1.64	1.61	1.57	1.53
	.95	2.21	2.18	2.11	2.03	1.98	1.94	1.89	1.84	1.79	1.73
	.975	2.59	2.54	2.44	2.33	2.27	2.21	2.15	2.08	2.01	1.94
	.99	3.09	3.03	2.89	2.74	2.66	2.58	2.49	2.40	2.31	2.21
	.999	4.50	4.39	4.14	3.87	3.74	3.59	3.45	3.29	3.14	2.97
30	.50	.961	.966	.978	.989	.994	1.00	1.01	1.01	1.02	1.02
	.75	1.35	1.34	1.32	1.30	1.29	1.28	1.27	1.26	1.24	1.23
	.90	1.79	1.77	1.72	1.67	1.64	1.61	1.57	1.54	1.50	1.46
	.95	2.13	2.09	2.01	1.93	1.89	1.84	1.79	1.74	1.68	1.62
	.975	2.46	2.41	2.31	2.20	2.14	2.07	2.01	1.94	1.87	1.79
	.99	2.91	2.84	2.70	2.55	2.47	2.39	2.30	2.21	2.11	2.01
	.999	4.11	4.00	3.75	3.49	3.36	3.22	3.07	2.92	2.76	2.59
40	.50	.956	.961	.972	.983	.989	.994	1.00	1.01	1.01	1.02
	.75	1.32	1.31	1.30	1.28	1.26	1.25	1.24	1.22	1.21	1.19
	.90	1.73	1.71	1.66	1.61	1.57	1.54	1.51	1.47	1.42	1.38
	.95	2.04	2.00	1.92	1.84	1.79	1.74	1.69	1.64	1.58	1.51
	.975	2.33	2.29	2.18	2.07	2.01	1.94	1.88	1.80	1.72	1.64
	.99	2.73	2.66	2.52	2.37	2.29	2.20	2.11	2.02	1.92	1.80
	.999	3.75	3.64	3.40	3.15	3.01	2.87	2.73	2.57	2.41	2.23
60	.50	.951	.956	.967	.978	.983	.989	.994	1.00	1.01	1.01
	.75	1.29	1.29	1.27	1.25	1.24	1.22	1.21	1.19	1.17	1.15
	.90	1.68	1.66	1.60	1.54	1.51	1.48	1.44	1.40	1.35	1.29
	.95	1.95	1.92	1.84	1.75	1.70	1.65	1.59	1.53	1.47	1.39
	.975	2.22	2.17	2.06	1.94	1.88	1.82	1.74	1.67	1.58	1.48
	.99	2.56	2.50	2.35	2.20	2.12	2.03	1.94	1.84	1.73	1.60
	.999	3.43	3.31	3.08	2.83	2.69	2.56	2.41	2.25	2.09	1.89
120	.50	.945	.950	.961	.972	.978	.983	.989	.994	1.00	1.01
	.75	1.27	1.26	1.24	1.22	1.21	1.19	1.18	1.16	1.13	1.10
	.90	1.62	1.60	1.55	1.48	1.45	1.41	1.37	1.32	1.26	1.19
	.95	1.87	1.83	1.75	1.66	1.61	1.55	1.50	1.43	1.35	1.25
	.975	2.10	2.05	1.95	1.82	1.76	1.69	1.61	1.53	1.43	1.31
	.99	2.40	2.34	2.19	2.03	1.95	1.86	1.76	1.66	1.53	1.38
	.999	3.12	3.02	2.78	2.53	2.40	2.26	2.11	1.95	1.76	1.54
∞	.50	.939	.945	.956	.967	.972	.978	.983	.989	.994	1.00
	.75	1.24	1.24	1.22	1.19	1.18	1.16	1.14	1.12	1.08	1.00
	.90	1.57	1.55	1.49	1.42	1.38	1.34	1.30	1.24	1.17	1.00
	.95	1.79	1.75	1.67	1.57	1.52	1.46	1.39	1.32	1.22	1.00
	.975	1.99	1.94	1.83	1.71	1.64	1.57	1.48	1.39	1.27	1.00
	.99	2.25	2.18	2.04	1.88	1.79	1.70	1.59	1.47	1.32	1.00
	.999	2.84	2.74	2.51	2.27	2.13	1.99	1.84	1.66	1.45	1.00

Entry is $\chi^2(a,\nu)$, where

$P[\chi^2(\nu) \leq \chi^2(a,\nu)] = a.$

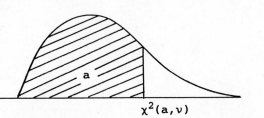

$\chi^2(a,\nu)$

df	a									
ν	.005	.010	.025	.050	.100	.900	.950	.975	.990	.995
1	$.0^439$	$.0^316$	$.0^398$	$.0^239$	.0158	2.71	3.84	5.02	6.63	7.88
2	.0100	.0201	.0506	.103	.211	4.61	5.99	7.38	9.21	10.60
3	.072	.115	.216	.352	.584	6.25	7.81	9.35	11.34	12.84
4	.207	.0297	.484	.711	1.064	7.78	9.49	11.14	13.28	14.86
5	.412	.554	.831	1.145	1.61	9.24	11.07	12.83	15.09	16.75
6	.676	.872	1.24	1.64	2.20	10.64	12.59	14.45	16.81	18.55
7	.989	1.24	1.69	2.17	2.83	12.02	14.07	16.01	18.48	20.28
8	1.34	1.65	2.18	2.73	3.49	13.36	15.51	17.53	20.09	21.96
9	1.73	2.09	2.70	3.33	4.17	14.68	16.92	19.02	21.67	23.59
10	2.16	2.56	3.25	3.94	4.87	15.99	18.31	20.48	23.21	25.19
11	2.60	3.05	3.82	4.57	5.58	17.28	19.68	21.92	24.73	26.76
12	3.07	3.57	4.40	5.23	6.30	18.55	21.03	23.34	26.22	28.30
13	3.57	4.11	5.01	5.89	7.04	19.81	22.36	24.74	27.69	29.82
14	4.07	4.66	5.63	6.57	7.79	21.06	23.68	26.12	29.14	31.32
15	4.60	5.23	6.26	7.26	8.55	22.31	25.00	27.49	30.58	32.80
16	5.14	5.81	6.91	7.96	9.31	23.54	26.30	28.85	32.00	34.27
17	5.70	6.41	7.56	8.67	10.09	24.77	27.59	30.19	33.41	35.72
18	6.26	7.01	8.23	9.39	10.86	25.99	28.87	31.53	34.81	37.16
19	6.84	7.63	8.91	10.12	11.65	27.20	30.14	32.85	36.19	38.58
20	7.43	8.26	9.59	10.85	12.44	28.41	31.41	34.17	37.57	40.00
21	8.03	8.90	10.28	11.59	13.24	29.62	32.67	35.48	38.93	41.40
22	8.64	9.54	10.98	12.34	14.04	30.81	33.92	36.78	40.29	42.80
23	9.26	10.20	11.69	13.09	14.85	32.01	35.17	38.08	41.64	44.18
24	9.89	10.86	12.40	13.85	15.66	33.20	36.42	39.36	42.98	45.56
25	10.52	11.52	13.12	14.61	16.46	34.38	37.65	40.65	44.31	46.93
26	11.16	12.20	13.84	15.38	17.29	35.56	38.89	41.92	45.64	48.29
27	11.81	12.88	14.57	16.15	18.11	36.74	40.11	43.19	46.96	49.64
28	12.46	13.56	15.31	16.93	18.94	37.92	41.34	44.46	48.28	50.99
29	13.12	14.26	16.05	17.71	19.77	39.09	42.56	45.72	49.59	52.34
30	13.79	14.95	16.79	18.49	20.60	40.26	43.77	46.98	50.89	53.67
40	20.71	22.16	24.43	26.51	29.05	51.81	55.76	59.34	63.69	66.77
50	27.99	29.71	32.36	34.76	37.69	63.17	67.50	71.42	76.15	79.49
60	35.53	37.48	40.48	43.19	46.46	74.40	79.08	83.30	88.38	91.95
80	51.17	53.54	57.15	60.39	64.28	96.58	101.9	106.6	112.3	116.3
100	67.33	70.06	74.22	77.93	82.36	118.5	124.3	129.6	135.8	140.2

$$\alpha = 0.05$$

ν	2	3	4	5	6	7	8	9	10
					p				
1	17.97	17.97	17.97	17.97	17.97	17.97	17.97	17.97	17.97
2	6.085	6.085	6.085	6.085	6.085	6.085	6.085	6.085	6.085
3	4.501	4.516	4.516	4.516	4.516	4.516	4.516	4.516	4.516
4	3.927	4.013	4.033	4.033	4.033	4.033	4.033	4.033	4.033
5	3.635	3.749	3.797	3.814	3.814	3.814	3.814	3.814	3.814
6	3.461	3.587	3.649	3.680	3.694	3.697	3.697	3.697	3.697
7	3.344	3.477	3.548	3.588	3.611	3.622	3.626	3.626	3.626
8	3.261	3.399	3.475	3.521	3.549	3.566	3.575	3.579	3.579
9	3.199	3.339	3.420	3.470	3.502	3.523	3.536	3.544	3.547
10	3.151	3.293	3.376	3.430	3.465	3.489	3.505	3.516	3.522
11	3.113	3.256	3.342	3.397	3.435	3.462	3.480	3.493	3.501
12	3.082	3.225	3.313	3.370	3.410	3.439	3.459	3.474	3.484
13	3.055	3.200	3.289	3.348	3.389	3.419	3.442	3.458	3.470
14	3.033	3.178	3.268	3.329	3.372	3.403	3.426	3.444	3.457
15	3.014	3.160	3.250	3.312	3.356	3.389	3.413	3.432	3.446
16	2.998	3.144	3.235	3.298	3.343	3.376	3.402	3.422	3.437
17	2.984	3.130	3.222	3.285	3.331	3.366	3.392	3.412	3.429
18	2.971	3.118	3.210	3.274	3.321	3.356	3.383	3.405	3.421
19	2.960	3.107	3.199	3.264	3.311	3.347	3.375	3.397	3.415
20	2.950	3.097	3.190	3.255	3.303	3.339	3.368	3.391	3.409
24	2.919	3.066	3.160	3.226	3.276	3.315	3.345	3.370	3.390
30	2.888	3.035	3.131	3.199	3.250	3.290	3.322	3.349	3.371
40	2.858	3.006	3.102	3.171	3.224	3.266	3.300	3.328	3.352
60	2.829	2.976	3.073	3.143	3.198	3.241	3.277	3.307	3.333
120	2.800	2.947	3.045	3.116	3.172	3.217	3.254	3.287	3.314
∞	2.772	2.918	3.017	3.089	3.146	3.193	3.232	3.265	3.294

$$\alpha = 0.01$$

ν	p								
	2	3	4	5	6	7	8	9	10
1	90.03	90.03	90.03	90.03	90.03	90.03	90.03	90.03	90.03
2	14.04	14.04	14.04	14.04	14.04	14.04	14.04	14.04	14.04
3	8.261	8.321	8.321	8.321	8.321	8.321	8.321	8.321	8.321
4	6.512	6.677	6.740	6.756	6.756	6.756	6.756	6.756	6.756
5	5.702	5.893	5.989	6.040	6.065	6.074	6.074	6.074	6.074
6	5.243	5.439	5.549	5.614	5.655	5.680	5.694	5.701	5.703
7	4.949	5.145	5.260	5.334	5.383	5.416	5.439	5.454	5.464
8	4.746	4.939	5.057	5.135	5.189	5.227	5.256	5.276	5.291
9	4.596	4.787	4.906	4.986	5.043	5.086	5.118	5.142	5.160
10	4.482	4.671	4.790	4.871	4.931	4.975	5.010	5.037	5.058
11	4.392	4.579	4.697	4.780	4.841	4.887	4.924	4.952	4.975
12	4.320	4.504	4.622	4.706	4.767	4.815	4.852	4.883	4.907
13	4.260	4.442	4.560	4.644	4.706	4.755	4.793	4.824	4.850
14	4.210	4.391	4.508	4.591	4.654	4.704	4.743	4.775	4.802
15	4.168	4.347	4.463	4.547	4.610	4.660	4.700	4.733	4.760
16	4.131	4.309	4.425	4.509	4.572	4.622	4.663	4.696	4.724
17	4.099	4.275	4.391	4.475	4.539	4.589	4.630	4.664	4.693
18	4.071	4.246	4.362	4.445	4.509	4.560	4.601	4.635	4.664
19	4.046	4.220	4.335	4.419	4.483	4.534	4.575	4.610	4.639
20	4.024	4.197	4.312	4.395	4.459	4.510	4.552	4.587	4.617
24	3.956	4.126	4.239	4.322	4.386	4.437	4.480	4.516	4.546
30	3.889	4.056	4.168	4.250	4.314	4.366	4.409	4.445	4.477
40	3.825	3.988	4.098	4.180	4.244	4.296	4.339	4.376	4.408
60	3.762	3.922	4.031	4.111	4.174	4.226	4.270	4.307	4.340
120	3.702	3.858	3.965	4.044	4.107	4.158	4.202	4.239	4.272
∞	3.643	3.796	3.900	3.978	4.040	4.091	4.135	4.172	4.205

Copied with permission from *Probability and Statistics for Engineers and Scientists,* Walpole, R.E. Myers, R.H., MacMillian, 1985

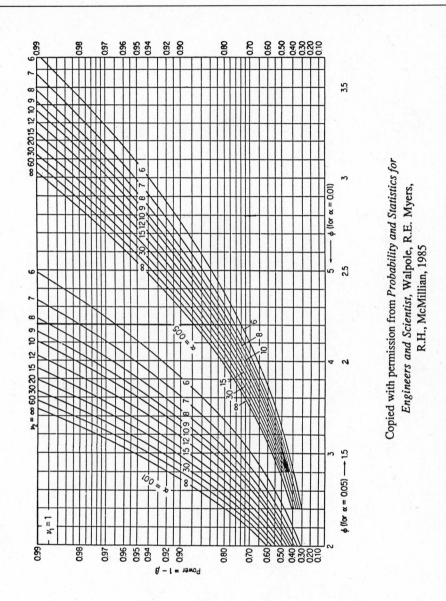

Copied with permission from *Probability and Statistics for Engineers and Scientist*, Walpole, R.E. Myers, R.H, McMillian, 1985

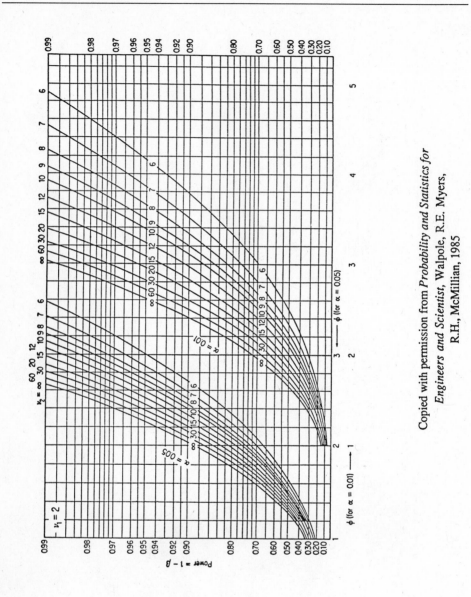

Copied with permission from *Probability and Statistics for Engineers and Scientist*, Walpole, R.E. Myers, R.H., McMillian, 1985

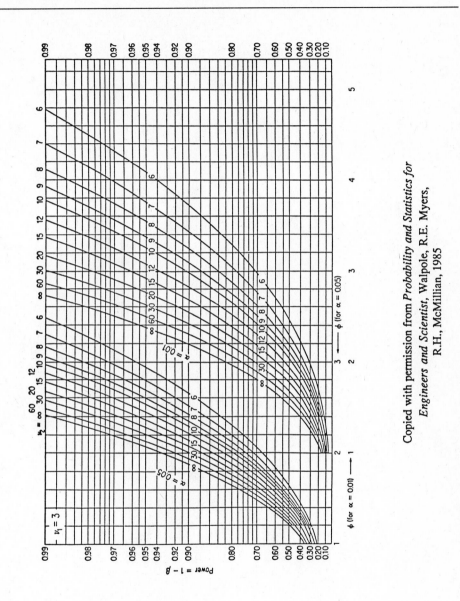

Copied with permission from *Probability and Statistics for Engineers and Scientist*, Walpole, R.E. Myers, R.H., McMillian, 1985

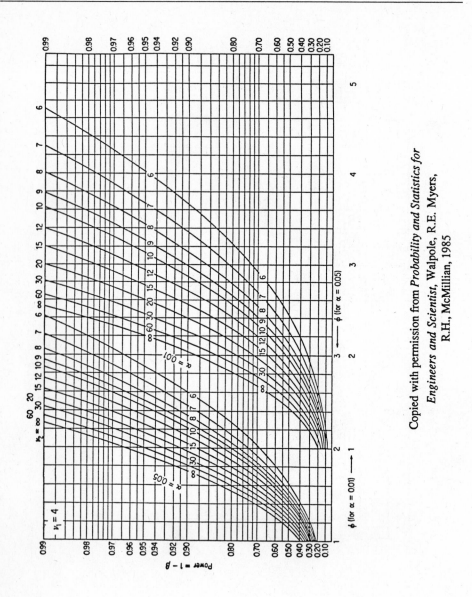

Copied with permission from *Probability and Statistics for Engineers and Scientist*, Walpole, R.E. Myers, R.H, McMillian, 1985

ANSWERS TO SELECTED PROBLEMS

Chapter 1

7. C_p = (Upper Spec − Lower Spec) / 6σ

      = (2100 − 1900) / 6(20)

      = 1.67

    C_{pk} = minimum (Upper Spec − $\bar{y}$, $\bar{y}$ − Lower Spec) / 3σ

      = minimum (60, 140) / 60

      = 60 / 60

      = 1.0

Chapter 2

1. b)

	RPM	Feed Rate	Tool Radius
1.	−	−	−
2.	−	−	+
3.	−	+	−
4.	−	+	+
5.	+	−	−
6.	+	−	+
7.	+	+	−
8.	+	+	+

2. a)

	(A) RPM	(B) Feed Rate	(C) Radius	AB	AC	BC	ABC
Avg(−)	131.9	66.9	107.2	110.0	134.1	119.1	108.4
Avg(+)	81.3	146.2	105.9	103.1	79.1	94.1	104.7
Δ	−50.6	79.3	−1.3	−6.9	−55.0	−25.0	−3.7
Δ/2	−25.3	39.7	−0.6	−3.5	−27.5	−12.5	−1.9

2. b)   The Pareto chart implies B, AC, A, and perhaps BC are important.

Chapter 3

5.   By folding over the R_{III} design, you can get a R_{IV} design.

6.   I = ABCDF = BCDEG = AEFG

13. a) I = ACD = BCE = ABDE

  A = CD = ABCE = BDE
  B = ABCD = CE = ADE
  C = AD = BE = ABCDE
  D = AC = BCDE = ABE
  E = ACDE = BC = ABD

  Alias problems are: B = CE
                       C = AD

16. a) α distance = $(n_F)^{1/4}$ = $(8)^{1/4}$ = 1.68

   b) Factor B: 10 to 80% (Assume this represents the maximum allowable range for the factors.)

design values	$-\alpha$	-1	0	1	α
real values	10	24.17	45	65.83	80

   c) $n_c = 4\sqrt{n_F + 1} - 2k = 6$

   d) Total number of runs: 8(factorial) + 6(centerpoints) + 6(axial) = 20

17. R^2_{AB} = .25, for A + B there is a 25% overlap between the two variables.

18. $_8C_2 = \dfrac{8!}{2! \ 6!} = 28$

Chapter 4

1.   MSE = .0415

a) 95% C.I. = $(4.10 - 3.86) \pm 2.306 \sqrt{.0415 \left(\frac{1}{5} + \frac{1}{5} \right)}$

   = [−.0571, .5371]

b) $t_0 = \dfrac{(4.1 - 3.86)}{\sqrt{.415 \left(\frac{1}{5} + \frac{1}{5} \right)}} = 1.862$

   $t_{\text{critical}} = 2.306$

c)

Source	SS	df	MS	f	P
Between	.1440	1	.1440	3.47	<0.05
Error	.3320	8	.0415		
Total	.4760	9			

2. $\Phi = 56\sqrt{n}$

n′	df = k(n′ − 1)	Φ	1 − B	Comments
15	28	2.17	0.82	too small
17	32	2.31	0.87	too small
18	34	2.38	0.89	close
19	36	2.44	0.92	correct

Use n = 19 for each group of bearings.

3. a) MSB = 17843
      MSE = 50.64
      F_0 = 352.3

$$\bar{d}_t = q_t \sqrt{\frac{MSE}{n_j}}$$

$$= 4.25 \sqrt{\frac{50.64}{7}} = 11.43$$

3. b) A_5 is the best choice for maximization.

4. a) $\bar{x}$ = 15.8, $\bar{y}$ = 9.6
      b_1 = .9415
      b_0 = −5.314

4. b) R^2 = .938

5. a)

Source	SS	df	MS	F_0
Temp (A)	1.439	2	.7195	3597.5
H_2SO_4	.0345	2	.0173	86.5
A × B	.1940	4	.0485	242.5
Error	8.9×10^{-4}	9	.0002	
Total	1.669	17		

Chapter 5

1. a)

Run #	ȳ	ln s
1	60.0	2.65
2	50.5	1.60
3	39.5	1.85
4	62.5	2.49
5	55.0	1.04
6	45.5	1.26
7	40.5	0.75
8	58.5	0.75

1. c) Setting A at the high setting appears to minimize the variance.

2. a) The prediction equation for the mean is:
$$y = 3.512 + 0.255(A) + 1.008(B) + .044(A)(B)$$

3. a)

	A	B	C	D	E	F	G
Avg (−)	72.26	77.77	77.39	79.80	78.18	77.30	77.22
Avg (+)	82.89	77.37	77.76	75.35	76.97	77.84	77.93
Δ	10.63	−.40	.37	−4.45	−1.21	.54	.71

3. b) To minimize, set A at high and D at low.
$$\hat{y} = 85.12$$

5.

ANALYSIS OF MEANS				
	A	B	C	D
Avg(−)	23.50	25.08	24.17	15.08
Avg(+)	26.17	24.58	25.50	34.58
Δ	2.67	−.50	1.33	19.50

Chapter 6

4.   Using S/N (nominal is best), the averages are:

	D	A	E	B
Avg(−)	109.9	107.9	105.9	104.9
Avg(+)	107.9	109.9	111.9	112.9
Δ	−2.0	2.0	6.0	8.0

5.   Using S/N (larger is better), the averages are:

	A	B	C	D	AB	AC	BC	D
Avg(−)	25.79	26.68	26.36	22.97	27.25	26.32	27.04	22.97
Avg(+)	27.41	26.52	26.84	30.23	25.95	26.88	26.16	30.23
Δ	1.62	−.16	.48	7.26	−1.3	.56	−.88	7.26

The best settings appear to be:
D at +1
A at +1
B at −1

Chapter 7

1.   You should use response surface methodology when you think the optimum is outside the experimental region.

2.   The gradient vector is the first derivative of the model with respect to each factor. It indicates which direction to travel to find the optimum.

4.   Coded values allow us to avoid multicollinearity problems in the analysis of the experimental data.

HISTORICAL PERSPECTIVE ON DOE

Early Paper on Latin Squares (Meziriac)	1st Statistical Hypothesis Test (Gauss)	Least Squares (Gauss) (Legendre)	Regression Concepts (Galton, Pearson)	Orthogonal Matrices (Hadamard)	t-test (Gosset)
1624	1812	1830	1888	1890	1908
Design of Experiments Developed (Fisher)	Hypothesis testing formalized (Neyman, Pearson)	Fisher's texts on DOE Blocking, Randomizing, Factorial Design, Anova (Fisher)	Incomplete randomized block designs (Yates)	2^k Factorial Designs (Yates)	
1920	1933	1925-35	1936	1937	
$L_{27}(2^{13})$ (Fisher)	Fractional Factorials (Finney, Rao)	Plackett-Burman Designs (Plackett-Burman)	$L_{18}(2^1 x 3^7)$ (Burman)	1st text on DOE applications (BrownLee)	
1945	1945	1945	1946	1948	
Taguchi develops his approach to DOE (Taguchi)	Multiple Regression ( )	Central Composite Design (Box,Wilson)	Optimal Designs (Elfving)	Multiple Comparisons (Tukey, Scheffe)	$L_{36}(2^3 x 3^{13})$ (Seiden)
1950	1950	1951	1952	1954	1954
RSM EVOP (Box)	Tukey Quick Test (Tukey)	Mixture Experiments (Scheffe)	Half Normal Plots (C.Daniel)	Optimal Designs (Kiefer, Wolfowitz)	Box-Behnken Designs (Box-Behnken)
1957	1958	1958	1959	1960	1960
Addelman Designs 2&3 Levels (Addelman)	2^{k-p} Fractional Factorials, Resolution (Box,Hunter)	Formalized Optimal Design Approach (Fedorov)	Detmax D-Optimal Algorithm (Mitchell)	1st Text on RSM (R.Myers)	Taguchi version of Loss Function & Robust Design (Taguchi)
1961	1961	1972	1974	1976	1980
K-Exchange Algorithum for D-Optimal (Johnson, Nachtsheim)	Residuals to find Variance Reducing Factors (Box,Meyer)	1st text to blend Taguchi and Classical (Schmidt/ Launsby)	D-Optimal Blocking techniques (Cox, Nachtsheim)	$\ln\dfrac{S^2+}{S^2-} \sim Z$ (Montgomery)	Sample Size Rules of Thumb for $\hat{S}$ Models (Schmidt/ Launsby)
1983	1986	1988	1989	1990	1991

Symbols and Abbreviations

LSL	Lower specification limit
USL	Upper specification limit
y	Some quality characteristic (response)
T	Target or nominal specification value
$\hat{\sigma} = s$	Standard deviation calculated from a sample
$\bar{y}$	Average response
CPI	Continuous process improvement
$\hat{\sigma}^2 = s^2$	Variance calculated from a sample
Σ	Mathematical summation (indicates "add them all up")
μ	Population mean
σ	Population standard deviation
σ^2	Population variance
UCL	Upper control limit (for control charts)
LCL	Lower control limit (for control charts)
dpm	Number of defects per million opportunities
C_p	Process capability index (assumes process is centered on the target)
C_{pk}	Process capability index (does not assume process in centered)
L	Loss
$- = -1$	Usually represents the low level of a factor
$+ = +1$	Usually represents the high level of a factor
0	Usually represents the middle level of a factor
$AB = A \times B$	2-way interaction of factors A and B
PAT	Process Action Teams
TQM	Total Quality Management
ANOVA	Analysis of Variance
Δ	Effect
$\Delta/2$	Half effect
$\hat{y}$	Predicted response
n_a	# of combinations associated with the axial points in a CCD
n_c	# of centerpoints in a CCD
n_f	# of combinations in the factorial portion of a CCD
df	Degrees of freedom
e_i	Error of prediction for the ith data point

F_0	Calculated F value
H_0	Null hypothesis
H_1	Alternative hypothesis
t_0	Calculated t values
t_c	Tabled t values
S/N	Signal-to-noise ratio
α	Probability of a Type I error (in Chapter 4)
α	Axial point distance from center (for CCD's in Chapters 3 & 7)
QFD	Quality Function Deployment
SST	Sum-of-squares total
SSR	Sum-of-squares due to regression
SSE	Sum-of-squares due to error
SSB	Sum-of-squares due to between group differences
MSB	Mean square between
MSE	Mean square error
$F_{critical}$	Tabled F value
R	Correlation coefficient
R^2	Squared correlation coefficient

RULES OF THUMB

In response to requests from several readers of the previous edition, we have added this appendix to provide quick and easy Rules of Thumb for use in designing and analyzing experimental tests. The following Rules of Thumb are based on sound experimental logic; however, they are not necessarily precise from a statistical point of view. The reader should only use these temporarily as a quick and easy substitute for some of the rigorous methods discussed in the actual text. The Rules of Thumb provided in this appendix will refer to the following:

1) Sample size per experimental condition or run (often referred to as the number of replicates and/or repetitions per experimental condition).

2) Quick test for detecting a significant shift in the average.

3) Quick test for detecting a significant shift in the standard deviation.

4) Two-level design choices for the number of factors (k) equal to 2, 3, 4, 5, and greater than 5.

5) Determining statistical significance for terms in the $\hat{y}$ model.

6) Determining statistical significance for terms in the $\hat{s}$ (or ln $\hat{s}$) model.

| Rule of Thumb (1) | Sample Size per Experimental Condition |

Historically, Rules of Thumb on sample size have been based on the amount of data required for only the $\hat{y}$ model. Since shifts in the variance are as important (if not more so) than shifts in the average, the following Rules of Thumb are based on the amount of data to adequately model § or ln §. The information is statistically derived and can be verified in *Engineering Statistics* [1]. The following Rules of Thumb on sample size also assume that you want to detect a 2.72 shift in the standard deviation, and the experiment is performed off line (i.e., factor lows and highs are chosen to be relatively far apart). For on-line experiments (i.e., factor lows and highs are relatively close together) you will want to multiply all tabled sample sizes by at least 10.

Percent Confidence that a term identified as significant, truly does belong in §	Percent chance of finding a significant term if one actually exists	Average number of (−) and (+) values in the experimental matrix columns			
		4	8	12	16
		Sample Size per Experimental Condition			
95% (α = .05)	40% (β = .60)	3	2	N/A	N/A
95% (α = .05)	75% (β = .25)	5	3	N/A	2
95% (α = .05)	90% (β = .10)	7	4	3	N/A
95% (α = .05)	95% (β = .05)	9	5	4*	3
95% (α = .05)	99% (β = .01)	11	6	5*	4*
99% (α = .01)	15% (β = .85)	3	2	N/A	N/A
99% (α = .01)	50% (β = .50)	5	3	N/A	2
99% (α = .01)	70% (β = .30)	7	4	3	N/A
99% (α = .01)	90% (β = .10)	9	5	4*	3
99% (α = .01)	96% (β = .04)	11	6	5*	4*

* For these sample sizes the percent chance of finding a significant term, if one actually exists is a rough approximation.

As an example consider a 16 run fractional factorial for the number of factors (k) = 5. Since each column in the 16 run matrix has a total of 16 (−) and (+) values, the ideal sample size per experimental condition (or run) is 3. This sample size produces 95% confidence in terms identified as significant and provides a 95% chance of finding a variance shifting factor if one exists. If we wanted to be 99% confident that a term identified as significant, truly does belong in §, there would be a 90% chance of finding an actual significant term.

Rule of Thumb (2)

**Quick Test for Detecting a
Significant Shift in the Average**

To determine if a significant average shift has occurred, a quick test developed by John Tukey in 1959, *Technometrics* [2], and popularized by Dorian Shainin, *World Class Quality* [3], is called the Tukey Quick Test or End Count Technique. To perform this test, arrange all of the data on a scale such that each of the two groups has a different label. Starting from the left-most data value, count the number of similar symbols until you reach an opposite symbol. Starting from the right, count the number of similar symbols until you reach an opposite symbol. Sum the two counts which is your end count. If both ends start with the same symbol your end count is zero.

The significance of an end count can be determined using the table below:

End Count	Significance	Confidence there exists a significant average shift
≥ 6	$\leq .10$	$\geq .90$
≥ 7	$\leq .05$	$\geq .95$
≥ 10	$\leq .01$	$\geq .99$
≥ 13	$\leq .001$	$\geq .999$

As an example demonstrating the simplicity of the End Count Technique is shown below on the same data from Chapter 4 where a t-test was used to determine whether or not a significant shift in the average occurred.

Strength Measurements											
Temp	y_1	y_2	y_3	y_4	y_5	y_6	y_7	y_8	y_9	$\bar{y}$	s^2
200°	2.8	3.6	6.1	4.2	5.2	4.0	6.3	5.5	4.5	4.6889	1.3761
300°	7.0	4.1	5.7	6.4	7.3	4.7	6.6	5.9	5.1	5.8667	1.1575

Composite Material Data (Low (L) = 200,° High (H) = 300°)

The hypotheses to be tested are

$$H_0: \quad \mu_L = \mu_H$$
$$H_1: \quad \mu_L \neq \mu_H$$

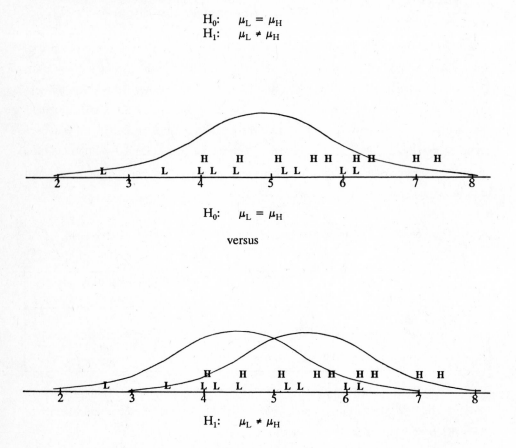

$$H_0: \quad \mu_L = \mu_H$$

versus

$$H_1: \quad \mu_L \neq \mu_H$$

Starting from the left, the number of L symbols before the first H is 3. Starting from the right, the number of H symbols before the first L is 4. Thus, the overall End Count is 7. Using the previous table with End Count values and associated significance values, our data indicates that we are approximately 95% confident in concluding $H_1: \mu_L \neq \mu_H$. This conclusion is identical to that of the t-test performed in Chapter 4.

| **Rule of Thumb (3)** | **Quick Test for Detecting a Significant Shift in the Standard Deviation** |

To detect a significant shift in standard deviation the proper statistical test is based on the ratio of sample variances, i.e., $F = \dfrac{s_-^2}{s_+^2}$.   In the text *Design and Analysis of Experiments* [4], the author states that $\ln \dfrac{s_-^2}{s_+^2}$ is approximately the standard normal, i.e., $\ln \dfrac{s_-^2}{s_+^2} \sim Z(0,1)$.   Consequently, a quick test can be developed as follows:

a)   Find the maximum of $\dfrac{s_-}{s_+}$ and $\dfrac{s_+}{s_-}$.

b)   If the maximum $\left[\dfrac{s_-}{s_+} , \dfrac{s_+}{s_-} \right]$ > 2.72 you are approximately 95% confidence that the two conditions have significantly different standard deviations, (i.e., a significant shift in standard deviation is highly probable).

Note 1:   This test is based on small sample sizes (i.e. $n_- = n_+ = 16$) and the desire to detect a 2.72 shift in standard deviation.

Note 2:   The value 2.72 is not a hard and fine line. You should use good judgement or more precise tests when max $\left[\dfrac{s_-}{s_+} , \dfrac{s_+}{s_-} \right]$ produces a value close to 2.72. (A gray zone occurs for values over the range 1.5 to 2.72. For values in this range you should conduct an F test.)

As an example for Rule of Thumb (3), consider the following:

$$n_- = n_+ = 16$$

$$s_- = 1.9 \qquad \max \left[\frac{1.9}{5.8} , \frac{5.8}{1.9} \right] = 3.053$$

$$s_+ = 5.8$$

Therefore, we are at least 95% confident that $\sigma_- \neq \sigma_+$.

Rule of Thumb (4)	**Two-Level Design Choices for the Number of Factors (k) = 2, 3, 4, 5, and greater than 5**

a) k = 2 : The only option is a full factorial where n = 2^2 = 4 experimental conditions (runs). According to Rule of Thumb (1) the ideal sample size per condition (run) is 9. Thus the total resources to conduct the ideal experiment of 2 factors at two-levels is 36.

b) k = 3 : Option 1 is to run a full factorial where n = 2^3 = 8 with 5 samples per condition (run). Thus, the total resources required would be 40.

Option 2 is a R_{III} half fraction where n = 2^{3-1} = 4 with 9 samples per condition (run). The total resources required is 36.

Note that Option 2 only saves four resources, yet would result in loss of all interactions. Therefore, most experimenters would prefer Option 1.

c) k = 4 : Option 1, the full factorial will have n = 2^4 = 16 runs, each with a sample size of 3. Total resources will be 48.

Option 2 is a R_{IV} ½ fraction where $n = 2^{4-1} = 8$ and the sample size per run is 5. Total resources is 40.

Option 2 will save us 8 resources compared to Option 1. Of the 6 total two-way interactions, Option 1 will evaluate them all and Option 2 will evaluate half of them. Which option to choose will typically depend on the cost per resource and the number of anticipated interactions.

d) $k = 5$: Option 1 is a full factorial with $n = 2^5 = 32$ runs each replicated 2 times for a total of 64 resources.

Option 2 is a R_V ½ fraction where $n = 2^{5-1} = 16$ runs each with 3 replications. Total resources will be 48.

In this case, Option 2 is usually the preferred choice because 16 resources are saved and all two-way interactions are still evaluated.

e) $k \geq 6$: In this case, the experimenter will usually conduct a screening experiment to identify important factors and then perform subsequent experimentation on these factors. Two-level screening design types and their ideal sample sizes are given below for various numbers of factors.

k	Design Type	Sample Size	Comments
6, 7	Taguchi L8	5	2 ways aliased with mains.
6 - 11	Taguchi L12	4	2 ways partially confounded with mains.
6 - 15	Taguchi L16	3	for $k \leq 8$, 2 ways unconfounded with mains; for $k > 8$, 2 ways aliased with mains.

Rule of Thumb (5)	**Determining Statistical Significance for Terms in the ŷ Model**

Using Analysis of Variance (ANOVA) produces a F value for each potential term to go into the ŷ model. Based upon the sample sizes discussed in Rule of Thumb (1), the critical F value will typically be between 4.0 and 6.0 assuming a confidence level of 95%. Therefore, a simple Rule of Thumb is:

    F < 4.0 implies insignificance (leave term out of ŷ model),

    F > 6.0 implies significance (leave term in ŷ model),

    4.0 < F < 6.0 is a gray zone;

Note: It is usually better to make an error placing the term in ŷ as opposed to leaving it out.

Multiple Regression usually results in a t value for each term where $t^2 = F$. Therefore, the same Rules of Thumb from ANOVA could apply to t^2. However, most regression software also produces a p(2-tail) value which represents the probability of placing an insignificant term in the ŷ model. Thus, the rule of thumb for p(2-tail) values is usually as follows:

    p(2-tail) < .05 term is significant, place in ŷ model.

    p(2-tail) > .10 term is insignificant, leave out of ŷ model.

    .05 < p(2-tail) < .10 is a gray zone where most experimenters will typically decide to place the term in ŷ.

Note: (1 − p(2-tail)) * 100% is your confidence that the term belongs in the model.

It should be noted, that these Rules of Thumb are just that; implying that the values presented are not hard and fast. Neither Rules of Thumb or rigorous statistical tests should ever replace common sense, size of effects and prior knowledge. Ideally, a blend of all these approaches works best.

Rule of Thumb (6)	**Determining Statistical Significance for Terms in the § or ln § Model**

Using $\ln \dfrac{s_+^2}{s_-^2} \sim Z(0,1)$ from Montgomery, cited in Rule of Thumb (3), you can use the following rules for determining significance when you have **2 level designs**:

- § models : $\left| \dfrac{\Delta}{2} \right|$ for any term should be $\geq \dfrac{\overline{s}}{2}$ (i.e., the term coefficient should be $\geq$ ½ of the constant).

- ln § models : $\left| \dfrac{\Delta}{2} \right| \geq .5$.

As previously stated with the $\hat{y}$ model Rules of Thumb, the reader should use several complimentary inputs such as common sense, prior knowledge and the above Rules of Thumb when making the final decision for significance. For example, any main factors with $\left| \dfrac{\Delta}{2} \right|$ close to $\dfrac{\overline{s}}{2}$ would be considered as significant. Whereas, a $\left| \dfrac{\Delta}{2} \right|$ close to $\dfrac{\overline{s}}{2}$ for a 3 way interaction is highly unlikely to be significant.

Appendix M Bibliography

1. Bowker, A.H. and Lieberman (1972), *Engineering Statistics*, Prentice-Hall Inc., Englewood Cliffs, New Jersey.

2. Tukey, J.W. (1959), *Technometrics*, 1, 31-48.

3. Bhote, K.R. (1988), *World Class Quality*, AMA Membership Publications Division, New York.

4. Montgomery, D.C. (1990), *Design and Analysis of Experiments*, 3rd Edition, Wiley, New York.

Glossary

2-LEVEL DESIGN - An experiment where all factors are set at one of two levels, denoted as low and high (-1, $+1$ or 1, 2).

3-LEVEL DESIGN - An experiment where all factors are set at one of three levels, denoted as low, medium and high (-1, 0, 1 or 1, 2, 3).

ALIASING - When two factors are set at the same levels throughout the experiment, i.e., the columns are 100% correlated.

ANOVA - Analysis of Variance. A statistical tool based on F ratios that measures if a factor contributes significantly to the variance of a response. Also determines the amount of variance due to pure error.

BALANCED DESIGN - A 2-level experimental design is balanced if each factor is run the same number of times at the high and low levels.

BLOCKING VARIABLE - A variable (factor) which cannot be randomized. The experiment is usually run in blocks for each level of the blocking variable and randomization is performed within blocks.

BOX-BEHNKEN DESIGN - A 3-level design used for quantitative factors and designed to estimate all main, quadratic, and 2-way interaction effects.

BRAINSTORMING - A group activity which generates a list of possible factors and levels, and the method by which the results may be evaluated.

CAUSALITY - The assertion that changes to an input factor will directly result in a specified change in an output.

CENTER POINTS - Experimental runs with all factor levels set half-way between the low and high settings.

CENTRAL COMPOSITE DESIGN - A 3-level design that starts with a 2-level fractional factorial and some center points. If needed, axial points can be tested to complete quadratic terms. Used typically for quantitative factors and designed to estimate all main effects plus desired quadratics and 2-way interactions.

CONFIDENCE INTERVAL - A range of values based on a sample mean and standard deviation that has a given probability of containing the true population parameter.

CONTROLLABLE FACTORS - Factors the experimenter has control of during all phases, i.e., experimental, production, and operational phases.

D-OPTIMAL DESIGN - An experimental design in which the minimum number or runs is based on degrees of freedom needed to analyze the desired effects. Not necessarily orthogonal or balanced, it does minimize the correlation between factors.

DEFINING RELATIONSHIP - A statement of one or more factor word equalities used to determine the aliasing structure in a fractional factorial design.

DEFINING WORDS - Factor word equalities in a defining relationship.

DEGREES OF FREEDOM (df) - The number of independent observations minus the number of parameters estimated.

DESIGN ARRAY - An array representing the experimental settings. Usually contains values ranging from -1 to 1, but could be wider if using CCD. The rows represent the runs and the columns represent the factors.

EIGENVALUES - With eigenvectors. Given a matrix multiplied by an eigenvalue, it will produce the same result as the related eigenvector multiplied by the same matrix. Used to describe stationary points on a surface.

EXPERIMENTAL DESIGN - Purposeful changes to the inputs (factors) to a process in order to observe corresponding changes in the outputs (responses).

F RATIO (F) - A ratio of two independent estimates of experimental error. If there isn't an effect, the ratio should be close to 1.

F TABLE - Provides a means for determining the significance of a factor to a specified level of confidence by comparing calculated F ratios to those from the F distribution (see Appendix). If the F ratio is greater than the table value, there is a significant effect.

FACTOR - An input to a process which can be manipulated during experimentation.

FISHBONE DIAGRAM - A wire diagram which a group can use to organize its thoughts during a brainstorming session. The backbone of the fish represents the response being measured. The "ribs" represent the types of factors that affect the response.

FOLDOVER DESIGN - A way to obtain a resolution IV design based on two designs of R_{III}. Used when the confirmation runs from a resolution III design differ substantially from their prediction and the experimenter desires to de-alias the two-way interaction from the main effects.

FRACTIONAL FACTORIALS - Instead of using a full factorial, some subset of it can be used if the experimenter can assume some interactions will not occur and he/she assigns a factor to that interaction column of the design.

FULL FACTORIAL - All possible combinations of the factors and levels. Given n factors all with two levels, there will be 2^n runs. If the factors have three levels, then 3^n runs would be needed.

GAUSSIAN DISTRIBUTION - The engineering name for the Normal Distribution.

GRADIENT - The slope at a point on a surface.

INNER ARRAY - Used in parameter design to identify the combinations of controllable factors to be studied in a designed experiment. Also called a "design array."

INTERACTION - Occurs when the combination of two or more interacting factors generates a result that is different from the result produced by the individual factors.

INTERACTION EFFECT - The influence of two or more interacting factors on the results when they are changed from one level to another.

LATIN SQUARES - A classical method used to generate orthogonal designs.

LEVEL - A setting or value of a factor.

LINEAR GRAPH - A tool used by Taguchi to identify sets of interacting columns in orthogonal arrays.

LOSS FUNCTION - A technique for quantifying loss due to product deviations from target values.

MAIN EFFECT - The influence a single factor has on the response when it is changed from one level to another. Also called the "mean response" or "average response."

MEAN SQUARE - Average squared deviation from the sample average used as an estimate of experimental error.

MEAN SQUARE ERROR - A weighted average of the sum of the variances for each run.

MULTICOLLINEARITY - The existence of strong correlations between input factors.

MULTIPLE REGRESSION - A Taylor series model using several independent variables to predict one dependent variable.

NOISE - Unexplained variability in the response.

SIGNAL-TO-NOISE (S/N) - A comparison of the influence of controllable factors (signals) to the influence of noise factors. The higher the S/N value, the better.

SIMPLE LINEAR REGRESSION - A model where one independent variable is used to predict one dependent variable.

SPECIFICATION (SPEC) LIMITS - The bounds of acceptable values for a given product or process.

STATIONARY POINT - The corollary in single variable calculus would be a critical point. A stationary point is a point where the gradient is zero, i.e., no slope at the point. The eigenvalues are used to identify a maximum, a minimum, a saddle point, or a ridge.

SUM OF SQUARES - The total of the squared differences from a set value, usually the mean.

SYSTEM DESIGN - The selection of materials, parts, products, factors, equipment, and process parameters.

T-TEST - A statistical test to measure if a significant difference exists between two sample mean values.

TAGUCHI METHODS - Experimental design thoughts and processes as developed by Taguchi. Based on the philosophy of the loss function, he uses Signal-to-Noise ratios as the primary analysis tool. Orthogonal design matrices are tabled and originate from fractional factorials, Plackett-Burman, and Latin Square designs.

TOLERANCE DESIGN - The specification of appropriate tolerances, product parameters, and process factors.

TRADITIONAL METHODS - Statistical experimental design thoughts and processes as originally developed by Fisher and others as early as the 1920's. Uses ANOVA as the primary analysis tool, along with orthogonal designs such as fractional factorials, Latin Squares, Plackett-Burman, Box-Behnken, Central Composite, and D-optimal.

TUKEY TEST - A statistical test to measure the difference between several mean values and tell the user which ones are statistically different from the rest.

UNCONTROLLABLE FACTORS - Factors that are difficult, undesirable, or impossible to change. Also called "noise factors." The uncontrollable factors included in a design matrix must be controllable in the experimental phase.

Index

NOTES

NOTES

NOTES

NOTES

NOTES

NOTES

NOTES

NOTES

NOTES

NOTES

NOTES

NOTES